T0184738

Mastering the Discrete Fourier Transform in One, Two or Several Dimensions: Pitfalls and Artifacts

Computational Imaging and Vision

This comprehensive book series embraces state-of-the-art expository works and advanced research monographs on any aspect of this interdisciplinary field.

Topics covered by the series fall in the following four main categories:

- Imaging Systems and Image Processing
- Computer Vision and Image Understanding
- Visualization
- Applications of Imaging Technologies

Only monographs or multi-authored books that have a distinct subject area, that is where each chapter has been invited in order to fulfill this purpose, will be considered for the series.

Volume 43

For further volumes:
http://www.springer.com/series/5754

Isaac Amidror

Mastering the Discrete Fourier Transform in One, Two or Several Dimensions

Pitfalls and Artifacts

 Springer

Isaac Amidror
School of Computer and Communication Sciences
Peripheral Systems Laboratory
Ecole Polytechnique Fédérale de Lausanne
Lausanne, Switzerland

ISSN 1381-6446
ISBN 978-1-4471-6131- 8   ISBN 978-1-4471-5167-8 (eBook)
DOI 10.1007/978-1-4471-5167-8
Springer London Heidelberg New York Dordrecht

Mathematics Subject Classification (2010): 65T50, 42A99, 42B99

Springer is part of Springer Science+Business Media (www.springer.com)

To my parents

*An expert is a man who has made all the mistakes,
which can be made, in a very narrow field.*

Niels Bohr [Mackay91 p. 35]

Front cover image: Aliasing due to insufficient sampling rate
may give unexpected shapes both in the signal domain and
in the spectral domain. See Fig. 5.17(k).

Contents

Appendices

Preface

The continuous Fourier transform (CFT) and its discrete counterpart, the discrete Fourier transform (DFT), are probably among the most widely used mathematical tools in science and engineering. Their fields of application are so diverse and numerous that it would be practically impossible to list them exhaustively in a short paragraph. It is not surprising, therefore, that almost anyone studying or working in a scientific or technological discipline had or will have the opportunity to use them one day or another.

The present book originates from the author's own experience as a frequent user of the DFT. Although several excellent textbooks already exist for teaching the DFT and its applications, it turns out that many practical issues of great importance to the users can still hardly be found in the existing literature. Most of the existing textbooks do not explain in sufficient detail the various pitfalls and artifacts that very often await users of the DFT, while in other books these topics are not easy to find, being dispersed between several sections and chapters. The situation in the multidimensional (MD) case is even worse, since many books defer this subject to a later chapter and only treat it there briefly as a straightforward extension of the one-dimensional case, without considering the particularities of the DFT pitfalls and artifacts in the MD case. Moreover, one can rarely find in the literature a detailed discussion on practical questions such as how to place the true unit values along the axes when plotting the DFT results, or similar practical issues that are often acquired only through the user's own experience, sometimes simply by trial and error. The aim of the present book is, therefore, to close these gaps in the existing literature.

This book is not intended to be a self-contained textbook for teaching the basics of the Fourier theory, a subject which is already covered by many excellent books. It rather intends to assist users who already have a basic knowledge of the DFT, but are not sufficiently familiar with its practical limitations, artifacts and pitfalls. The present book can be used, therefore, as a useful additional reference for students and researchers in various branches of science and engineering who have to work on applications of DFT. For this reason we have preferred throughout this book an informal style based on a pictorial, intuitive approach over a rigorous mathematical treatment, and we have intentionally avoided a purist's approach based on theorems and proofs. And just like the author's previous books on the theory of the moiré phenomenon, this volume, too, contains a large number of illustrative examples and figures, some of which are visually fascinating and even spectacular.

The material covered in this book includes a wide range of subjects, ranging from the aliasing and leakage phenomena to less widely known topics such as symmetry related issues and the DFT artifacts that may result thereof, or the influence of the various possible types of data extension on the DFT results. It is our aim in this book to attract the reader's attention to the many possible pitfalls and sources of error in the use of DFT, including potential errors due to the different CFT or DFT definitions being used in different books and software packages, the need for input and output data reorganizations, etc. And most importantly, we show the users how to correctly interpret the DFT results they obtain, and how to distinguish between true spectral contents and the various artifacts that are only due to DFT.

This book is intended for students, scientists and engineers wishing to widen their knowledge of the DFT and its practical use. In particular, it will be very useful for "naive" users from various scientific or technical disciplines who have to use the DFT for their respective applications. The reader will find in this book not only a theoretical explanation of the DFT artifacts and pitfalls in question, but also practical recipes accompanied by many examples for the correct use of DFT in one or several dimensions. The prerequisite mathematical background is limited to an elementary familiarity with calculus and with the continuous and discrete Fourier theory.

This concentrated treatment of the DFT artifacts and pitfalls in a single volume is, indeed, new, and it aims to make this book a valuable source of information for the widest possible range of DFT users. We hope this book will not only help to "demystify" this subject, but also encourage readers to further deepen their interest in the Fourier theory, which is undoubtedly one of the most beautiful and enlightening branches of mathematics.

The material in this book is based on the author's personal research at the EPFL (*Ecole Polytechnique Fédérale de Lausanne*). This work would have never been possible without the support and the excellent research environment provided by the EPFL. The author wishes to express his gratitude to Prof. Roger D. Hersch, the head of the Peripheral Systems Laboratory of the EPFL, for his encouragement throughout the different stages of this work. Many thanks are also due to the publishers for their continued helpfulness and availability throughout the publishing cycle.

Chapter 1

Introduction

1.1 The discrete Fourier transform

The discrete Fourier transform (DFT) is the discrete-world counterpart of the continuous Fourier transform (CFT). The DFT is widely used as a practical and efficient computing tool for calculating numerically the Fourier transform (i.e. the frequency spectrum) of functions or signals [Brigham88 pp. xiv, 1–3, 98]. In many circumstances the values of our given signal are only known on a discrete grid (for example, if the signal values have been measured at discrete intervals or obtained by a digital computer). In such cases using DFT is the natural way for computing the Fourier transform of the given data. But even when the given signal is a continuous function whose analytical expression is fully known, DFT often remains the most convenient way for getting a visual glimpse at its spectrum, especially when the analytic calculation of the continuous Fourier transform proves to be too laborious or impractical.

One has to remember, however, that although the use of DFT as a numerical approximation to the continuous Fourier transform is extremely simple and attractive, there is a price to pay for this convenience: The DFT results are often "soiled" by some artifacts which are not part of the underlying continuous Fourier transform, and which may be quite hard to detect. Moreover, the use of DFT also requires caution due to some inherent pitfalls that may await an inexperienced user. A basic discussion on these artifacts and pitfalls can be found dispersed in the literature, and yet, most of these subjects are usually learned by DFT users through their own experience, sometimes simply by trial and error, and often they remain in the form of personal notes, recipes or "tricks". Moreover, these questions have been treated in the literature mostly for the one dimensional (1D) case. But in many disciplines, including applications in optics and in image processing, it is rather the multidimensional (MD) case, and in particular the two dimensional (2D) case, that is being used. Although most DFT artifacts and pitfalls in the MD case can be explained theoretically as a multidimensional extension of their 1D counterparts, there still exist many practical issues that are specifically of interest in the 2D or MD case.[1]

Our aim in this book is twofold: To show how to avoid such DFT artifacts, whenever possible; and if they cannot be avoided, at least to show how to identify them in time, in order not to interpret them erroneously as spectral properties of the given function. Our

---

[1] Note that the multidimensional Fourier theory is indeed more rich, versatile and interesting than its 1D counterpart. For example, some 2D Fourier theorems, such as the rotation theorem, the shear theorem, the separable product theorem and the projection-slice theorem [Bracewell95 pp. 157–159, 166, 499], have no 1D equivalents.

explanations are given from the *user's* point of view rather than from the mathematical point of view of a DFT theoretician. We favour a pictorial, intuitive approach rather than a purely theoretic one, since an intuitive understanding may often be extremely helpful in practical engineering situations. We mainly mention the 1D and 2D cases, due to their particular importance in applications, and also because they lend themselves more easily to a pictorial presentation that is the key for intuitive understanding. But our discussion covers the general multidimensional case, too.

Remark 1.1: In most applications, DFT is implemented using one of the numerous variants of the very efficient Fast Fourier Transform (FFT) algorithm [Briggs95, Chapter 10; Brigham88, Chapter 8; Nussbaumer82, Chapters 4–5], but it can be also computed using other methods (see, for example, [Kammler07, Sec. 6.1]). Because the DFT results are basically independent of the specific algorithm being used for their computation, we prefer to use here the generic term DFT, although our discussion is also applicable to all cases where an FFT algorithm is being used.   ∎

1.2 A brief historical background

The early seeds of what we call today the Fourier theory can be traced back to 18th-century mathematicians such as Euler, D. Bernoulli, Lagrange, and others [Briggs95 pp. 2–4]. Nevertheless, it is widely accepted that the first landmark contribution to this theory can be attributed to Jean Baptiste Joseph Fourier in his famous paper on the propagation of heat in solid bodies, which was presented to the French Academy of Sciences (*Institut de France*) in December 1807. This paper claimed, for the first time, that any arbitrary function, whether continuous or discontinuous, can be represented as an infinite series of sines and cosines. In spite of the skepticism and controversy aroused by this paper at that time, an upgraded version thereof finally earned Fourier the Academy's grand prize in January 1812, yet with some reservations. But it was only a decade later that Fourier's work was first published in his book *Théorie Analytique de la Chaleur* [Fourier22] (meaning in English: Analytic Theory of Heat). His prize-winning paper was finally also published by the Academy, in two parts, in 1824 and 1826 [Fourier24, Fourier26].

Fourier's work was pursued in the 19th and 20th centuries by famous mathematicians such as Poisson, Dirichlet, Riemann, Lesbegue and others, who further extended its scope but also established it on a more rigorous mathematical basis by addressing challenging questions such as convergence, integration, etc.

But as already mentioned, the continuous Fourier transform cannot always be computed analytically, either due to the nature of the problem at hand or due to the complexity of the integrals, which cannot always be expressed in closed form. And indeed, a discrete version of the Fourier theory, which offers a more practical numerical approach, has been developed in parallel to its continuous counterpart. Eighteenth century mathematicians such as Euler and Lagrange already used the discretized approach in their work, but it seems that the earliest explicit formulas of the DFT appear in a paper by Clairaut published in 1754 [Briggs95 p. 4] (although Clairaut's work was based on cosines alone, and can be considered therefore as a discrete cosine transform).

The use of the Fourier theory expanded rapidly over the years and found many new fields of application. But even the computation of the more practical discrete approach proved to be a formidable task when done by hand. Special-purpose mechanical analog computers were invented for this purpose at the end of the 19th century (see, for example, [Kammler07 pp. 86–87, 294]), and when the first digital computers appeared in the middle of the 20th century they were soon utilized for numeric DFT calculations, too.

It is important to note, however, that even when powerful computers are available, a straightforward application of the DFT formulas incurs a significant computation cost, since the number of operations required for an N-point DFT is proportional to N^2. As a simple example, a personal computer that performs 10^7 operations per second will do a 1000-point DFT in one tenth of a second, which seems fast. But if we need to apply such a DFT to a 2000×2000-point matrix or image, which is a routine task in modern image processing, this would already require some minutes. Obviously, before the advent of modern digital computers such a computation would simply be prohibitive.

This illustrates, indeed, the overwhelming importance of the scientific revolution that was brought about by the fast Fourier transform (FFT), a very efficient algorithm for the computation of the DFT which was published in 1965 by J. W. Cooley and J. W. Tukey [Cooley65]. Using this algorithm, the number of operations required for an N-point DFT is reduced from N^2 to $N\log_2 N$ (up to some constant factors), which is a very significant improvement. For example, a 1024-point FFT may require about 10,000 operations rather than about 1,000,000 operations, which is an improvement by a factor of 100; and furthermore, this factor itself ($N/\log_2 N$) continuously increases with N. The dramatic reduction in computation cost thanks to the FFT algorithm made it possible to do Fourier analysis on a digital computer, even when large-scale data arrays are involved.

Ironically, however, it turns out that the FFT algorithm, one of the most influential computational tools of the 20th century, was already used by Gauss in a paper dated to 1805, and which appeared in 1866 after his death in his collected works [Heideman85; Briggs95 pp. 5, 381; Kammler07 p. 295].

Today, the Fourier theory and in particular its discrete counterpart are used in virtually any imaginable branch of science and technology. For example, one could mention applications in fields as far apart as optics, electricity, acoustics, communications, signal processing, imaging, biological engineering, hydrodynamics, heat propagation, mechanics, geophysics, spectroscopy, statistics, mathematics, and the list is still long.

A more detailed historical account on the Fourier transform and on the DFT can be found, for example, in [Prestini04]. [Brigham88] gives a survey of various applications of the DFT as well as an extensive bibliography on the subject. An overview on the life of

Fourier and his main mathematical achievements can be found, for example, in [Bracewell86 pp. 462–464], in Chapter 1 of [Prestini04], or in dedicated books such as [Herivel75].

1.3 The scope of the present book

This book is written for people who already know and use the discrete Fourier transform, but are not yet sufficiently familiar with its practical limitations, pitfalls and artifacts. For this reason, no attempt is made here to explain the fundamentals of the Fourier theory and of the DFT. Many excellent textbooks already exist on these subjects, and readers who wish to review basic notions of the Fourier theory are encouraged to refer to any of these books. To mention just a few, one may consider classical works such as Bracewell's or Gaskill's books [Bracewell86; Bracewell95; Gaskill78], and for the discrete case Brigham's and Briggs' books on the DFT [Brigham88; Briggs95].

In particular, we do not explain here the implementation of DFT by the very efficient FFT algorithm; this well-known subject can be found in many existing textbooks such as [Nussbaumer82], [Loan92], [Brigham88, Chapter 8] or in the excellent overview provided in Chapter 10 of [Briggs95]. All our discussions in this book are valid for the DFT in general, independently of the algorithm being used to compute it.

Because this book is addressed to a wide audience in various scientific and engineering branches, a pictorial, intuitive approach supported by mathematics is preferred over a rigorous mathematical treatment. In many cases we give informal demonstrations rather than formal proofs, or defer detailed derivations to an appendix.

Furthermore, it is beyond the scope of this book to state rigorously pure mathematical issues such as the precise conditions under which each formula holds, the precise class of functions which satisfy the required convergence conditions, etc. In fact, a large number of functions encountered in engineering problems possess Fourier transforms only in a generalized sense (e.g. the sine and cosine functions, the step function, etc.). Following the practice adopted by [Bracewell86], [Gaskill78], and many other authors, such generalized Fourier transforms are manipulated in the same fashion as ordinary transforms, and we make no distinction between the two: It is to be understood that the term "Fourier transform" means "generalized Fourier transform" if we are dealing with functions that violate the existing conditions (see, for example, [Gaskill78 p. 184]). Some good references are given, though, to help the interested readers find a more detailed mathematical discussion on these questions.

Finally, it should be noted that although we occasionally use questions related to image processing to illustrate our discussion, this book has not been written with any specific DFT application in mind. In fact, our principal aim is to present the material in a general, application-independent way. Consequently, a full discussion on the various applications of the DFT remains beyond the scope of the book. A survey on the different fields of application of the DFT in science and technology can be found in [Brigham88, pp. 1–3]. The short, incomplete list at the end of Sec. 1.2 above may give the reader a small glimpse into the diversity of the various fields in which the DFT has found applications.

1.4 Overview of the following chapters

Chapter 2 provides the background and the basic notions for the entire book. It starts with an overview of the various definitions and notations that are being used in the literature for the continuous Fourier transform and for its discrete counterpart. Although the different variants used in the literature are essentially equivalent in terms of the Fourier theory, each of them gives slightly different results, and one should be aware of this fact to avoid any possible confusion, especially when working with several textbooks, references or Fourier software packages. After sorting out the notations issue, Chapter 2 reviews the basic rules which govern, respectively, the continuous and the discrete Fourier transforms, both in the 1D case and in the 2D and MD cases. Then it proceeds with a brief presentation of the graphical development of the DFT, and finally it reviews the main potential pitfalls that may await us when using the DFT as an approximation to the CFT, and which are the subject of the following chapters.

Chapter 3 deals with the data reorganizations that may be needed before or after the application of the DFT or the inverse DFT (IDFT). As we will see, the discrete input or output sequences are not necessarily organized the way we would expect them to be in the continuous world. Therefore, in order to avoid misinterpretations or other potential pitfalls and errors, one may need to reorganize these discrete sequences correctly before interpreting or displaying them. This is explained first for the 1D case and then for the 2D and MD cases, and illustrated by several examples and figures.

Chapter 4 explains how to interpret the DFT results in terms of true units along the axes. The input and output arrays of the DFT consist of a list of values (real or complex numbers), that are indexed by an integer running for example between $0,...,N-1$ (or, in the MD case, by M such indices). But when presenting or plotting the data we would like to mark along the axes the true values in terms of the units that are being used in the original continuous function (meters, seconds, etc.) or in its spectrum (cycles per unit, or Hz). Furthermore, there also exists the question of how to assign the true units along the vertical axis, i.e. what is the connection between the *values* obtained in the output array of the DFT and the true heights in the spectrum of our original function. Each of these questions is treated separately, and various examples and figures are provided to illustrate the discussion.

Chapter 5 is dedicated to the most important and well-known DFT artifact, aliasing. Aliasing is one of the main sources of discrepancy between the DFT results and the continuous Fourier transform. As in the previous chapters we start with the simpler 1D case, and only then we proceed to the more general (and more interesting) MD case.

Many illustrative examples and figures are also provided, some of which are quite surprising and sometimes even spectacular.

In Chapter 6 we focus our attention to the second major source of discrepancy between the DFT results and the continuous Fourier transform, the leakage artifact. Once again we start with the 1D case, and only then we proceed to the more general MD case. Both cases are accompanied by examples and figures to illustrate the discussion.

Then, in Chapter 7 we explain how to choose optimal values for the various parameters that are related to the resolution and the range of the data, both in the signal and in the spectral domains. We discuss the significance of each of these parameters, and show the interrelations between them.

Finally, in Chapter 8 we group together some further questions and potential pitfalls that may await the unaware DFT users, but which do not fit into any of the previous chapters. These include, among other topics, the representation of discontinuities in the input and output arrays of the DFT; issues that are related to the phase when using the polar (i.e. magnitude and phase) representation of the complex-valued spectrum; symmetry related issues and the DFT artifacts that may result thereof; artifacts that may occur due to sampling and reconstruction of continuous-world signals (jaggies and sub-Nyquist artifacts); and the various displaying considerations that should be taken into account in order to obtain an optimal graphic representation of the DFT results.

The main body of the book is accompanied by several appendices:

In Appendix A we briefly review the mathematical concept of an impulse, both in the continuous and in the discrete worlds, and we explain how impulses should be treated in the discrete world in order to obtain a good approximation to the continuous Fourier transform.

Appendix B examines the most straightforward ways to extend the input data when higher precision is required, and explains how each of these ways influences the DFT results (including the possible effects on the DFT artifacts). It is shown that some of these methods are indeed useful and advantageous, depending on one's aims and on the nature of the data, while others are rather misleading and may introduce significant artifacts.

In Appendix C we study in more detail the discrete-world properties of cases involving periodic functions. In particular, we discuss the discrete counterparts of the period and the frequency, explain their roles and their interconnections in the signal and spectral domains, and provide their multidimensional generalization.

In Appendix D we group together various issues, including the derivations of several results that we preferred for different reasons not to include in the main body of the work.

And finally, Appendix E provides a glossary of the most important terms that have been used in the present book.

1.5 About the graphic presentation of sampled signals and discrete data

Because graphic presentation of discrete data is essential to the understanding of many issues throughout this book, we find it necessary to give here a short introductory section on this subject, before delving into our actual work in the following chapters.

Many different ways exist for presenting graphically discrete data such as sampled signals, but in this section we mainly focus on the methods being used in this book. We start with the 1D case, where we have to plot the discrete function $y_k = g(x_k)$ with constant x_k increments of Δx. Then we continue with the 2D case, where we have to plot the discrete function $z_{k,l} = g(x_k, y_l)$ with constant x_k and y_l increments of Δx and Δy, respectively, and finally we proceed to the MD case, in which we have to plot the discrete function $z_{\mathbf{k}} = g(\mathbf{x_k})$, $\mathbf{x_k}$ being an M-dimensional discrete vector with constant increments along each of its M components. Further details can be found in references on the visual display of quantitative data such as [Cleveland93; Cleveland94; Wright07], or in the user manuals of mathematical software packages such as *Mathematica*®, *Matlab*®, *Maple*®, etc.

1.5.1 Graphic presentations in the 1D case

In the 1D case, the most obvious way to plot a discrete function $y_k = g(x_k)$ consists of the *symbol plot*, where each successive point (x_k, y_k) is graphically represented by a symbol (such as a dot or any other shape) at the corresponding coordinates; see Fig. 1.1(a).[2] This straightforward plotting method gives good visual results when the discrete data to be plotted varies slowly, so that the eye can easily follow the successive dots in the plot. However, when the given discrete data varies more rapidly, as shown on the right-hand side of Fig. 1.1(a), the eye may have some difficulties in following the successive dots in their right order. In such cases it may be useful to add to the symbol plot straight line segments that connect between the successive points, giving a *connected symbol plot* as in Fig. 1.1(b). Furthermore, if the successive points (x_k, y_k) are to be drawn densely, it may be advantageous to drop the dots altogether and only draw the connecting lines, resulting in a *connected line plot*, as in Fig. 1.1(c). This can give an illusion of a continuous plot (which may be similar to the continuous graph of the underlying continuous unsampled function, inasmuch as such a function exists[3]). Although this illusion could mislead one to believe that the plotted data is continuous, this should not really be a problem when we are aware that the graph in question represents a discrete signal. And indeed, this drawing method is widely used throughout this book, and dots are only used when confusion may arise. Note, however, that in this pseudo-continuous plotting method, when the points (x_k, y_k) to be plotted are not sufficiently dense, abrupt jumps in the signal may be rendered by slanted ramps rather than by sharp vertical transitions, so that an isolated impulse or a sharp-edged pulse may look, respectively, as an isosceles triangle or trapezium (Fig. 1.2).

---

[2] Note that we reserve the term *point* to the element (x_i, y_i) itself or its location in the plot, and use the term *dot* for the actual symbol that represents this point graphically.

[3] Such an underlying continuous function containing all the in-between data may exist, for example, if our data represents the temperature y_k sampled at the discrete moments x_k; but it does not exist if our data represents, for instance, the number of cars sold per agent or per country.

Another useful graphic improvement of the symbol plot consists of adding to the latter a *vertical line plot*, where each of the points (x_k,y_k) is connected to the horizontal axis by a thin vertical line (Fig. 1.1(d)); it is also possible to omit the dots altogether and draw the vertical line plot alone (Fig. 1.1(e)).[4] These successive vertical lines tend to fill the area below the graph of the discrete signal and guide the eye along the successive (x_k,y_k) points, without really connecting them by a continuous line. Yet another possible variant consists of replacing the vertical lines by adjacent vertical rectangles or boxes, like in a histogram. However, this is not recommended for reasons that we will see in Sec. 8.5.

1.5.2 Graphic presentations in the 2D case

We now proceed to the 2D case, where we have to plot the discrete function $z_{k,l} = g(x_k,y_l)$ with constant x_k and y_l increments. Here the situation is indeed more interesing, since each discrete point must be represented by three values: its spatial coordinates x_k and y_l and its value $z_{k,l}$. Basically, there are two families of methods which allow us to graphically represent such cases: (1) planar presentations, and (2) pseudo-3D presentations.

(1) In the family of planar presentation methods the paper's surface represents a 2D matrix or grid whose elements correspond to the spatial locations (coordinates) x_k and y_l of our discrete points. The values $z_{k,l}$ are then represented in the plot either by using gray levels (or colour tones), giving a *density plot*, or alternatively by drawing contour lines like in a topographical map, giving a *contour plot*.

In a *density plot*, each point x_k,y_l is represented in the plot by a small square pixel (corresponding to one matrix element or grid element) whose gray level represents its value $z_{k,l}$. The gray level scale can be chosen in various ways, but usually black represents the lowest possible value, white represents the highest possible value, and intermediate gray levels represent all the in-between values. Thus, if $z_{k,l}$ can only take values between 0 and 1 (as is often the case in the image or signal domain), black may represent the value 0 and white then represents the value 1. But if the values of $z_{k,l}$ can take both positive and negative values (as is often the case when drawing the real or imaginary valued parts of the spectral domain), black may represent the most negative

Figure 1.1: Comparison of various graphic methods for plotting 1D discrete data, using the signal $y_k = \cos(2\pi f x_k)$, $x_k = k/8$, $k \in \mathbb{Z}$ with $f = 0.5$ (left-hand column) and with $f = 3.75$ (right-hand column). Both of these two signals are plotted in each row using a different graphic presentation method: (a) Symbol plot. (b) Connected symbol plot (i.e. combination of symbol plot and connected line plot). (c) Connected line plot. (d) Combination of symbol plot and vertical line plot. (e) Vertical line plot.

---

[4] In the present book vertical line plots (without dots) are used, for example, in Fig. B.1.

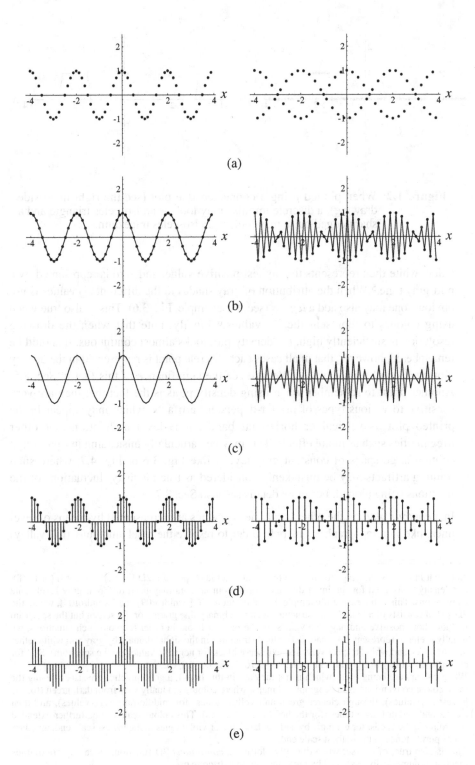

(a)

(b)

(c)

(d)

(e)

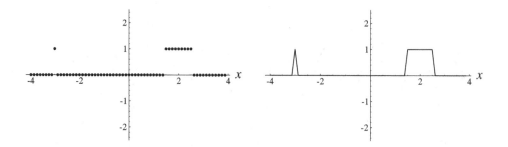

Figure 1.2: When plotted using a connected line plot (see the right-hand side drawing), a discrete impulse may look as an isosceles triangle and a sharp-edged pulse may look as an isosceles trapezium.

value, white then represents the highest positive value, and zero is represented by a mid-gray tone.[5] When the attribution of gray shades to the different $z_{k,l}$ values is not obvious, one may also add a *legend* (see, for example, Fig. 3.6). This is also true when using colours to represent the $z_{k,l}$ values.[6] Finally, note that when the drawing resolution is sufficiently high, the density plot looks almost continuous. It should be remembered, however, that in all cases each discrete point is represented in the density plot by a small square pixel.[7] We will see the significance of this fact in Sec. 8.5. Another point to remember when using density plots is the fact that they are very sensitive to various types of printer-dependent artifacts, which may appear in the printed plot as vertical or horizontal bands, gray-level undulations, or other irregularities such as moiré effects. This may be particularly misleading in figures that contain large spans of constant gray levels, like Fig. 3.6 or Fig. 4.7, where such printing artifacts may be mistakenly considered as true intensity fluctuations of the functions being plotted. For more details see also Sec. 8.7.

In a *contour plot*, on the other hand, the $z_{k,l}$ values are represented by means of level lines, like in a topographical map. In order to resolve the relief ambiguity (hill/valley)

---

[5] Note that in figures showing side by side the input and the output of a 2D Fourier transform (or DFT), the density plots used for the input data and for the output data may need different gray level scale conventions. This is the case, for example, in many figures in [Amidror09] and [Amidror07], where the 2D signal consists of a black/white structure such as a binary line grating or dot screen but the spectrum includes both positive and negative values. In the present book, in order to avoid such situations, we usually prefer to represent 2D binary 0/1-valued structures in the signal domain by gray and white rather than by black and white, respectively, and to reserve black for negative values (in the spectrum). See, for example, the binary line gratings in Figs. 5.14–5.16, 5.18, 8.6–8.9, etc.

[6] When colours are being used (which is not the case in this book), a good practice consists of using the same colour conventions as in geographical maps, where colours gradually vary from dark green (for the lowest $z_{k,l}$ values), through clearer green and yellow tones (for middle-range $z_{k,l}$ values), and then clearer and darker brown tones (for the heighest $z_{k,l}$ values). This colour scale can be further extended according to the needs, for example by adding dark blue or violet tones in the lowest scale end, and dark red or purple tones in the highest scale end.

[7] This is also true, of course, when digitally plotting a *continuous* 2D function, since any continuous function is numerically evaluated by the computer on a discrete grid.

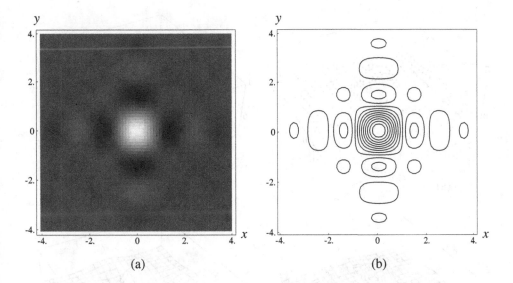

Figure 1.3: Comparison of two planar presentation methods for plotting 2D discrete data, using the signal $z_{k,l} = \text{sinc}(x_k)\,\text{sinc}(y_l)$, $x_k = k/8$, $y_l = l/8$, $k,l \in \mathbb{Z}$. (a) Density plot. (b) Contour plot (note that the level lines are obtained by interpolation, and they are drawn with continuous curves although the data is discrete).

in such a contour plot we can superpose it on top of the corresponding density plot, which better conveys the scale of the $z_{k,l}$ values. Note that the level lines are obtained by interpolations, and they are drawn with continuous curves even if they represent in reality discrete data.

(2) The second family of methods for the graphic presentation of 2D discrete functions consists of pseudo-3D presentations. In this case, the spatial locations x_k and y_l of our discrete points are no longer represented along the standard x and y axes of the paper surface. Instead, one tries to draw, using the rules of perspective, an oblique view of a 3D rigid model which corresponds to the given data, where the $z_{k,l}$ values are represented by heights along the model's z axis rather than by gray or colour shades. Here, too, there exist several possible variants: For example, one may represent the $z_{k,l}$ values by thin vertical lines (impulses) of the corresponding heights that emanate from the model's x,y plane, as in Fig. 1.4(b); by using 3D vertical boxes rather than vertical lines (like in a 3D histogram), as in Fig. 1.4(c); or by a continuous mesh or wireframe that represents the surface generated by the points $(x_k, y_l, z_{k,l})$ in the 3D space, as in Fig. 1.4(d). These variants are, in fact, simple 2D generalizations of methods that we have already discussed in the 1D case. Note, however, that the 2D generalization of the symbol plot (i.e. representing each point $(x_k, y_l, z_{k,l})$ by a single dot in the 3D space) is not really effective, since such a drawing does not clearly convey the 3D location of each point (see Fig. 1.4(a)).

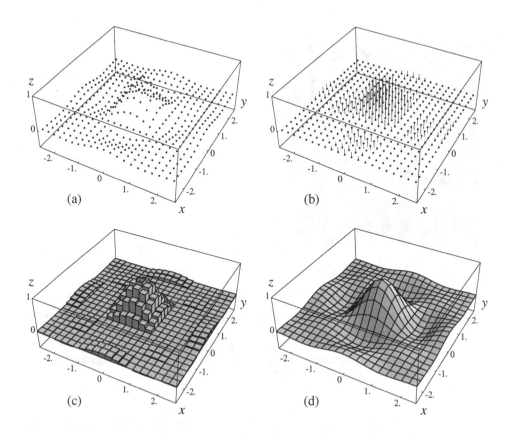

Figure 1.4: Comparison of various pseudo-3D presentation methods for plotting
2D discrete data, using the signal $z_{k,l} = \text{sinc}(x_k)\,\text{sinc}(y_l)$, $x_k = k/4$,
$y_l = l/4$, $k,l \in \mathbb{Z}$. (a) Symbol plot. (b) Vertical line plot (nailbed); note
that a dot has been added here at the *basis* of each vertical line, to
clearly show its location in the *x,y* plane as well as its vertical direction
(positive or negative). (c) Box plot. (d) Mesh or wireframe plot.

Pseudo-3D presentations have the advantage of being very intuitive, but on the other
hand they are limited to relatively small data sets, since beyond a certain element
density the plot may become too dark or even completely black. Moreover, they often
suffer from difficulties in showing the data in the "rear" side of the 3D model,
depending on the oblique viewing angle being used.

In the present book we usually represent 2D discrete data by simple gray-level density
plots consisting of juxtaposed square gray-level pixels. It should be stressed, however, as
already mentioned in point (1) above, that the attribution of gray levels to the $z_{k,l}$ values
may vary according to the case. Furthermore, as explained later in Sec. 8.7, when drawing

Fourier or DFT spectra it is commonplace to truncate the range of $z_{k,l}$ values being actually plotted, so that the entire range of gray levels can be used for representing in detail the most interesting part of the $z_{k,l}$ range.

1.5.3 Graphic presentations in the MD case

Proceeding now to higher dimensions, it is clear that discrete functions $z_k = g(x_k)$ of three or more dimensions cannot be directly plotted in a single drawing. Rather, they can be represented by two or more drawings showing various projections, i.e. partial dimensional representations of the MD data (just as we do, for example, when we plot a 2D complex-valued function by drawing separately its real-valued and its imaginary-valued parts).

1.6 About the exercises and the internet site

Having presented the graphic methods we use to depict discrete data, we conclude this introductory chapter with some further practical information about the present book.

First of all, for the benefit of the readers we have added at the end of each chapter a section containing a few problems and exercises. Some of these are not merely routine exercises, but rather intriguing or even challenging problems. Their aim is to aid the assimilation of the material covered by the chapter, but also to develop new insights beyond it. We therefore highly encourage readers to dedicate some time for reviewing these problems.

This book is also accompanied by an internet site, which can be found at the following address:

http://lspwww.epfl.ch/publications/books/dft/

This site will provide useful complementary information on the book, including various updates or even printable versions of some of the key figures. This site will also include a corrigendum section, where all the typos and errors to be discovered in this volume will be posted along with their respective corrections.

$$*\qquad*\qquad*$$

Finally, a word about our notations. Throughout this book we adopt the following notational conventions:

Sec. 3.2   —   Section 2 of Chapter 3.

Sec. A.2   —   Section 2 of Appendix A.

Fig. 3.2   —   Figure 2 of Chapter 3.

Fig. A.2   —   Figure 2 of Appendix A.

(3.2)   —   Equation or formula 2 of Chapter 3.

(A.2)   —   Equation or formula 2 of Appendix A.

Similar conventions are also used for enumerating tables, examples, propositions, remarks, etc.; for instance, Example 3.2 is the second example of Chapter 3.

The mathematical symbols and notations used in the present volume are listed at the end, following the list of the main relations. A glossary of the main terms is provided in Appendix E.

Chapter 2
Background and basic notions

2.1 Introduction

In this chapter we review the main definitions and notations that are being used in the literature for the continuous Fourier transform and for its discrete counterpart. Unfortunately, no standard definition exists for the Fourier transform, and one may find in the literature several slightly different variants. Although these variants are essentially equivalent in terms of the Fourier theory, each of them gives slightly different results, and one should be aware of this fact to avoid any possible confusion, especially when working with several textbooks, references or Fourier software packages.

We start in Sec. 2.2 with the continuous Fourier transform (CFT), and then we proceed in Sec. 2.3 to the discrete Fourier transform (DFT). In both cases we review some of the most commonly used definitions, and present the notations that will be used in the sequel. In Sec. 2.4 we list the basic rules which govern the CFT and the DFT, respectively, in the 1D, 2D and MD cases. Then, in Sec. 2.5 we briefly present the graphical development of the DFT, and in Sec. 2.6 we review the potential pitfalls that may await us when using the DFT as an approximation to the CFT, and which are the subject of the following chapters. Finally, in Sec. 2.7 we provide a first glimpse into cases having impulsive spectra.

2.2 The continuous Fourier transform: definitions and notations

Let us start with a brief reminder on the definition of the continuous Fourier transform. The continuous Fourier transform can be defined in several slightly different ways, all of which are in use in the literature (see, for example, [Bracewell86 pp. 6–8; Gaskill78 pp. 181–182; Brigham88 pp. 22–23; Champeney73 pp. 8–10; Cartwright90 p. 98; Kammler07 pp. 63–64]). According to the common convention used in optics, the Fourier transform of a possibly complex-valued function $g(x)$ of a real variable x is defined by:

$$G(u) = \int_{-\infty}^{\infty} g(x)\, e^{-i2\pi ux}\, dx \qquad\qquad -\infty < u < \infty \qquad\qquad (2.1)$$

where $G(u)$, too, is a possibly complex-valued function of a real variable u. The inverse transform is given by:

$$g(x) = \int_{-\infty}^{\infty} G(u)\, e^{i2\pi ux}\, du \qquad\qquad -\infty < x < \infty \qquad\qquad (2.2)$$

Here, the variable x may represent a quantity such as distance or time, and it is measured accordingly in terms of units such as meters, seconds, etc. The variable u represents frequency, and it is measured in terms of cycles per unit of x, such as cycles per meter, cycles per second (also called Hertz), etc. In the following we will often use "Hertz" (Hz) as a generic frequency unit, i.e. as a shorthand notation for "cycles per unit of x".

Remark 2.1: It is important to note that in order for the Fourier transform and its inverse to exist, i.e. in order that the integrals converge, the functions in question must satisfy some conditions; for example, they must be absolutely integrable on the entire range of integration. Nevertheless, it turns out that a large number of functions encountered in engineering problems do not strictly satisfy these conditions, and they possess Fourier transforms only in a generalized sense (we may mention, for example, the sine and cosine functions, the step function, etc.). Following the practice adopted by [Bracewell86], [Gaskill78], and many other authors, such generalized Fourier transforms are manipulated in the same fashion as ordinary transforms, and we make no distinction between the two: It is to be understood that the term "Fourier transform" means "generalized Fourier transform" if we are dealing with functions that violate the existing conditions (see, for example, [Gaskill78 p. 184] or [Bracewell86 pp. 8–13]). In particular, periodic functions are treated like any other function, with the understanding that their continuous-world spectrum consists, in fact, of impulses (whose amplitudes reflect the coefficients of the Fourier series decomposition). A rigorous mathematical discussion on issues such as convergence conditions, generalized Fourier transforms, etc. is beyond the scope of this book; the interested readers can find more detailed information in references such as [Kammler07 pp. 48–58 and Chapter 7], [Champeney87 Chapters 6–9], [Champeney73 Sec. 2.3], [Papoulis62 Chapter 2], etc. ∎

In an alternative definition, the factor 2π appearing in the exponential part is lumped together with u, giving the following definition of the Fourier transform and of its inverse:

$$G(\omega) = \int_{-\infty}^{\infty} g(x)\, e^{-i\omega x}\, dx \qquad\qquad -\infty < \omega < \infty \qquad\qquad (2.3)$$

$$g(x) = \frac{1}{2\pi} \int_{-\infty}^{\infty} G(\omega)\, e^{i\omega x}\, d\omega \qquad\qquad -\infty < x < \infty \qquad\qquad (2.4)$$

For the sake of symmetry, some authors prefer to split the constant factor $\frac{1}{2\pi}$ of Eq. (2.4) and to distribute it equally between the direct and the inverse transforms, giving:

$$G(\omega) = \frac{1}{\sqrt{2\pi}} \int_{-\infty}^{\infty} g(x)\, e^{-i\omega x}\, dx \qquad\qquad -\infty < \omega < \infty \qquad\qquad (2.5)$$

$$g(x) = \frac{1}{\sqrt{2\pi}} \int_{-\infty}^{\infty} G(\omega)\, e^{i\omega x}\, d\omega \qquad\qquad -\infty < x < \infty \qquad\qquad (2.6)$$

Note that while in the definition given by Eqs. (2.1)–(2.2) u stands for the spectral frequency measured in cycles per unit of x (meter, second, etc.), in the definitions given by Eqs. (2.3)–(2.4) or (2.5)–(2.6) ω is the angular frequency $\omega = 2\pi u$ measured in radians per unit of x [Gaskill78 p. 182; Cartwright90 p. 98].

Finally, each of these definitions has also a variant in which the signs in the exponential part of the direct transform and of its inverse are interchanged.

It is clear that each of these definitions gives for the same function $g(x)$ a slightly different Fourier transform; but all of the definitions work equally well — provided that one always sticks to the same definition. Conversion rules from one convention to another can be found, for example, in [Bracewell86 p. 7] or in [Champeney73 Sec. 2.2]. In the present work we will use the definition provided by Eqs. (2.1)–(2.2), unless explicitly mentioned otherwise.

All of these definitions can be also extended to the multidemensional case (see, for example, [Champeney73 pp. 40–41]). Thus, in the 2D case Eqs. (2.1)–(2.2) become:

$$G(u,v) = \int_{-\infty}^{\infty} \int_{-\infty}^{\infty} g(x,y)\, e^{-i2\pi(ux+vy)}\, dx\, dy \qquad -\infty < u,v < \infty \qquad (2.7)$$

$$g(x,y) = \int_{-\infty}^{\infty} \int_{-\infty}^{\infty} G(u,v)\, e^{i2\pi(ux+vy)}\, du\, dv \qquad -\infty < x,y < \infty \qquad (2.8)$$

and in the general MD case they become (see, for example, [Marks09 Sec. 8.4]):

$$G(\mathbf{u}) = \int_{-\infty}^{\infty} g(\mathbf{x})\, e^{-i2\pi \mathbf{u}\cdot\mathbf{x}}\, d\mathbf{x} \qquad\qquad (2.9)$$

$$g(\mathbf{x}) = \int_{-\infty}^{\infty} G(\mathbf{u})\, e^{i2\pi \mathbf{u}\cdot\mathbf{x}}\, d\mathbf{u} \qquad\qquad (2.10)$$

where $\mathbf{x}, \mathbf{u} \in \mathbb{R}^M$, $\mathbf{u}\cdot\mathbf{x}$ is their scalar product, and the integration between $-\infty$ and ∞ is done along each of the M dimensions.

2.3 The discrete Fourier transform: definitions and notations

Returning now to the discrete case, the DFT and the inverse DFT (IDFT) are often defined as follows [Nussbaumer82 p. 81]:[1]

$$G_n = \sum_{k=0}^{N-1} g_k\, e^{-i2\pi kn/N} \qquad\qquad n = 0, \dots, N-1 \qquad (2.11)$$

$$g_k = \frac{1}{N} \sum_{n=0}^{N-1} G_n\, e^{i2\pi kn/N} \qquad\qquad k = 0, \dots, N-1 \qquad (2.12)$$

where g_k and G_n are sequences of N possibly complex-valued numbers. Note that unlike their continuous counterparts, the discrete definitions are valid in all cases, and they are not subject to any particular conditions [Kammler07 p. 38].[2]

But just as in the continuous Fourier transform, in the discrete case, too, many slightly different definitions can be found in the literature. For example, in some references (such

---

[1] The constant factor $\frac{1}{N}$ is required for the sake of reversibility, i.e. in order to get back by the IDFT exactly the original input array that was fed to the DFT [Nussbaumer pp. 80–81], [Smith03 p. 577]. The reason it appears can be also understood from Remark 2.2 below.

[2] This does not yet mean, however, that Eqs. (2.11)–(2.12) can approximate the continuous Fourier transform in all cases, with no conditions on the original underlying function $g(x)$. We will return to this question in Sec. 2.6 below.

as [Bracewell86 p. 358] or [Kammler07 p. 196]) the constant $\frac{1}{N}$ appears in the DFT rather than in the IDFT, while in other references it is replaced in a more symmetric way by the constant $\frac{1}{\sqrt{N}}$ in both DFT and IDFT (like in [Smith07 p. 109]). In each of these variants the signs in the exponential parts of the DFT and of the IDFT can be also interchanged. And even the range of the summation may be defined in various ways (for example, in [Bracewell86 p. 362] all indices vary between $-N/2$ and $N/2{-}1$, while in [Briggs95 p. 23] the indices vary between $-N/2{+}1$ and $N/2$).[3]

Also, in order to stress the connection between the discrete transforms and their continuous counterparts,[4] many authors prefer to use $g(k)$ and $G(n)$, i.e. the discrete (sampled) version of the continuous functions $g(x)$ and $G(u)$, rather than g_k and G_n:

$$G(n) = \sum_{k=0}^{N-1} g(k)\, e^{-i2\pi kn/N} \qquad\qquad n = 0, \dots, N{-}1 \qquad\qquad (2.13)$$

$$g(k) = \frac{1}{N}\sum_{n=0}^{N-1} G(n)\, e^{i2\pi kn/N} \qquad\qquad k = 0, \dots, N{-}1 \qquad\qquad (2.14)$$

The input step of the DFT (i.e. the sampling interval Δx) and the frequency step of the DFT, Δu, may be explicitly inserted into the definition, as follows:

$$G(n\Delta u) = \sum_{k=0}^{N-1} g(k\Delta x)\, e^{-i2\pi kn/N} \qquad\qquad n = 0, \dots, N{-}1$$

$$g(k\Delta x) = \frac{1}{N}\sum_{n=0}^{N-1} G(n\Delta u)\, e^{i2\pi kn/N} \qquad\qquad k = 0, \dots, N{-}1$$

And if we note that the frequency step Δu of the DFT output is given by $\Delta u = \frac{1}{N\Delta x}$ (see Eq. (4.4)), we obtain, like in [Brigham88 p. 97]:

$$G(\tfrac{n}{N\Delta x}) = \sum_{k=0}^{N-1} g(k\Delta x)\, e^{-i2\pi kn/N} \qquad\qquad n = 0, \dots, N{-}1 \qquad\qquad (2.15)$$

$$g(k\Delta x) = \frac{1}{N}\sum_{n=0}^{N-1} G(\tfrac{n}{N\Delta x})\, e^{i2\pi kn/N} \qquad\qquad k = 0, \dots, N{-}1 \qquad\qquad (2.16)$$

In the two last equation pairs the integer k in $g(\ )$ has been multiplied by the step Δx (the sampling interval), while in $G(\ )$ the index n has been multiplied by the corresponding frequency step $\Delta u = \frac{1}{N\Delta x}$. The advantage in doing so is that in this notation the discrete abscissa of $g(\ )$, $k\Delta x$, is no longer a simple integer index, but consists of true sampling points along the axis, measured in terms of the true units (seconds, meters, etc.); and similarly, the discrete abscissa of $G(\ )$, $n\Delta u$ or equivalently $\frac{n}{N\Delta x}$, is not merely an integer index, but indeed consists of frequencies measured in cycles per second, meter, etc. [Bracewell86 p. 362].[5] This notation may be less convenient, however, in other circumstances, for example when expressing the discrete counterparts of the classical Fourier theorems; and indeed, when discussing the DFT theorems $k\Delta x$ and $\frac{n}{N\Delta x}$ are usually replaced by k and n for convenience of notation (see [Brigham88 p. 107]).

---

[3] On the various possible indexing methods, their equivalence, and the reasons for choosing one method rather than another, see for example [Briggs95 p. 24 and Sec. 3.1].
[4] This assumed connection will be justified in Remark 2.2 and Secs. 2.5, 2.6 below.
[5] A fuller and more precise discussion on this subject is provided in Sec. 4.3; see in particular Eq. (4.11) and the paragraph which follows it.

If we further denote by $x_k = k\Delta x$ the k-th sampling point and by $u_n = n\Delta u = \frac{n}{N\Delta x}$ the n-th frequency sample measured in cycles per unit we obtain yet another variant:

$$G(u_n) = \sum_{k=0}^{N-1} g(x_k)\, e^{-i2\pi x_k u_n} \qquad\qquad n = 0, \dots ,N-1 \qquad\qquad (2.17)$$

$$g(x_k) = \frac{1}{N}\sum_{n=0}^{N-1} G(u_n)\, e^{i2\pi x_k u_n} \qquad\qquad k = 0, \dots ,N-1 \qquad\qquad (2.18)$$

In some other references (such as [Smith07 p. 2]) the factor 2π appearing in the exponential part is lumped together with u_n, giving the following definition of the DFT and of its inverse (where $x_k = k\Delta x$ is the k-th sampling point and $\omega_n = \frac{2\pi n}{N\Delta x}$ is the n-th frequency sample measured in radians per unit):

$$G(\omega_n) = \sum_{k=0}^{N-1} g(x_k)\, e^{-ix_k \omega_n} \qquad\qquad n = 0, \dots ,N-1 \qquad\qquad (2.19)$$

$$g(x_k) = \frac{1}{N}\sum_{n=0}^{N-1} G(\omega_n)\, e^{ix_k \omega_n} \qquad\qquad k = 0, \dots ,N-1 \qquad\qquad (2.20)$$

Obviously, just as in the continuous case, each of these definitions gives for the same input a slightly different DFT; but all of the definitions work equally well — provided that one always sticks to the same definition. Conversion rules from one convention to another can be easily established, like in the continuous case.

In the present work we will usually use the definition provided by Eqs. (2.13)–(2.14), unless mentioned otherwise. Note, however, that this notation does not explicitly distinguish between the original continuous functions g, G and their discrete counterparts. If the distinction is not sufficiently clear from the context, for example when it is desired to compare the value of the original function G to the value of its DFT approximation at the same point, we will denote the discrete counterparts of the functions g, G by $\tilde{g}$, $\tilde{G}$, like in [Brigham88 p. 95]. Alternatively, one may use the sequence notation g_k, G_n as in Eqs. (2.11)–(2.12), like in [Briggs95 pp. 22–23, xv]. Yet another possibility consists of using square brackets for the discrete case: $g[k]$, $G[n]$, like in [Kammler07 p. 5].

Remark 2.2: Note that the DFT can be obtained from the continuous Fourier transform as a numerical approximation to the integrals in Eqs. (2.1)–(2.2), using the rectangular rule for approximating integrals (see Sec. D.9 in Appendix D or [Press02 p. 508]):[6]

$$G(n\Delta u) \approx \Delta x \sum_{k=-N/2}^{N/2-1} g(k\Delta x)\, e^{-i2\pi k \Delta x n \Delta u} \qquad n = -\frac{N}{2}, \dots ,\frac{N}{2}-1 \qquad (2.21)$$

$$g(k\Delta x) \approx \Delta u \sum_{n=-N/2}^{N/2-1} G(n\Delta u)\, e^{i2\pi k \Delta x n \Delta u} \qquad k = -\frac{N}{2}, \dots ,\frac{N}{2}-1 \qquad (2.22)$$

Because $\Delta u = \frac{1}{N\Delta x}$ (as we will see below in Eq. (4.4)), this means, indeed, using Eqs. (2.15)–(2.16), that the DFT and the IDFT approximate their continuous counterparts up to a constant factor of Δx or $\frac{1}{\Delta x}$, respectively:[6]

---

[6] The choice of the index ranges here is explained in Secs. 2.5 and D.9 (see in particular Remark D.2). As we will see, due to the N-element periodicity of the input and output arrays of the DFT and IDFT, all index ranges of N consecutive integers are equivalent, so once the best match with the underlying CFT is obtained we can change the index ranges and choose the most convenient ones depending on the case.

$$G(n\Delta u) \approx \Delta x\, \widetilde{G}(n\Delta u), \quad \text{where} \quad \widetilde{G}(n\Delta u) = \sum_{k=-N/2}^{N/2-1} g(k\Delta x)\, e^{-i2\pi kn/N} \qquad n = -\frac{N}{2}, \dots, \frac{N}{2}-1$$

$$g(k\Delta x) \approx \frac{1}{\Delta x}\, \widetilde{g}(k\Delta x), \quad \text{where} \quad \widetilde{g}(k\Delta x) = \frac{1}{N} \sum_{n=-N/2}^{N/2-1} G(n\Delta u)\, e^{i2\pi kn/N} \qquad k = -\frac{N}{2}, \dots, \frac{N}{2}-1 \quad \blacksquare$$

Remark 2.3: For the sake of completeness, it should be mentioned here that the DFT can also be represented in an elegant way using matrix notation. Since its input and output are arrays of N elements (real or complex numbers), they can be considered as N-element vectors; and thanks to the linearity property of the DFT (see rule 1 in Sec. 2.4.3 below), it can be represented as a product of such an N-element vector and an $N\times N$ matrix. The matrix representation of the DFT is explained, for example, in [Briggs95 pp. 31–33] or in [Smith07 Sec. 6.12]. This matrix representation gives a new algebraic insight into the properties of the DFT as a linear operator, but we will not need it in the present book. Note that just like the explicit DFT formulas that we have seen above, the matrix representation, too, may be subject to some slight form variations. ■

All of the variants of the DFT definition can be easily extended to the multidimensional case, too (see, for example, [Rosenfeld82 pp. 21–22], [Brigham88 Sec. 11.2] or [Bracewell95 pp. 167–168]). For example, in the 2D case Eqs. (2.13)–(2.14) become:[7]

$$G(m,n) = \sum_{k=0}^{N-1}\sum_{l=0}^{N-1} g(k,l)\, e^{-i2\pi(mk+nl)/N} \qquad m, n = 0, \dots, N-1 \qquad (2.23)$$

$$g(k,l) = \left(\frac{1}{N}\right)^2 \sum_{m=0}^{N-1}\sum_{n=0}^{N-1} G(m,n)\, e^{i2\pi(mk+nl)/N} \qquad k, l = 0, \dots, N-1 \qquad (2.24)$$

and in the general MD case they become (see, for example, [Marks09 Sec. 8.6.3]):

$$G(\mathbf{n}) = \sum_{\mathbf{k}} g(\mathbf{k})\, e^{-i2\pi \mathbf{n}\cdot\mathbf{k}/N} \qquad n_0, \dots, n_{M-1} = 0, \dots, N-1 \qquad (2.25)$$

$$g(\mathbf{k}) = \left(\frac{1}{N}\right)^M \sum_{\mathbf{n}} G(\mathbf{n})\, e^{i2\pi \mathbf{n}\cdot\mathbf{k}/N} \qquad k_0, \dots, k_{M-1} = 0, \dots, N-1 \qquad (2.26)$$

where $\mathbf{k} = (k_0, \dots, k_{M-1})$, $\mathbf{n} = (n_0, \dots, n_{M-1})$, $\mathbf{n}\cdot\mathbf{k}$ is their scalar product, and the summations go between $0, \dots, N-1$ for each element of $\mathbf{k}$ or $\mathbf{n}$, respectively. More generally, if each dimension j has a different length (number of elements) N_j, then the MD case becomes:

$$G(\mathbf{n}) = \sum_{\mathbf{k}} g(\mathbf{k})\, e^{-i2\pi \mathbf{k}\cdot(\mathbf{n}/\mathbf{N})} \qquad n_j = 0, \dots, N_j-1, \quad j = 0, \dots, M-1 \qquad (2.27)$$

$$g(\mathbf{k}) = \frac{1}{N_0 \cdots N_{M-1}} \sum_{\mathbf{n}} G(\mathbf{n})\, e^{i2\pi \mathbf{n}\cdot(\mathbf{k}/\mathbf{N})} \qquad k_j = 0, \dots, N_j-1, \quad j = 0, \dots, M-1 \qquad (2.28)$$

where the summations go between $0, \dots, N_j-1$ for each element of $\mathbf{k}$ or $\mathbf{n}$, $\mathbf{N} = (N_0, \dots, N_{M-1})$, and the vector division is performed element-wise: $\mathbf{v}/\mathbf{w} = (v_0/w_0, \dots, v_{M-1}/w_{M-1})$.

Finally, for the benefit of the DFT users, we list below the DFT and IDFT definitions as they are implemented in some widely used mathematical software packages. We only give

---

[7] For the sake of simplicity we assume here that the number of elements is N along each of the dimensions. But the definitions given here can be also extended to the more general case where the number of elements along each dimension is different, as shown below for the MD case.

here the 1D forms, the MD definitions being straightforward extensions as shown above. For the sake of clarity, we stick here to the same letter conventions as above, even if the letters being used in the respective package documentations are different than ours.

• In Mathematica® [Wolfram96 p. 868] the definitions being used are:

$$G_n = \frac{1}{\sqrt{N}} \sum_{k=1}^{N} g_k \, e^{i2\pi(k-1)(n-1)/N} \qquad\qquad n = 1, \dots, N \qquad\qquad (2.29)$$

$$g_k = \frac{1}{\sqrt{N}} \sum_{n=1}^{N} G_n \, e^{-i2\pi(k-1)(n-1)/N} \qquad\qquad k = 1, \dots, N \qquad\qquad (2.30)$$

• In Matlab® [Matlab02 p. 1-48] the definitions are:

$$G(n+1) = \sum_{k=0}^{N-1} g(k+1) \, e^{-i2\pi kn/N} \qquad\qquad n = 0, \dots, N-1 \qquad\qquad (2.31)$$

$$g(k+1) = \frac{1}{N} \sum_{n=0}^{N-1} G(n+1) \, e^{i2\pi kn/N} \qquad\qquad k = 0, \dots, N-1 \qquad\qquad (2.32)$$

• In Maple® [Maplesoft] the definitions used by default are:

$$G_n = \frac{1}{\sqrt{N}} \sum_{k=1}^{N} g_k \, e^{-i2\pi(k-1)(n-1)/N} \qquad\qquad n = 1, \dots, N \qquad\qquad (2.33)$$

$$g_k = \frac{1}{\sqrt{N}} \sum_{n=1}^{N} G_n \, e^{i2\pi(k-1)(n-1)/N} \qquad\qquad k = 1, \dots, N \qquad\qquad (2.34)$$

The reason for the index shift of +1 or −1 in the respective cases is that the index range associated with an array in these software packages is $1,\dots,N$ while the DFT indices are supposed to run between $0,\dots,N-1$.

2.4 Rules for deriving new Fourier transforms from already known ones

The most straightforward way to find the Fourier transform of a given function is by using directly the definition formula, for example Eq. (2.1). However, in many cases the computation of the integral is quite tedious. Fortunately, things may become much simpler thanks to some general rules which tell us how we may obtain new Fourier transforms from already known ones or from their combinations. These rules are well known in the literature and they can be found along with their proofs in virtually any book on the Fourier theory, usually under the name of *Fourier transform properties* or *Fourier transform theorems*. Many of these rules subsist also in the discrete case, while others must be adapted accordingly; but there are also rules which only apply to the continuous case and rules which only apply to the discrete case. For the sake of convenience we provide here (without proofs) a list of some of the most useful rules that we have collected from various sources as cited below. We start with the continuous 1D case, then we proceed to the more interesting 2D and MD cases, and finally we give their 1D, 2D and MD counterparts for the discrete case. Note, however, that these lists are non-exhaustive, and further rules can be found in the cited references. Whenever required, we have added remarks to draw the reader's attention to some particularities or potential sources of error.

2.4.1 Rules for the 1D continuous Fourier transform

Assume that the functions $g(x)$ and $h(x)$ have the Fourier transforms $G(u)$ and $H(u)$, respectively, and that a, b and f are real numbers. Then we have the following general rules (see, for example, [Bracewell86 Chapter 6], [Gaskill78 Sec. 7.3], [Brigham88 Chapter 3] and [Chu08 pp. 175–180]):

Rule name	Function	Fourier transform	Remarks
1. Linearity	$ag(x) + bh(x)$	$aG(u) + bH(u)$	
1a. Scaling	$ag(x)$	$aG(u)$	
1b. Addition	$g(x) + h(x)$	$G(u) + H(u)$	
2. Dilation	$g(ax)$	$\frac{1}{\|a\|}G(u/a)$	$a \neq 0$
2a. Frequency dilation	$\frac{1}{\|a\|}g(x/a)$	$G(au)$	$a \neq 0$
3. Reflection	$g(-x)$	$G(-u)$	
4. Inversion	$G(x)$	$g(-u)$	
5. Complex conjugation	$g^*(x)$	$G^*(-u)$	
6. Shift (or translation)	$g(x - a)$	$e^{-i2\pi au}\, G(u)$	
6a. Frequency shift	$e^{i2\pi fx}\, g(x)$	$G(u - f)$	
7. Convolution	$g(x)*h(x)$	$G(u)H(u)$	
7a. Product	$g(x)h(x)$	$G(u)*H(u)$	
8. Cross correlation	$g(x)\star h(x)$ $= g(x)*h(-x)$	$G(u)H(-u)$	(a)
8a. Autocorrelation	$g(x)\star g(x)$ $= g(x)*g(-x)$	$\|G(u)\|^2$	
9. Real part	$\mathrm{Re}[g(x)]$	$\frac{1}{2}G(u) + \frac{1}{2}G^*(-u)$	
9a. Imaginary part	$\mathrm{Im}[g(x)]$	$\frac{1}{2}G(u) - \frac{1}{2}G^*(-u)$	

Remarks:

(a) The cross-correlation rule deserves particular attention in order to avoid confusion. First of all, it should be remembered that cross correlation is not commutative, meaning that $h(x) \star g(x)$ generally differs from $g(x) \star h(x)$. There also exists in the literature an alternative convention for the definition of cross correlation, in which the roles of g and h are inversed (see, for example, [Coulon84 pp. 81–82]): $g(x) \star h(x) = g(-x) * h(x)$. In this case the resulting Fourier transform becomes $G(-u)H(u)$. Unfortunately, these two conventions are not compatible, and one should be aware of this source of confusion when using different references or software packages. Furthermore, in the case of *complex-valued* functions $g(x)$ and $h(x)$ one can also define their *complex cross correlation* to be: $g(x) \star h(x) = g(x) * h^*(-x)$, in which case the resulting Fourier transform is $G(u)H^*(u)$. Note that if $h(x)$ is a real-valued function then $H(u)$ is Hermitian [Bracewell86 p. 15] and therefore $H^*(u) = H(-u)$, so that the complex cross correlation indeed reduces into real-valued cross correlation. But in the complex case, too, some authors use the alternative convention in which the roles of g and h are inversed (see, for example, [Weisstein99 p. 352]): $g(x) \star h(x) = g^*(-x) * h(x)$; in this case the Fourier transform becomes $G^*(u)H(u)$. For further details see, for example, [Gaskill78 pp. 172–173, 200], [Amidror07 pp. 413–414].

Note that the rules given above are adapted to the Fourier transform as defined in Eqs. (2.1)–(2.2); but as one could expect, some of these rules may undergo slight changes when other conventions are being used. For example, the rules which correspond to the definition based on angular frequency $\omega = 2\pi u$ (Eqs. (2.5)–(2.6)) can be found in references such as [Chu08 Sec. 5.7], [Poularikas96 Sec. 2.2] or [Champeney73 Sec. 2.4]. Moreover, the names of the rules may also vary between different sources; for example, our *dilation* rule [Kammler07 p. 138] is called *similarity* in [Bracewell86 p. 101], while in [Brigham88 pp. 32–33] it is called *scaling*.

A similar list of rules can be also formulated for the ICFT. For example, it follows from rules 7 and 7a above that the ICFT of $G(u)*H(u)$ is $g(x)h(x)$, and that the ICFT of $G(u)H(u)$ is $g(x)*h(x)$. If one uses an alternative definition of the CFT such as the one based on angular frequency, the ICFT rules, too, should be adapted accordingly.

2.4.2 Rules for the 2D continuous Fourier transform

Assume that the functions $g(x,y)$ and $h(x,y)$ have the Fourier transforms $G(u,v)$ and $H(u,v)$, respectively, and that a, b, f_1 and f_2 are real numbers. Then we have the following rules (see, for example, [Bracewell95 pp. 154–166], [Gaskill78 Sec. 9.3]):

Rule name	Function	Fourier transform	Remarks
1. Linearity	$ag(x,y) + bh(x,y)$	$aG(u,v) + bH(u,v)$	
1a. Scaling	$ag(x,y)$	$aG(u,v)$	
1b. Addition	$g(x,y) + h(x,y)$	$G(u,v) + H(u,v)$	
2. Dilation	$g(ax,by)$	$\frac{1}{|ab|}G(u/a,v/b)$	$a,b \neq 0$
2a. Frequency dilation	$\frac{1}{|ab|}g(x/a,y/b)$	$G(au,bv)$	$a,b \neq 0$

3. Reflection	$g(-x,-y)$	$G(-u,-v)$
4. Inversion	$G(x,y)$	$g(-u,-v)$
5. Complex conjugation	$g^*(x,y)$	$G^*(-u,-v)$
6. Shift (or translation)	$g(x-a, y-b)$	$e^{-i2\pi(au+bv)}G(u,v)$
6a. Frequency shift	$e^{i2\pi(f_1 x+f_2 y)}g(x,y)$	$G(u-f_1, v-f_2)$
7. Convolution	$g(x,y)**h(x,y)$	$G(u,v)H(u,v)$
7a. Product	$g(x,y)h(x,y)$	$G(u,v)**H(u,v)$
8. Cross correlation	$g(x,y)\star\star h(x,y)$ $= g(x,y)**h(-x,-y)$	$G(u,v)H(-u,-v)$ (a)
8a. Autocorrelation	$g(x,y)\star\star g(x,y)$ $= g(x,y)**g(-x,-y)$	$\|G(u,v)\|^2$
9. Real part	$\mathrm{Re}[g(x,y)]$	$\tfrac{1}{2}G(u,v)+\tfrac{1}{2}G^*(-u,-v)$
9a. Imaginary part	$\mathrm{Im}[g(x,y)]$	$\tfrac{1}{2}G(u,v)-\tfrac{1}{2}G^*(-u,-v)$
10. Separable product	$g(x)h(y)$	$G(u)H(v)$ (b),(c)
11. Rotation	$g(x\cos\alpha-y\sin\alpha,$ $x\sin\alpha+y\cos\alpha)$	$G(u\cos\alpha-v\sin\alpha,$ $u\sin\alpha+v\cos\alpha)$ (b)
12. Shear	$g(x+by, y)$	$G(u, v-bu)$ (b)

Remarks:

(a) The same remark about cross correlation as in the 1D case applies here, too.

(b) These 2D rules have no non-trivial 1D counterparts.

(c) See an example in Sec. D.10 of Appendix D.

As in the 1D case, the rules given above are adapted to the Fourier transform as defined in Eqs. (2.7)–(2.8); when using other conventions some of the rules may undergo respective minor changes.

Further rules for the 2D case (such as the compound shear rule, the affine rule, etc.) can be found in [Bracewell95 pp. 158–166] and [Gaskill78 Sec. 9-3]. Note that the 2D case is indeed richer and more interesting than its 1D counterpart: Not only it allows for more flexibility with the 1D rules, which can be now applied along any of the axes or along both; but it also offers new 2D rules which have no equivalents in the 1D case, like rules 10–12 above.

2.4.3 Rules for the MD continuous Fourier transform

The generalization of the 2D rules to the MD case is usually straightforward. Assume that the functions $g(\mathbf{x})$ and $h(\mathbf{x})$ have the Fourier transforms $G(\mathbf{u})$ and $H(\mathbf{u})$, respectively, where $\mathbf{x} = (x_0, \dots, x_{M-1})$, $\mathbf{u} = (u_0, \dots, u_{M-1}) \in \mathbb{R}^M$; that a, b are real numbers; that $\mathbf{a} = (a_0, \dots, a_{M-1})$, $\mathbf{f} = (f_0, \dots, f_{M-1}) \in \mathbb{R}^M$; and that $\mathbf{v} \cdot \mathbf{w}$ is the scalar product of the two vectors in question. Then we have the following rules (see [Marks09, Sec. 8.4; Stein71, Chapter 1]):

Rule name	Function	Fourier transform	Remarks
1. Linearity	$ag(\mathbf{x}) + bh(\mathbf{x})$	$aG(\mathbf{u}) + bH(\mathbf{u})$	
1a. Scaling	$ag(\mathbf{x})$	$aG(\mathbf{u})$	
1b. Addition	$g(\mathbf{x}) + h(\mathbf{x})$	$G(\mathbf{u}) + H(\mathbf{u})$	
2. Dilation	$g(a_0 x_0, \dots, a_{M-1} x_{M-1})$	$\frac{1}{\|a_0 \cdots a_{M-1}\|} G(u_0/a_0, \dots, u_{M-1}/a_{M-1})$	
2a. Frequency dilation	$\frac{1}{\|a_0 \cdots a_{M-1}\|} g(x_0/a_0, \dots, x_{M-1}/a_{M-1})$	$G(a_0 u_0, \dots, a_{M-1} u_{M-1})$	
3. Reflection	$g(-\mathbf{x})$	$G(-\mathbf{u})$	
4. Inversion	$G(\mathbf{x})$	$g(-\mathbf{u})$	
5. Complex conjugation	$g^*(\mathbf{x})$	$G^*(-\mathbf{u})$	
6. Shift (or translation)	$g(\mathbf{x} - \mathbf{a})$	$e^{-i2\pi \mathbf{a} \cdot \mathbf{u}} G(\mathbf{u})$	
6a. Frequency shift	$e^{i2\pi \mathbf{f} \cdot \mathbf{x}} g(\mathbf{x})$	$G(\mathbf{u} - \mathbf{f})$	
7. Convolution	$g(\mathbf{x}) * h(\mathbf{x})$	$G(\mathbf{u})H(\mathbf{u})$	
7a. Product	$g(\mathbf{x})h(\mathbf{x})$	$G(\mathbf{u}) * H(\mathbf{u})$	
8. Cross correlation	$g(\mathbf{x}) \star h(\mathbf{x})$ $= g(\mathbf{x}) * h(-\mathbf{x})$	$G(\mathbf{u})H(-\mathbf{u})$	(a)
8a. Autocorrelation	$g(\mathbf{x}) \star g(\mathbf{x})$ $= g(\mathbf{x}) * g(-\mathbf{x})$	$\|G(\mathbf{u})\|^2$	
9. Real part	$\mathrm{Re}[g(\mathbf{x})]$	$\frac{1}{2}G(\mathbf{u}) + \frac{1}{2}G^*(-\mathbf{u})$	
9a. Imaginary part	$\mathrm{Im}[g(\mathbf{x})]$	$\frac{1}{2}G(\mathbf{u}) - \frac{1}{2}G^*(-\mathbf{u})$	
10. Separable product	$g(x_0, \dots, x_{j-1})h(x_j, \dots, x_{M-1})$	$G(u_0, \dots, u_{j-1})H(u_j, \dots, u_{M-1})$	
10a. Separable products	$g_0(x_0)\dots g_{M-1}(x_{M-1})$	$G_0(u_0)\dots G_{M-1}(u_{M-1})$	(b)
11. Affine	$g(A\mathbf{x})$	$\frac{1}{\|A\|}G(A^{-T}\mathbf{u})$	(c)

Remarks:

(a) The same remark about cross correlation as in the 1D case applies here, too.

(b) See an example in Sec. D.10 of Appendix D.

(c) This rule is an MD generalization of the 2D rules 2, 11, 12 and of all similar 2D rules having the form $g(a_1x+b_1y, a_2x+b_2y)$. Here, A is an $M \times M$ matrix, $|A|$ is its determinant, and A^{-T} is the transpose of the inverse matrix. Of course, $\mathbf{x}$ and $\mathbf{u}$ should be treated here as column vectors in order for the products $A\mathbf{x}$ and $A^{-T}\mathbf{u}$ to make sense.

Note that the MD rules given above are adapted to the Fourier transform as defined in Eqs. (2.9)–(2.10). The MD rules which correspond to the definition based on angular frequency $\omega = 2\pi u$ can be found, for example, in [Cartwright90 pp. 115–117], [Poularikas96 pp. 149–155] or [Champeney73 Sec. 3.6].

2.4.4 Rules for the 1D discrete Fourier transform

Many of the rules of the 1D continuous Fourier transform subsist in the discrete case, too; but some of them must be slightly amended or even completely redefined in order to be adapted to the discrete case. For example, the convolution, cross correlation and autocorrelation rules may remain unchanged if we agree to replace the respective operations by their cyclical discrete counterparts (see, for example, [Brigg95 pp. 78–89]). Other rules, such as the dilation rule, must be entirely redefined; but there also exist new rules which have no equivalent in the continuous case, and vice versa (see Problem 2-5).

Note that just as in the continuous case, the discrete rules may depend on the choice of the DFT definition. The rules presented in the list below are adapted to the convention used in Eqs. (2.13)–(2.14); but when using the convention in which the constant $\frac{1}{N}$ appears in the DFT definition rather than in the IDFT definition, like in [Bracewell86 p. 358] or [Kammler07 p. 196], some of the rules must be modified by factor N. When the rules obtained in this alternative convention are different, we give them in the "Alt. DFT" column. Note that a third convention also exists, which uses $\frac{1}{\sqrt{N}}$ in both DFT and IDFT definitions [Smith07 p. 109]. In this case, too, an appropriate change should be made in those rules which depend on the DFT definition.

Assume that the sequences $g(k)$ and $h(k)$ have the discrete Fourier transforms $G(n)$ and $H(n)$, respectively, that a and b are real numbers, and that j is an integer. Then we have the following general rules (see, for example, [Brigham88 Sec. 6.5], [Smith07 Sec. 7.4], [Briggs95 Secs. 3.2–3.3]):

Rule name	Function	DFT	Alt. DFT	Remarks
1. Linearity	$ag(k) + bh(k)$	$aG(n) + bH(n)$	Same	
1a. Scaling	$ag(k)$	$aG(n)$	Same	
1b. Addition	$g(k) + h(k)$	$G(n) + H(n)$	Same	

2. Dilation	Complicated; see, for example, [Kammler07 pp. 205–208]		
3. Reflection	$g(-k)$	$G(-n)$	Same
4. Inversion	$G(k)$	$Ng(-n)$	$\frac{1}{N}g(-n)$
5. Complex conjugation	$g^*(k)$	$G^*(-n)$	Same
6. Shift (or translation)	$g(k-j)$	$e^{-i2\pi jn/N}G(n)$	Same
6a. Frequency shift	$e^{i2\pi jk/N}g(k)$	$G(n-j)$	Same
7. Cyclic convolution	$g(k)*h(k)$	$G(n)H(n)$	$NG(n)H(n)$
7a. Product	$g(k)h(k)$	$\frac{1}{N}G(n)*H(n)$	$G(n)*H(n)$
8. Cross correlation	$g(k)\star h(k)$ $= g(k)*h(-k)$	$G(n)H(-n)$	$NG(n)H(-n)$ (a)
8a. Autocorrelation	$g(k)\star g(k)$ $= g(k)*g(-k)$	$\lvert G(n)\rvert^2$	$N\lvert G(n)\rvert^2$
9. Real part	$\mathrm{Re}[g(k)]$	$\frac{1}{2}G(n)+\frac{1}{2}G^*(-n)$	Same
9a. Imaginary part	$\mathrm{Im}[g(k)]$	$\frac{1}{2}G(n)-\frac{1}{2}G^*(-n)$	Same
10. Periodicity	$g(k+jN)$	$G(n)$	Same (b)
10a. Spectral periodicity	$g(k)$	$G(n+jN)$	Same (b)
11. Stretch (zero packing)	$\mathrm{Stretch}_L[g(k)]$	$\mathrm{Repeat}_L[G(n)]$	$\frac{1}{L}\mathrm{Repeat}_L[G(n)]$ (c)
11a. Repeat	$\mathrm{Repeat}_L[g(k)]$	$L\mathrm{Stretch}_L[G(n)]$	$\mathrm{Stretch}_L[G(n)]$ (c)

Remarks:

(a) The same remark about cross correlation as in the continuous case applies here, too.

(b) In other words, the periodicity rules say that $g(k+jN)=g(k)$ and $G(n+jN)=G(n)$ for any integer j. These rules have no equivalent in the continuous case, where the functions $g(x)$ and $G(u)$ are not necessarily periodic.

(c) These rules have no equivalent in the list given above for the continuous case. The discrete operator $\mathrm{Stretch}_L[g(k)]$ (also known as zero packing) receives a sequence $g(k)$ of length N and generates a sequence of length LN by adding $L-1$ zeroes after each element of $g(k)$. Similarly, the discrete operator $\mathrm{Repeat}_L[g(k)]$ receives a sequence $g(k)$ of length N and generates a sequence of length LN by repeating the original sequence $g(k)$ $L-1$ more times. A full explanation of these rules can be found, for example, in [Smith07 pp. 157–158] or for the alternative DFT definition in [Bracewell86 p. 370], [Kammler07 pp. 201–203].

Just as in the continuous case, here, too, the names of the rules may vary between different sources. For example, our *reflection* rule [Kammler07 p. 199] is called in [Bracewell86 p. 366] *reversal* while in [Nussbaumer82 p. 82] it is called *symmetry*.

Various other rules which are valid for the discrete case alone can be found, for example, in [Smith07 Sec. 7.4], [Briggs95 Sec. 3.3] or [Kammler07 pp. 201–212].

Note that a similar list of rules can be also deduced for the IDFT. For example, it follows from rules 7 and 7a above that the IDFT of $G(n)*H(n)$ is $Ng(k)h(k)$, and that the IDFT of $G(n)H(n)$ is $g(k)*h(k)$ [Čížek86 Sec. 4.2]. If one uses an alternative definition of the DFT (such as the definition in which the constant $\frac{1}{N}$ appears in the DFT rather than in the IDFT), the IDFT rules, too, should be adapted accordingly.

2.4.5 Rules for the 2D discrete Fourier transform

The extension of most of the 1D DFT rules to the 2D case is straightforward. Assume that the sequences $g(k,l)$ and $h(k,l)$ have the discrete Fourier transforms $G(m,n)$ and $H(m,n)$, respectively, that a and b are real numbers, and that j_1 and j_2 are integers. Then the following general rules can be stated (see, for example, [Briggs95 pp. 149, 173]):[8]

Rule name	Function	DFT	Alt. DFT	Remarks
1. Linearity	$ag(k,l) + bh(k,l)$	$aG(m,n) + bH(m,n)$	Same	
1a. Scaling	$ag(k,l)$	$aG(m,n)$	Same	
1b. Addition	$g(k,l) + h(k,l)$	$G(m,n) + H(m,n)$	Same	
2. Dilation	Complicated			
3. Reflection	$g(-k,-l)$	$G(-m,-n)$	Same	
4. Inversion	$G(k,l)$	$N^2 g(-m,-n)$	$\frac{1}{N^2}g(-m,-n)$	
5. Complex conjugation	$g^*(k,l)$	$G^*(-m,-n)$	Same	
6. Shift (or translation)	$g(k - j_1, l - j_2)$	$e^{-i2\pi(j_1 m + j_2 n)/N} G(m,n)$	Same	
6a. Frequency shift	$e^{i2\pi(j_1 k + j_2 l)/N} g(k,l)$	$G(m - j_1, n - j_2)$	Same	
7. Cyclic convolution	$g(k,l)**h(k,l)$	$G(m,n)H(m,n)$	$N^2 G(m,n)H(m,n)$	
7a. Product	$g(k,l)h(k,l)$	$\frac{1}{N^2}G(m,n)**H(m,n)$	$G(m,n)**H(m,n)$	

---

[8] For the sake of simplicity we assume here that the number of elements is N along each of the two dimensions, but the rules given here can be also extended to the more general case where the number of elements along each dimension is different.

| 8. Cross correlation | $g(k,l)\star\star h(k,l)$ $= g(k,l)**h(-k,-l)$ | $G(m,n)H(-m,-n)$ | $N^2G(m,n)H(-m,n)$ | (a) |

| 8a. Autocorrelation | $g(k,l)\star\star g(k,l)$ $= g(k,l)**g(-k,-l)$ | $|G(m,n)|^2$ | $N^2|G(m,n)|^2$ | |

| 9. Real part | $\text{Re}[g(k,l)]$ | $\tfrac{1}{2}G(m,n) + \tfrac{1}{2}G^*(-m,-n)$ | Same | |

| 9a. Imaginary part | $\text{Im}[g(k,l)]$ | $\tfrac{1}{2}G(m,n) - \tfrac{1}{2}G^*(-m,-n)$ | Same | |

| 10. Separable product | $g(k)h(l)$ | $G(m)H(n)$ | Same | |

| 11. Periodicity | $g(k+j_1N, l+j_2N)$ | $G(m,n)$ | Same | (b) |

| 11a. Spectral periodicity | $g(k,l)$ | $G(m+j_1N, n+j_2N)$ | Same | (b) |

| 12. Stretch | $\text{Stretch}_L[g(k,l)]$ | $\text{Repeat}_L[G(m,n)]$ | $\tfrac{1}{L^2}\text{Repeat}_L[G(m,n)]$ | (c) |

| 12a. Repeat | $\text{Repeat}_L[g(k,l)]$ | $L^2\text{Stretch}_L[G(m,n)]$ | $\text{Stretch}_L[G(m,n)]$ | (c) |

Remarks:

(a) The same remark about cross correlation as in the continuous case applies here, too.

(b) In other words, the periodicity rules say that $g(k+j_1N, l+j_2N) = g(k,l)$ and $G(m+j_1N, n+j_2N) = G(m,n)$ for any integers j_1, j_2. These rules have no equivalent in the continuous case, where the functions $g(x,y)$ and $G(u,v)$ are not necessarily periodic.

(c) These rules have no equivalent in the list given above for the 2D continuous case. The discrete operator $\text{Stretch}_L[g(k,l)]$ (also known as zero packing) receives a 2D matrix $g(k,l)$ of $N\times N$ elements and generates a 2D matrix of $LN\times LN$ elements by replacing each element of the original matrix by an $L\times L$ matrix having the original element at the first position and zeroes everywhere else. Similarly, the discrete operator $\text{Repeat}_L[g(k,l)]$ receives a 2D matrix $g(k,l)$ of $N\times N$ elements and generates a 2D matrix of $LN\times LN$ elements by repeating the original matrix to both directions into a block consisting of $L\times L$ copies of the original matrix.

2.4.6 Rules for the MD discrete Fourier transform

The generalization of the 2D rules to the MD case is usually straightforward. Assume that the MD sequences $g(\mathbf{k})$ and $h(\mathbf{k})$ have the Fourier transforms $G(\mathbf{n})$ and $H(\mathbf{n})$, respectively, where $\mathbf{k} = (k_0, \dots, k_{M-1})$, $\mathbf{n} = (n_0, \dots, n_{M-1}) \in \mathbb{Z}^M$; that a, b are real numbers; that $\mathbf{a} = (a_0, \dots, a_{M-1})$, $\mathbf{j} = (j_0, \dots, j_{M-1}) \in \mathbb{Z}^M$; and that $\mathbf{v}\cdot\mathbf{w}$ is the scalar product of the two vectors in question. Then we have the following rules:[9]

---

[9] For the sake of simplicity we assume here that the number of elements is N along each of the dimensions, but the rules given here can be also extended to the more general case where the number of elements along each dimension is different.

Rule name	Function	DFT	Alt. DFT	Remarks
1. Linearity	$ag(\mathbf{k}) + bh(\mathbf{k})$	$aG(\mathbf{n}) + bH(\mathbf{n})$	Same	
1a. Scaling	$ag(\mathbf{k})$	$aG(\mathbf{n})$	Same	
1b. Addition	$g(\mathbf{k}) + h(\mathbf{k})$	$G(\mathbf{n}) + H(\mathbf{n})$	Same	
2. Dilation		Complicated		
3. Reflection	$g(-\mathbf{k})$	$G(-\mathbf{n})$	Same	
4. Inversion	$G(\mathbf{k})$	$N^M g(-\mathbf{n})$	$\frac{1}{N^M}g(-n)$	
5. Complex conjugation	$g^*(\mathbf{k})$	$G^*(-\mathbf{n})$	Same	
6. Shift (or translation)	$g(\mathbf{k} - \mathbf{j})$	$e^{-i2\pi \mathbf{j}\cdot\mathbf{n}/N}\, G(\mathbf{n})$	Same	
6a. Frequency shift	$g(\mathbf{k})\, e^{i2\pi \mathbf{j}\cdot\mathbf{k}/N}$	$G(\mathbf{n} - \mathbf{j})$	Same	
7. Cyclic convolution	$g(\mathbf{k}){*}h(\mathbf{k})$	$G(\mathbf{n})H(\mathbf{n})$	$N^M G(\mathbf{n})H(\mathbf{n})$	
7a. Product	$g(\mathbf{k})h(\mathbf{k})$	$\frac{1}{N^M}G(\mathbf{n}){*}H(\mathbf{n})$	$G(\mathbf{n}){*}H(\mathbf{n})$	
8. Cross correlation	$g(\mathbf{k}){\star}h(\mathbf{k})$ $= g(\mathbf{k}){*}h(-\mathbf{k})$	$G(\mathbf{n})H(-\mathbf{n})$	$N^M G(\mathbf{n})H(-\mathbf{n})$	(a)
8a. Autocorrelation	$g(\mathbf{k}){\star}g(\mathbf{k})$ $= g(\mathbf{k}){*}g(-\mathbf{k})$	$\lvert G(\mathbf{n})\rvert^2$	$N^M\lvert G(\mathbf{n})\rvert^2$	
9. Real part	$\mathrm{Re}[g(\mathbf{k})]$	$\frac{1}{2}G(\mathbf{n}) + \frac{1}{2}G^*(-\mathbf{n})$	Same	
9a. Imaginary part	$\mathrm{Im}[g(\mathbf{k})]$	$\frac{1}{2}G(\mathbf{n}) - \frac{1}{2}G^*(-\mathbf{n})$	Same	
10. Separable product	$g(k_0, \dots ,k_{j-1})$ $h(k_j, \dots ,k_{M-1})$	$G(n_0, \dots ,n_{j-1})$ $H(n_j, \dots ,n_{M-1})$	Same	
10a. Separable products	$g_0(x_0)\dots g_{M-1}(x_{M-1})$	$G_0(u_0)\dots G_{M-1}(u_{M-1})$	Same	
11. Periodicity	$g(\mathbf{k} + \mathbf{j}N)$	$G(\mathbf{n})$	Same	(b)
11a. Spectral periodicity	$g(\mathbf{k})$	$G(\mathbf{n} + \mathbf{j}N)$	Same	(b)
12. Stretch (zero packing)	$\mathrm{Stretch}_L[g(\mathbf{k})]$	$\mathrm{Repeat}_L[G(\mathbf{n})]$	$\frac{1}{L^M}\mathrm{Repeat}_L[G(\mathbf{n})]$	(c)
12a. Repeat	$\mathrm{Repeat}_L[g(\mathbf{k})]$	$L^M\mathrm{Stretch}_L[G(\mathbf{n})]$	$\mathrm{Stretch}_L[G(\mathbf{n})]$	(c)

Remarks:

(a) The same remark about cross correlation as in the continuous case applies here, too.

(b) In other words, the periodicity rules say that $g(\mathbf{k} + \mathbf{j}N) = g(\mathbf{k})$ and $G(\mathbf{n} + \mathbf{j}N) = G(\mathbf{n})$ for any integer vector $\mathbf{j}$. These rules have no equivalent in the continuous case, where the functions $g(\mathbf{x})$ and $G(\mathbf{u})$ are not necessarily periodic.

(c) The discrete operator $\text{Stretch}_L[g(\mathbf{k})]$ (also known as zero packing) receives an MD sequence $g(\mathbf{k})$ of N elements to each direction and generates an MD sequence of LN elements to each direction by replacing each element of the original sequence by a sequence of L elements to each direction having the original element at the first position and zeroes everywhere else. Similarly, the discrete operator $\text{Repeat}_L[g(\mathbf{k})]$ receives an MD sequence $g(\mathbf{k})$ of N elements to each direction and generates an MD sequence of LN elements to each direction by repeating the original sequence to all directions into an MD block with L copies of the original sequence to each direction.

2.5 Graphical development of the DFT — a three-stage process

As already mentioned in Remark 2.2, the DFT definition can be derived as a numerical approximation to the definition of the continuous Fourier transform. But in fact, the DFT can be developed in many different ways. For example, Chapter 2 in [Briggs95] shows in detail how the DFT can be obtained as an approximation to the Fourier transform, as an approximation to the Fourier series coefficients of a periodic function, as a least-squares approximation to N given data pairs by a sum of sines and cosines, and as the Fourier transform of an impulse train. There also exists in the literature a very appealing graphical approach illustrating the development of the DFT from the CFT (see, for example, [Brigham88 Chapter 6], [Gonzalez87 pp. 95–97], [Bergland69 pp. 45–47] or [Cochran67 pp. 47–48]). Although this graphical approach may be considered less rigorous, it is most beneficial from the user's point of view since it offers an intuitive understanding of the theory, without recurring to detailed mathematical expressions and formulations. In the present section we will review this graphical approach very briefly; further details can be found, for example, in the references mentioned above.

Consider Fig. 2.1. The first row of this figure schematically shows a "suitably well behaved" function $g(x)$[10] and its continuous Fourier transform $G(u)$; and the last row of the figure shows their discrete counterparts, namely the sequence $\tilde{g}(k\Delta x)$ and its DFT $\tilde{G}(n\Delta u)$. The in-between rows of the figure illustrate the intermediate steps which lead us from the continuous case to the discrete one. As shown in the figure, the DFT approximation is obtained from the original CFT by means of three simple operations:

1. Sampling the original continuous function.

2. Truncating the sampled function into a finite range.

3. Sampling the resulting continuous spectrum.

---

[10] Namely, a function whose continuous Fourier transform can be expected to be reasonably well approximated by DFT; see Sec. 2.6.

The first stage is illustrated by rows (a)–(c) of our figure. Row (a) shows the original function $g(x)$ and its CFT $G(u)$. Row (b) shows a sampling impulse train with sampling interval of Δx and impulse height of 1, along with its CFT, which is an impulse train with impulse interval of $1/\Delta x$ and impulse height of $1/\Delta x$ (see, for example, [Brigham88 p. 25]). As shown in row (c), the sampled version of the original signal $g(x)$ is obtained by multiplying the signals in rows (a) and (b). The resulting spectrum (CFT) is, therefore, according to the convolution theorem, the convolution of the spectra in rows (a) and (b), i.e. a periodic repetition of replicas of $G(u)$, as shown indeed in row (c). Note that due to this convolution the resulting spectrum in row (c) is now periodic with period $F = 1/\Delta x$, unlike the original spectrum $G(u)$ in row (a) which is not necessarily periodic. The possible overlapping between consecutive replicas in this spectrum (which is, in reality, additive) is the source of the *aliasing* artifact of the DFT, as we will see in Chapter 5.

The second stage is illustrated by rows (c)–(e) of the figure. Row (d) shows a rectangular truncation window of height 1 and length $R = N\Delta x$ (i.e. N sampling intervals) and its CFT, which is a narrow sinc function with amplitude R (see, for example, [Brigham88 p. 24]). As shown in row (e), the truncated version of the sampled signal is obtained by multiplying the signals in rows (c) and (d). The resulting spectrum (CFT) is again, according to the convolution theorem, the convolution of the spectra in rows (c) and (d); as we can see in row (e), this spectrum is a somewhat distorted version of the spectrum of the sampled signal shown in row (c). This distortion is the source of the *leakage* artifact of the DFT, that will be discussed in detail in Chapter 6.

We now proceed to the third stage, which is illustrated by rows (e)–(g) of our figure. So far our spectral-domain function is still continuous, as shown in row (e), and we therefore need to sample it. Row (f) shows in the spectral domain a sampling impulse train with sampling interval of $\Delta u = 1/R$ (see point (iii) below) and impulse height of 1, along with its inverse CFT in the signal domain, which is an impulse train with impulse interval of $1/\Delta u = R$ and impulse height of R. As shown in row (g), the sampled version of the CFT of row (e) is obtained by multiplying the spectra in rows (e) and (f). The resulting signal is, according to the convolution theorem, the convolution of the signals in rows (e) and (f), namely a periodic repetition of replicas of the sampled and truncated signal, as shown indeed in the signal domain of row (g). Note that due to this signal-domain convolution the resulting signal in row (g) is now periodic with period $R = N\Delta x$, unlike the original function $g(x)$ in row (a) which is not necessarily periodic. And as we have already seen above, the resulting DFT, shown in the spectral domain of row (g), is periodic with period $F = 1/\Delta x$. Although the signal-domain period R and the spectral-domain period F are not necessarily equal, each of them consists of exactly N sampling intervals (respectively, $R = N\Delta x$ and $F = N\Delta u$; see point (iii) below).

A few remarks concerning this process are now in order:

(i) Formally, all the rows of Fig. 2.1 subsist in the continuous world (including the impulse trains, the sampled functions and the sampled spectra), and the Fourier transform used in all rows is the CFT. But row (g) can be also interpreted as being discrete, by simply ignoring the continuous zero-valued intervals between the individual impulse locations along the x and u axes (see, for example, [Granlund95 p. 192]). Similarly, row (c) can be also interpreted as a continuous Fourier transform of a discrete signal[11] [Kammler07 p. 11; Smith07 pp. 207–209]; or, considering this row in the opposite direction, as a Fourier series development of the periodic signal shown in the right-hand column.

(ii) Although the original function $g(x)$ and its CFT $G(u)$ in row (a) are not necessarily periodic, their resulting discrete counterparts shown in row (g) are periodic. As we have seen, this happens due to the replications caused by the spectral-domain convolution in row (c) and the signal-domain convolution in row (g). If $g(x)$ and/or $G(u)$ *are* periodic, their periods may or may not coincide with the DFT periodicity; the consequences will be discussed later, for example in Chapter 6 and Sec. 7.4.

(iii) The spectral-domain sampling interval in row (f) is not chosen arbitrarily. We take it to be $\Delta u = 1/R$ in order that the impulse interval in the signal domain of row (f), and hence the period of the discrete signal in row (g), be exactly equal to the length of the truncation window, $1/\Delta u = R$. This choice also guarantees that each period of the DFT contains exactly N discrete elements, just like each period of the discrete signal $g(k)$ (since one spectral period, $F = 1/\Delta x$, contains N intervals of length $\Delta u = 1/R = 1/N\Delta x$).

(iv) The second step (signal-domain truncation) is required here since the length of the input array is limited, meaning that when we sample the original continuous-world signal $g(x)$ we must stop the sampling operation at the input array borders. Note, however, that although the length of the output array is limited, too, a dual spectral-domain truncation is not included here. The reason is that unlike the input sampling process, the DFT operation does not necessarily stop at the borders of the output array; rather, if higher frequencies are present, they will be folded over cyclically to the other end of the output array, causing *spectral domain aliasing* (see Chapter 5).[12]

(v) On the flexibility in the order we perform the operations in the figure, see Sec. D.8.

(vi) We have used here Fig. 2.1 as a didactic tool for illustrating graphically the relationship between the CFT and the DFT. Note, however, that in this figure the original function $g(x)$ is symmetric about the origin, which suggests a symmetric truncation in row (d). As we can see in row (g), this implies a centered DFT definition where the integer summation index varies between $-N/2$ to $N/2-1$. But when $g(x)$ is *causal* (i.e. $g(x) = 0$ for $x < 0$ [Smith07 p. 133]), it may be more natural to use in row

---

[11] Also called in the literature *Discrete-time Fourier transform* [Smith03 p. 145; Smith07 p. 208].

[12] From another point of view, our second step (signal-domain truncation) is used here to avoid overlapping of far-away tails of the original signal with the signal's replicas (that are generated by the signal-domain convolution in stage 3), a problem known as *signal-domain aliasing* (see Sec. 5.7). In other words, without the signal-domain truncation in row (e) of Fig. 2.1 we would have in the signal domain of row (g) an overlapping problem analogous to the spectral-domain overlapping in row (c). If we also wish to avoid the overlapping problem in the spectral domain of row (c), known as *spectral-domain aliasing*, a dual truncation step (spectral-domain truncation) may be added before step 1 (see end of Sec. 5.2). Adding this spectral-domain truncation step to our 3-step process would make this process more symmetric; but this new step is not completely harmless (see Sec. 6.7), and the decision whether or not to add it is left to the users. If this spectral-domain truncation of $G(u)$ is indeed desired, it will be typically done by low-pass filtering the continuous input signal $g(x)$ prior to its sampling.

(d) a truncation window going between $x = 0$ and $x = R$, as shown in Fig. 2.2. This leads naturally to a non-centered DFT definition where the integer summation index varies between 0 and $N-1$, as in Eqs. (2.13)–(2.14) above.[13] In the most general case, where $g(x)$ is neither symmetric about the origin nor causal (see Fig. 2.3), there is no natural reason to prefer either of the truncation windows; we will therefore stick to the non-centered one, which leads to the indexing convention used in Eqs. (2.13)–(2.14). In fact, this non-centered indexing is the standard convention used by virtually all software implementations of the DFT (see, for example, the partial list given at the end of Sec. 2.3); furthermore, this convention is used by such DFT implementations in *all* circumstances, including for symmetric functions $g(x)$. The practical implications of this fact are very important, and they will be discussed at length in Chapter 3.

(vii) A final remark concerns the amplitudes of the resulting signals in the last row. Although we started in the first row with a signal $g(x)$ and a CFT whose maximum amplitudes are c and d, respectively, we see that the corresponding amplitudes in the

Figure 2.1: A schematic drawing illustrating graphically the development of the DFT from the CFT. (a) The original function $g(x)$ and its CFT, $G(u)$. (b) A sampling impulse train with sampling interval of Δx and impulse height of 1, and its CFT, which is an impulse train with impulse interval of $1/\Delta x$ and impulse height of $1/\Delta x$. (c) The sampled version of the original signal $g(x)$ is obtained by multiplying the signals in rows (a) and (b). The resulting spectrum (CFT) is, therefore, the convolution of the spectra in rows (a) and (b), i.e. a periodic repetition of replicas of $G(u)$. (d) A rectangular truncation window of height 1 and length $R = N\Delta x$ (i.e. N sampling intervals) and its CFT, which is a narrow sinc function with amplitude R. (e) The truncated version of the sampled signal is obtained by multiplying the signals in rows (c) and (d); the resulting spectrum (CFT) is the convolution of the spectra in rows (c) and (d). This spectrum is a somewhat distorted version of the spectrum of the sampled signal shown in (c). (f) A spectral-domain sampling impulse train with sampling interval of $\Delta u = 1/R$ and impulse height of 1, along with its inverse CFT (in the signal domain), which is an impulse train with impulse interval of $1/\Delta u = R$ and impulse height of R. (g) The sampled version of the CFT of row (e) is obtained by multiplying the spectra in rows (e) and (f). The resulting signal in the signal domain is the convolution of the signals in rows (e) and (f), which is a periodic repetition of replicas of the sampled and truncated signal. Row (g) shows indeed the discrete counterparts of the function $g(x)$ and of its CFT $G(u)$ from row (a), namely, the corresponding discrete sequence $\tilde{g}(k\Delta x)$ and its DFT. The two pairs of vertical dotted lines going down throughout the figure indicate the extent of the input and output arrays of the DFT. Note that this figure is provided here for didactic reasons only; most software implementations of the DFT use non-centered truncation and indexing, as shown in Figs. 2.2 and 2.3.

---

[13] Note, however, that in this case the CFT of the truncation window has a non-zero imaginary part due to the shift theorem (see point 6 in Sec. 2.4.1 above); this does not have a major impact on the results (see [Brigham88 pp. 94, 100]) since it only affects the phase of the spectrum, but not its magnitude.

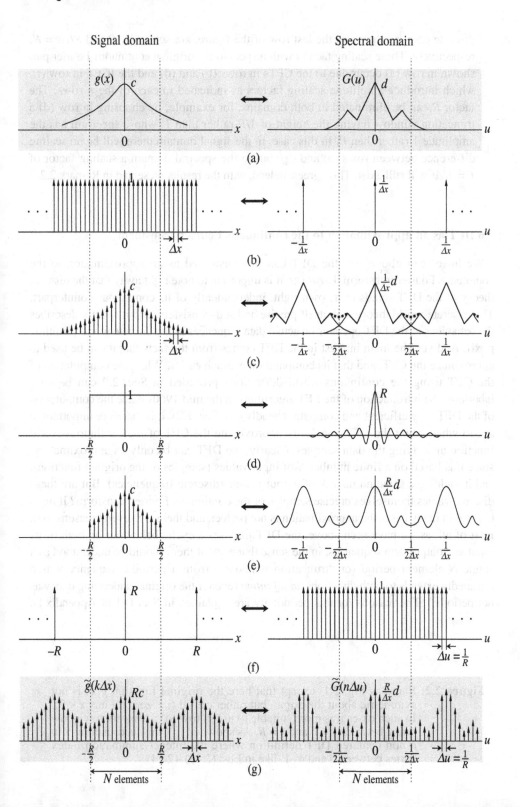

Signal domain   Spectral domain

(a)

(b)

(c)

(d)

(e)

(f)

(g)

discrete case, as shown in the last row of the figure, are scaled by R and $R/\Delta x = N$, respectively. These scaling factors with respect to the original continuous Fourier pair shown in row (a) occur due to the CFTs in rows (b) and (d) and the ICFT in row (f), which introduce amplitude scaling factors as indicated in each of these rows. The factor R can be eliminated in both domains, for example, by choosing in row (d) a truncation window having the height of $1/R$ rather than 1, whose spectrum has the amplitude 1 rather than R. In this case, in the signal domain there will be no scaling difference between rows (a) and (g), but in the spectral domain a scaling factor of $F = 1/\Delta x$ will still exist. This agrees, indeed, with the results presented in Remark 2.2.

2.6 DFT as an approximation to the continuous Fourier transform

We have seen above that the DFT can be considered as an approximation to the continuous Fourier transform. However, it is important to note right away that the discrete theory of the DFT stands in its own right, independently of its continuous counterpart. This discrete-world theory is in itself precise and self-consistent, and it exactly describes the behaviour of the DFT operator on actual data sequences in the discrete world [Smith07 p. xi]. And yet, the main interest in the DFT comes from the facts that it can be used to approximate the CFT, and that its computation is much easier. While the computation of the CFT using the continuous-world definitions provided in Sec. 2.2 can be very laborious, the introduction of the FFT algorithms in the mid 1960s made the computation of the DFT very efficient and computer friendly (see Sec. 1.2). It is therefore important to see to what extent the DFT can really approximate the CFT of the continuous-world function underlying the data samples. Clearly, the DFT can be only an approximation, since it is based on a finite number N of input values (samples of the original function), and it yields the same number N of output values (discrete frequencies). But are these discrete values themselves precise samples of the continuous Fourier transform? It turns out that in many cases the approximation is not perfect, and there are several reasons to it. First of all, as we have seen above, the DFT imposes a cyclic behaviour on both of its input and output data sequences, in the sense that each of them should be understood as a single N-element period (or "truncation window") from a periodic sequence which repeatedly extends to both directions *ad infinitum* (even if the original underlying data was not periodic). The reasons for this periodicity are explained in Sec. D.1 of Appendix D.

Figure 2.2: Same as Fig. 2.1, except that here the original function $g(x)$ is not symmetric about the origin, but rather causal (i.e. zero for any $x < 0$). In such cases it is more suitable to use in row (d) a truncation window going between $x = 0$ and $x = R$. As shown in (g), this leads naturally to a non-centered DFT definition where the integer summation index varies between 0 and $N-1$, like in Eqs. (2.13)–(2.14).

Signal domain   Spectral domain: Magnitude

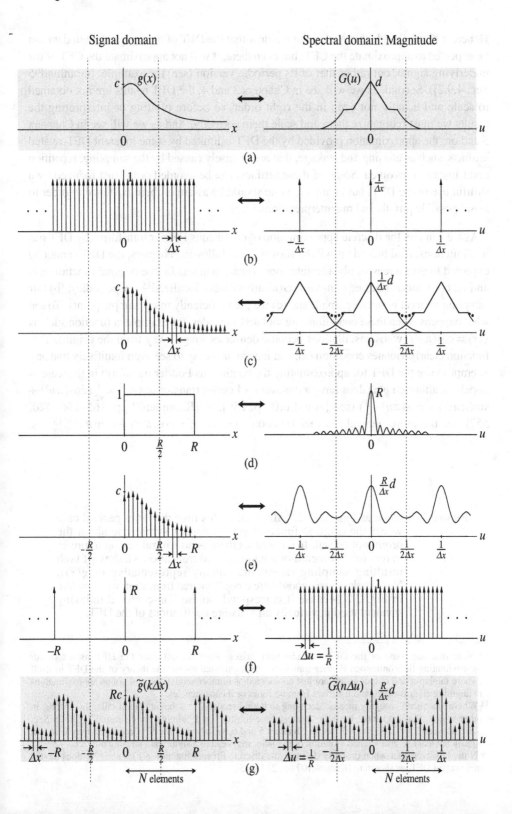

Hence, it is only inside this N-element window that the DFT of the sampled input data can be expected to approximate the CFT; but even there, it will not approximate the CFT of the underlying signal, but rather that of its periodic version (see, for example, [Granlund95 Sec. 4.9.2]). Second, as we will see in Chapters 3 and 4, the DFT results are not obtained to scale and usually not even in the right order, so before plotting or interpreting the results we must reorganize them and scale them correctly. And as we will see in Chapters 5 and on, the approximation provided by the DFT is limited by some inherent DFT-related artifacts, such as aliasing and leakage, that are precisely caused by the sampling, repetition and truncation involved. Some of these artifacts can be avoided or at least reduced by a skillful use of the DFT, but in any case one should be aware of their existence in order to avoid possible pitfalls and misinterpretations.[14]

As we can see, the discrete approximation of continuous Fourier transforms by DFT has its limitations. And indeed, as will be shown in the following chapters, the DFT cannot be expected to give a reasonably accurate approximation unless (a) the original function $g(x)$ and its continuous Fourier transform $G(u)$ are suitably localized[15] (or periodic); (b) our sampling interval is suitably small; and (c) we use sufficiently many sample points. To see what happens when these conditions are violated, consider, for example, a function such as $g(x) = \cos(\pi x^2)$ (whose oscillations become denser as we go away from the origin). This function clearly violates condition (a), and indeed, it is easy to see even intuitively that any attempt to use the DFT for approximating the continuous Fourier transform in this case is hopeless, although $g(x)$ does have a continuous Fourier transform: $G(u) = \frac{1}{\sqrt{2}}(\cos(\pi u^2) + \sin(\pi u^2)) = \cos(\pi u^2 - \frac{\pi}{4})$ (see [Amidror09 p. 259], or [Kammler07 pp. 165–166, 420, 557] for the more general Fresnel function $g(x) = e^{i\pi x^2} = \cos(\pi x^2) + i\sin(\pi x^2)$).[16] As

Figure 2.3: Same as Figs. 2.1–2.2, illustrating this time the more general case where the original function $g(x)$ is neither symmetric about the origin nor causal. In such cases there is no natural reason to prefer in row (d) a centered or a non-centered truncation window (if both resulting sampling ranges are equally representative of $g(x)$). Nevertheless, we choose here a non-centered truncation window like in Fig. 2.2 because it corresponds to the non-centered indexing standard used in virtually all software applications of the DFT.

---

[14] Note that the issue of the DFT artifacts only arises when we consider the DFT as a tool for approximating the continuous Fourier transform. As mentioned above, the theory of the DFT in itself (where the input and output arrays are just considered as number sequences, and not as approximations to anything else) is exact, and it suffers from no flaws or inconsistencies.

[15] Where "localized" roughly means "decaying to both directions", i.e. having small tails; this means, in other words, that $g(x)$ must be both "almost time-limited" and "almost band-limited" [Chu08 Sec. 5.8.1; Zayed93 p. 9]. As we will see in Chapters 5 and 6, violation of the first condition may lead to a significant leakage artifact, and violation of the latter may lead to a significant aliasing artifact.

[16] Note, however, that when properly defined, the discrete Fresnel function $g(k) = e^{i2\pi k^2/N}$ does have a meaningful DFT, as shown in [Kammler07 pp. 214–215].

Signal domain   Spectral domain: Magnitude

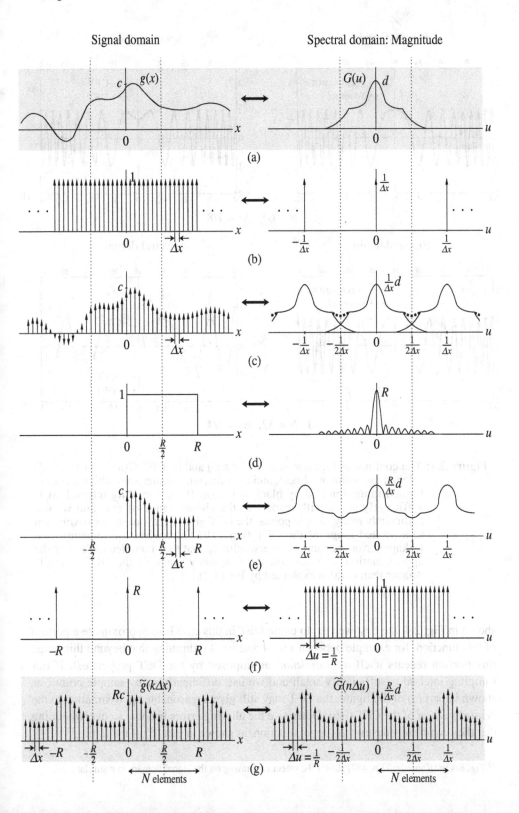

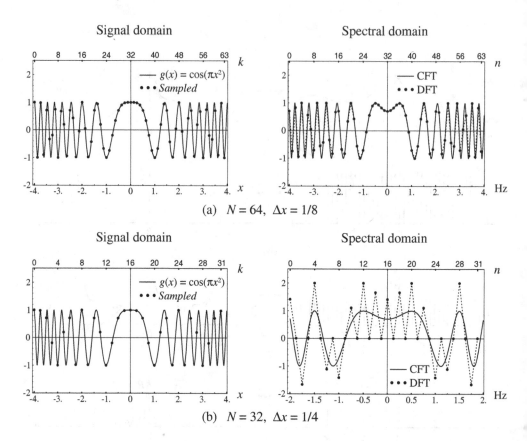

Figure 2.4: The continuous function $g(x) = \cos(\pi x^2)$ and its CFT $G(u) = \cos(\pi u^2 - \frac{\pi}{4})$ (plotted in continuous lines), and their discrete counterparts within a certain limited range (plotted by black dots). (a) If the sampling interval Δx is sufficiently small with respect to the chosen range of $g(x)$ and we use sufficiently many sample points, the DFT gives a reasonable approximation to a limited range of the CFT $G(u)$. (b) Otherwise, the aliasing (and leakage) errors become so overwhelming that they completely destroy the approximation. (The reason the frequency range of the DFT in (b) is smaller than in (a) is explained by Eq. (7.2).)

shown in Fig. 2.4, the best we can do using DFT in this case is to approximate a portion of this function, for example between $x = -4$ and $x = 4$ (admitting that beyond this range this portion repeats itself *ad infinitum*, as imposed by the DFT properties). If our sampling interval is sufficiently small and we use sufficiently many sample points, as shown in part (a) of the figure, the DFT may still give a reasonable approximation to the *central portion* of the CFT $G(u)$. Otherwise the aliasing errors become so overwhelming that they completely destroy the approximation, as shown in part (b) of the figure.[17]

---

[17] Fig. 8.4 in [Kammler07 p. 487] shows the effect of aliasing on the signal domain in a similar case.

Mathematical estimations of the errors between the continuous Fourier transform and its DFT approximation can be found in the literature (see, for example, Sec. 2.7 and Chapter 6 in [Briggs95] or Chapters 12 and 13 in [Morrison94]). Here, however, we will rather be interested in understanding the nature and the origin of these errors. Our aim in the following chapters is not to review the theory of the DFT, a subject about which the literature already abounds, but rather to point out some of the subtle issues that may arise when using the DFT as an approximation to the Fourier transform (including in the special case of periodic functions, as explained in more detail in the following section). An understanding of these subtle issues is necessary to guarantee a correct utilization of this important tool, and a correct interpretation of its results.

2.7 The use of DFT in the case of periodic or almost-periodic functions

So far, we have considered the use of DFT as a digital approximation to the continuous Fourier transform. It should be mentioned, however, that DFT can be also used to obtain a digital approximation to the Fourier series development of a periodic function. This should not be surprising, since the special case of periodic functions is already included in the continuous Fourier theory: Simply, the spectrum of a periodic function is not continuous, but rather consists of an impulse train where the distance between successive impulses equals the frequency $f = 1/T$ of the periodic function, and the impulse amplitudes correspond (up to a possible constant scaling factor) to the Fourier series coefficients (see, for example, [Papoulis68 p. 107] or [Amidror09 p. 377]). Therefore, once we get by DFT an approximation to the impulsive spectrum of our given periodic function, we can use the resulting impulse amplitudes (up to a constant scaling factor) as an approximation to the corresponding Fourier series coefficients.

The main condition for this to work properly (i.e. for the DFT to correctly approximate the impulse train which makes up the continuous-world spectrum of our periodic function) is that the sampling range length $R = N\Delta x$ must be exactly equal to one period of our given function [Smith07 pp. 109, 210–212], or to an integer number of periods. As we will see in Chapter 6, this condition guarantees that the impulses of the continuous-world spectrum fall exactly on sampling points of the DFT spectrum, so that no *leakage* occurs. A further condition is that the given periodic function should be band limited, and the sampling rate must be at least twice the maximum frequency of our periodic function; as we will see in Chapter 5, this condition guarantees that the resulting impulses are not corrupted by *aliasing*. Under these conditions the resulting impulse amplitudes in the DFT spectrum should be essentially identical (up to a certain scaling factor and a small discretization error[18]) to the original continuous-world impulse amplitudes. This is explained in detail, along with several examples, in Sec. D.9 of Appendix D and in

---

[18] Note that a discretization error is unavoidable here (just as in the DFT approximation to CFT), since the DFT approximates the integrals of the Fourier coefficients in the same way as numerical integration does; this is explained in detail in [Briggs95 pp. 53–54 and Sec. 2.4].

Appendix A. Note that for any real-valued periodic function, the maximum number of impulses (Fourier series coefficients) that can be thus approximated by the DFT cannot exceed $N/2$, where N is the array length; the other $N/2$ elements of the output belong to the negative frequencies, and they are simply a Hermitian-symmetric copy of the positive frequencies. In the most general case, where the given periodic function is complex-valued and has no particular symmetry properties, so that its negative spectral frequencies are independent of its positive frequencies, the maximum number of impulses (Fourier series coefficients) that can be approximated by the DFT cannot exceed N.

A more detailed and rigorous discussion on the DFT approximation to Fourier series coefficients can be found in [Brigham88, Sec. 5.2], and in particular in [Briggs95, Sec. 2.4]. This last reference also clearly explains why if the given periodic function contains discontinuities (even discontinuities that fall between the two end points of the sampled data), the DFT input at such points must be the average of the values to both sides of the discontinuity (see also Sec. 8.2 below).

A detailed discussion on the errors in the approximation of the Fourier series coefficients by DFT can be found in [Briggs95 pp. 180–192]. A discussion on the errors in the inverse DFT that reconstructs the original periodic function from the Fourier series coefficients can be found in [Briggs95 pp. 212–215].

Finally, note that DFT can be also used to approximate the spectrum of an almost-periodic function (i.e. a function whose continuous-world spectrum is purely impulsive but its impulse locations are not commensurable; see [Amidror09, Appendix B]). However, in this case it is impossible for all the impulses in the continuous-world spectrum to fall exactly on sampling points of the DFT spectrum: Otherwise, the impulse locations would be commensurable and the given function would be periodic. This means that leakage cannot be avoided here and the DFT approximation cannot be as good as in the periodic case.

PROBLEMS

2-1. *The different Fourier transform definitions.* As we have seen in Sec. 2.2, the continuous Fourier transform can be defined in several slightly different ways, all of which are frequently used in the literature. Some of the differences between these definitions are just a matter of notation (such as the different letters being used for the function names or for the variables), but other differences are more substantial and may give, indeed, slightly different Fourier transforms for the same original function. Such are, for example, the choice of the constant factors that precede the integral in the definition, and the location of the minus sign in the exponential part (it may appear either in the direct transform or in the inverse transform).

(a) Make a list of the main definitions you can find in various sources, and try to see what are the advantages and the shortcomings of each of them. What could be the reasons that lead authors to prefer one definition over the others? (See, for example,

[Bracewell86 pp. 6–8; Gaskill78 pp. 181–182; Brigham88 pp. 22–23; Champeney73 pp. 8–10, Cartwright90 p. 98].)

(b) Is it possible and practical to impose a universally agreed convention, in order to eliminate once and for all the inconvenience caused by the coexistance of several different and non-compatible versions in the literature?

2-2. *Conversion rules between the different Fourier transform definitions.* Suppose you have found in the literature a table containing various functions $g(x)$ and their Fourier transforms $G(u)$. Indeed, classical references such as [Erdélyi54], [Champeney73 pp. 20–39], [Bracewell86, Chapter 21], [Oberhettinger90], [Gradshteyn07, Sec. 17.2], etc. provide a vast selection of such Fourier pairs, which could be used as a good starting point when trying to find the analytic CFT expression of a given function. But before using such a table, you should always check under which conventions it has been established. It often happens that the conventions used in the reference in question are not the same as those you are using; such a mismatch can be, indeed, a source of many pitfalls and errors for an unaware user. To avoid such pitfalls, identify which of the different Fourier transform variants you are using, and always stick to the same convention. It is also useful to establish the conversion rules for going back and forth between your favorite convention and the other main conventions; a small selection of such conversion rules can be found, for example, in [Bracewell68 p. 7] or in [Champeney73 pp. 8–10]. Such conversion rules will allow you to easily convert Fourier transforms from one convention to another, and to safely use Fourier transform tables that are provided in various sources by different authors.

2-3. *Conversion rules between the different DFT definitions.* The same is also true for the Discrete Fourier transform. Unfortunately, here, too, no standard definition exists, and one may find in the literature several slightly different variants. Although these variants are essentially equivalent in terms of the Fourier theory, each of them may give slightly different results, and one should be aware of this fact to avoid any possible confusion, especially when working with several textbooks, references or Fourier software packages. For this end, just as you did in the previous problem, identify which variant of the DFT you are using, and always stick to the same convention. Then, establish the conversion rules for going back and forth between your favorite convention and the other main conventions. This will allow you to easily convert discrete Fourier transforms from one convention to another, to work with different DFT software packages, and to use DFT tables that are provided in the literature by different authors. Also, make sure that the DFT variant you are using is compatible with your favorite CFT variant; for example, if you have chosen a CFT definition with the minus sign in the exponential part of the *inverse* transform, it would be wise (if possible) to use the same convention in the DFT software, too.

2-4. What would you expect to get if you cannot follow the advice given at the end of the previous problem, and you must use a DFT software package having a different exponent-sign convention than your CFT? (See also the end of Sec. 3.6 and Footnote 7 in Sec. 4.4).

2-5. *Transformation and combination rules for the continuous and for the discrete Fourier transforms.* The Fourier theory benefits from a certain number of rules which tell us how we may obtain new Fourier transforms from already known ones or from their combinations (see Sec. 2.4). These transformation and combination rules prove to be very useful, since they allow us to find the Fourier transforms of many new functions quite easily, without having to use the explicit definition formula to compute each Fourier transform from scratch. As we have seen in Sec. 2.4.4, many of these rules for the continuous Fourier transform subsist in the discrete case, too; but some of them must be slightly amended or even completely redefined in order to be adapted to the discrete case. Furthermore, there also exist discrete-world rules which have no equivalent in the continuous case, and continuous-world rules that cannot be extended to the discrete case. Can you mention a few rules from each of these categories?

2-6. *Transformation and combination rules for the continuous and for the discrete Fourier transforms (continued)*. One example of a continuous-world rule that cannot be easily transferred to the discrete world is the dilation rule. Can you explain the difficulty? (see, for example, [Kammler07 pp. 205–208]). Another well-known 2D example is the rotation theorem, which says that when a 2D function $g(x,y)$ is rotated by angle θ, its CFT $G(u,v)$, too, undergoes a rotation by the same angle [Bracewell95 p. 157]. Why doesn't this rule have a discrete-world equivalent?

2-7. *Symmetry rules for the continuous and for the discrete Fourier transforms*. An extremely nice feature of both continuous and discrete Fourier transforms is that they possess many symmetry properties, that prove to be very helpful. For example, we give below some of the most useful symmetry rules for the 1D CFT:

(1) If $g(x)$ is real-valued and even, then its CFT $G(u)$, too, is real-valued and even.

(2) If $g(x)$ is real-valued and odd, then its CFT $G(u)$ is imaginary-valued and odd.

(3) If $g(x)$ is even, then its CFT $G(u)$, too, is even.

(4) If $g(x)$ is odd, then its CFT $G(u)$, too, is odd.

(5) If $g(x)$ is real-valued, then its CFT $G(u)$ is Hermitian.

(6) If $g(x)$ is imaginary-valued, then its CFT $G(u)$ is Anti-Hermitian.

The main symmetry rules for the 1D continuous case can be found, for example, in [Bracewell86 pp. 14–16]. Their discrete counterparts can be found in [Bracewell86 p. 366] or in [Briggs95 pp. 74–78]. Note that sometimes the symmetry rules are listed together with the transformation and combination rules (see, for example, the table provided in [Briggs95 pp. 88–89]).

(a) Are there any difficulties in transferring these symmetry rules from the continuous world to the discrete world or vice versa?

(b) Can we reformulate these symmetry rules using "if and only if" rather than "if"? *Hint*: See, for example, [Cooley69 p. 79] or [Cooley70 p. 317].

(c) How can these symmetry rules be extended to the 2D or MD cases? Can you think of any difficulties in transferring such 2D or MD symmetry rules from the continuous world to the discrete world? *Hint:* See Sec. 8.4.

2-8. Do the transformation and combination rules (see Problem 2-5 and Sec. 2.4) vary depending on the Fourier transform definition being used, or are they independent of the definitions? Do the symmetry rules mentioned in the previous problem vary depending on the chosen Fourier transform definition? Would you give the same answers in the discrete case, too?

2-9. *Analog vs. digital*. It is often said that the real world is analog, while computations are digital. In particular, analog signals and spectra belong to the real world, while their digital counterparts are the results of computation.

(a) Is this always true?

(b) Is it possible to approach continuous-world signals and spectra at will by sufficiently increasing the resolution of digital computation?

2-10. What is the difference between *digital* and *discrete*? and what is the difference between *analog* and *continuous*? *Hint*: See in the glossary.

Chapter 3

Data reorganizations for the DFT and the IDFT

3.1 Introduction

While the continuous Fourier transform takes as its input a function $g(x)$ and returns as its output another function $G(u)$, the discrete Fourier transform takes as its input a sequence of N values and returns as its output another sequence of N values. However, if we consider the input and output sequences of the DFT as a digital approximation to the continuous-world functions $g(x)$ and $G(u)$, we must be aware of an important difference between the continuous and discrete cases, which may be a source of many pitfalls and errors. While the continuous functions $g(x)$ and $G(u)$ are generally defined throughout a continuous domain along their x or u axes, the input and output sequences of the DFT are only defined on a discrete domain consisting of N consecutive points, and outside this domain they are considered to behave periodically. Furthermore, for reasons we will see below the discrete values which correspond to $x = 0$ or $u = 0$ are not necessarily located in the center of the input or output sequences of the DFT. Therefore, in order to present or plot these discrete sequences the way they would be presented in the continuous world, one may need to reorganize them, i.e. to reorder the sequence elements differently (or more precisely, to shift the N-element sequence along its periodic replications until the elements corresponding to $x = 0$ or $u = 0$ fall where they are expected to be with respect to their continuous-world counterparts). This is also true in the 2D and MD cases, where the same reasoning applies along each dimension individually.

In this chapter we explain the data reorganizations that may be required for understanding the DFT and IDFT results, both in the 1D and in the MD cases. We start in Sec. 3.2 with the reorganization of the output data of the DFT. Then in Sec. 3.3 we proceed to the reorganization of the input data of the DFT, and in Sec. 3.4 we discuss the required reorganizations in the case of the IDFT. Several illustrative examples are given in Sec. 3.5, and finally Sec. 3.6 provides some concluding remarks.

3.2 Reorganization of the output data of the DFT

A typical 1D DFT algorithm receives as input a 1D array of N complex numbers (say, samples of a given continuous complex- or real-valued function $y = g(x)$), and returns as its output a 1D array of N complex numbers indexed by $n = 0,...,N-1$, which contains the resulting discrete frequency spectrum. However, the frequency values in the resulting output array are not ordered in the conventional way we use to plot the spectrum: The first element in the array ($n = 0$) corresponds to the frequency 0 (that is traditionally called DC because it represents in electrical transmission theory the direct current component

[Smith03 pp. 152, 633]); the second element ($n = 1$) corresponds to f_1, the smallest frequency above 0 in the discrete spectrum (see Sec. 4.3); and the next elements represent the following frequencies of the discrete spectrum, i.e. $2f_1$, $3f_1$, etc. However, this is only true until $n = N/2-1$; the output array elements starting from $n = N/2$ and until the last element, $n = N-1$, correspond to the negative frequencies of the spectrum. Thus, the last element of the output array, $n = N-1$, corresponds to the negative frequency $-f_1$, which should conventionally be plotted just before the frequency 0; its predecessor, the element $n = N-2$, corresponds to the frequency $-2f_1$; and finally, the element $n = N/2$ corresponds to $-(N/2)f_1$, the most negative frequency in the spectrum.[1] This is illustrated in Fig. 3.1(b) and in Fig. 3.2(a). It should be mentioned that this complication only occurs when using a DFT definition with *non-centered* indices, like Eqs. (2.13)–(2.14), in which the indices vary between 0 and $N-1$ rather than between $-N/2$ and $N/2-1$ [Briggs95 p. 67]. But because this is the standard convention used in virtually all software applications of the DFT (see, for example, the partial list given at the end of Sec. 2.3), we have to accept this fact and get used to it. The reason for this ordering of the output is explained in detail in Sec. D.2 of Appendix D. It follows, therefore, that in order to plot the resulting spectrum in the conventional way, with the zero frequency in the center and the negative frequencies to its left, we have to *reorganize* the output array of the DFT by interchanging (swapping) its first half (the segment containing the elements $n = 0,...,N/2-1$) and its second half (the segment containing the elements $n = N/2,...,N-1$), as shown in Figs. 3.1(c) and 3.2(a).[2] This reorganization of the spectrum is not usually done by the DFT algorithm [Brigham88 p. 169], and it is up to the user to perform it when plotting the DFT results.[3]

Of course, whether or not to reorganize the plotted data in the conventional way may be a question of personal preference. As long as we remember that the results for $n > N/2$ actually relate to negative frequencies, we should encounter no interpretation problems; but in order to avoid misunderstanding of the DFT results by people who may be unaware of this non-conventional ordering, it is usually recommended to reorganize the data whenever plotting the DFT results. If one prefers not to do so, a possible alternative solution is to plot a true frequency scale beneath the scale of the output index n, as done, for example, in Fig. 9.1 in [Brigham88 pp. 168–169]. For more details on plotting the true units along the axes, see Chapter 4.

---

[1] In the vast majority of DFT implementations N is a power of two, or at least an even number. In these cases the output array returned by the DFT consists of the frequency 0, followed by $N/2-1$ positive frequencies ($f_1, ... ,(N/2-1)f_1$), and then by $N/2$ negative frequencies ($-(N/2)f_1, ... ,-f_1$). Thus, the output array contains one more negative frequency than positive frequencies (although thanks to the cyclic nature of the DFT the element $n = N/2$ corresponds in fact to both of the positive and negative frequencies $\pm(N/2)f_1$ [Press02 p. 510]). In the rare cases where N is odd the number of positive and negative frequencies in the output array of the DFT is identical, and the reorganization of the array should be done accordingly [Briggs95 Sec. 3.1].

[2] Note that this is also equivalent to a cyclic shift of the array (i.e. a shift with wraparound) through $N/2$ positions in either direction.

[3] For example, the Matlab® mathematical software package provides a dedicated routine called fftshift that can be manually called, if required, for reordering the output of the routines fft and fft2 [Matlab02 p. 1-50]. But many other applications do not even provide this manual tool, and the reordering operation must be explicitly programmed by the user himself.

A 2D DFT algorithm is the 2D generalization of its 1D counterpart: it receives as input a 2D array of $N{\times}N$ complex numbers (say, samples of a given continuous 2D complex- or real-valued function $z = g(x,y)$), and returns as its output a 2D array of $N{\times}N$ complex numbers, usually indexed by $m = 0,...,N{-}1$ and $n = 0,...,N{-}1$, which contains the discrete 2D frequency spectrum. The 2D DFT algorithm works by applying the 1D DFT in two steps, first along each row in the horizontal direction, and then once again along each of the resulting columns in the vertical direction. Therefore, the reorganization required in the 2D case consists of swapping the first half and the second half of each horizontal row, and then swapping the first half and the second half of each of the resulting vertical columns. This results in a reshuffle of the four quadrants of the output matrix provided by the 2D DFT algorithm as shown in Fig. 3.2(b) [Brigham88 pp. 244–247; Bracewell86 p. 380].[4]

This can be easily extended to any MD DFT, if we remember that the MD DFT algorithm is based on successive applications of the 1D DFT along each of the M dimensions. The required MD reorganization consists therefore of swapping the first half and the second half of each row along the first dimension, then along the second dimension, and so on until the M-th dimension.

3.3 Reorganization of the input data of the DFT

While the reorganization of the output data (the discrete frequency values) obtained by the DFT is widely discussed in the literature, it is less commonly mentioned that sometimes a reorganization is also required in the input data. This is, again, a consequence of the non-centered DFT definition (i.e. the DFT definition with non-centered indices) which is used by virtually all DFT software applications. Indeed, just as a non-centered DFT gives the frequency 0 in the first element of its output array, so does it expect to get the sample belonging to $x = 0$ (the origin) in the first element of its input array (see Eqs. (2.13)–(2.14)). But if we wish to sample a function $g(x)$ within a symmetric range $-r...r$ and feed it to the DFT, the sample belonging to $x = 0$ will not be located in the first position of the input array ($k = 0$), but rather in the position $k = N/2$. Therefore, in such cases a reorganization (or rather *unreorganization*[5]) of the input array must be performed before feeding it to the DFT, as illustrated in Fig. 3.3.[6] More generally, if the sample

---

[4] As pointed out in [Brigham88 p. 244], this shuffle consists of a right circular shift of the four quadrants through two quadrants. It should be stressed, however, that no rotation is involved, and each quadrant remains in its original orientation.

[5] Note that in this case the original input array is already organized (i.e. ready to be plotted), so that we rather need to *unreorganize* it (i.e. to adapt it to the DFT convention). But because unreorganization involves precisely the same swapping as reorganization, the term reorganization can be used in both cases. For the sake of clarity, we will also use the more explicit but non-standard terms "DFT-organization" and "display-organization" whenever confusion may arise.

[6] Without this reorganization, the input function will be considered by the DFT as being laterally shifted, resulting in a corresponding change in the spectrum according to the shift theorem (see point 6 in Sec. 2.4.4 or [Brigham88 p. 107, 114]). But when the complex-valued spectrum is expressed in terms of magnitude (absolute value) and phase, this lateral shift will only affect the phase, while the magnitude will remain unchanged; see Sec. 8.3.

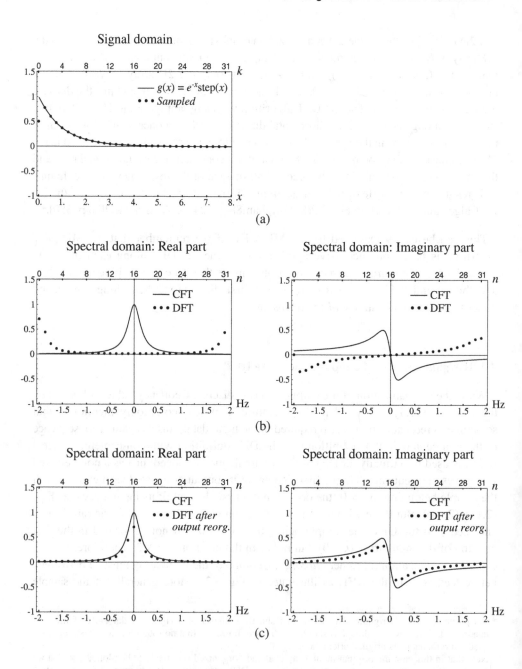

Figure 3.1: Example illustrating the need for DFT output reorganization when plotting the DFT results. (a) The continuous function $g(x) = e^{-x}$ step(x), represented here by a continuous line, and its discrete counterpart, represented by black dots, which is obtained by sampling $g(x)$ within the range 0...8 with a step of $\Delta x = 0.25$, so that $N = 32$. (b) The CFT (continuous Fourier

Spectral domain: Real part   Spectral domain: Imaginary part

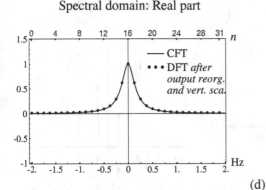
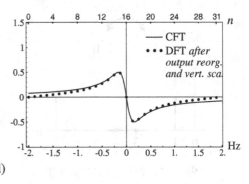

(d)

transform) of $g(x)$, $G(u) = 1/(1+[2\pi u]^2) - 2\pi ui/(1+[2\pi u]^2)$ [Bracewell86 p. 418], represented here by a continuous line, and its discrete counterpart, represented by black dots, as obtained by applying DFT to the discrete sequence shown in (a), without any corrections. (c) Same as (b), after reorganization of the DFT output. (d) Same as (c), after having applied to the DFT output the required vertical scaling correction, too, as explained in Sec. 4.4; note the close match between the corrected DFT and the CFT. In each of the plots the horizontal axis at the bottom corresponds to the continuous function being plotted, and the horizontal axis at the top corresponds to the integer index of the discrete data sequence that is represented by the dots. (Note that in (a) the value of the sampled function at the discontinuity point of $g(x)$, $x = 0$, is neither 0 nor 1, but rather the midvalue 0.5; see Sec. 8.2.)

belonging to $x = 0$ appears in any other position of the input array, the array should be cyclically shifted so as to bring $x = 0$ to the position $k = 0$; this may be considered as a *non-standard reorganization*. But if the sample belonging to $x = 0$ is already located in the first element of the input array, like in Fig. 3.1, then no reorganization of the input array is necessary. This could be the case, for example, when the input array contains a signal that was obtained from measurements that can be considered as starting from $x = 0$ [Smith07 p. 133], or when sampling a *causal* signal, i.e. a signal that is only non-zero for $x \geq 0$, such as the function $g(x) = e^{-x}$ step(x) of Fig. 3.1 [Brigham88 pp. 167–170]. The output array of the DFT, on its part, should always be reorganized, as explained above, for display purposes. Further examples illustrating the need for input and output reorganizations in 1D and 2D cases are given in Sec. 3.5. One may also see Example 9.3 and Fig. 9.5 in [Brigham88 pp. 175–177], or for the 2D case, Example 11.7 and Fig. 11.9 in [Brigham88 pp. 248–250]. It is important to note, however, that while a wrongly reorganized DFT output only gives a wrongly *displayed* output, a wrongly reorganized DFT input gives indeed a *wrong DFT output* (see also Sec. 3.6).

Before reorganization   After reorganization

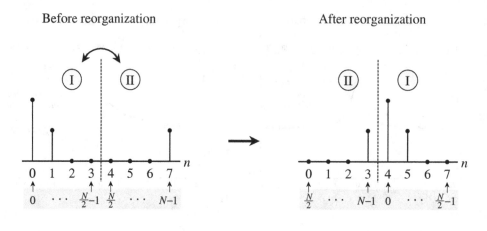

(a) The 1D case

Before reorganization   After reorganization

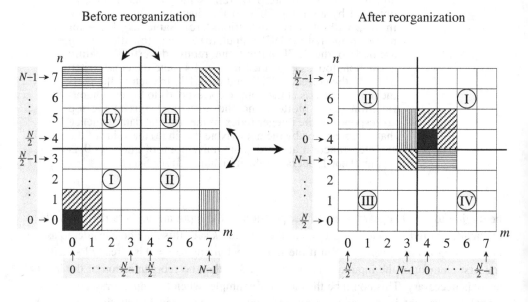

(b) The 2D case

Figure 3.2: Graphical illustration of the output reorganization operation: (a) in the 1D case, and (b) in the 2D case. Both cases are drawn here at a very low resolution, with $N = 8$, so that the individual elements (dots in (a) and square pixels in (b); see Sec. 1.5) can be clearly seen. In the 1D case the reorganization consists of swapping the first and second halves of the array (which is equivalent to a cyclical shift of $N/2$). In higher dimensions the reorganization consists of swapping half spaces along each dimension (which is equivalent to a cyclical shift of $N/2$ along each dimension).

Finally, note that if we are only interested in the *power spectrum* or in the *amplitude spectrum* of the given data (which is the case, for instance, when our input consists of random or statistical data) then no input reorganizations are necessary, since the input reorganization (being a cyclical shift of the DFT input array) only affects the *phase* of the spectrum, in accordance with the discrete counterpart of the shift theorem (see point 6 in Sec. 2.4.4 or [Brigham88 pp. 107, 114]). But this change in the phase *does* affect, of course, the real-valued part and the imaginary-valued part of the spectrum.

3.4 Data reorganizations in the case of IDFT

We now arrive to the inverse DFT (IDFT), which is used to retrieve the signal-domain data back from the frequency-domain data. Because the IDFT is computed with the same basic algorithm as the DFT (see, for example, [Press02 pp. 511–513] for the 1D case and [Press02 pp. 527–529] for the MD case), the reorganization issue may also arise when using the IDFT [Brigham88 Sec. 9.4]. Let us try to clarify this point.

Obviously, due to the reversibility of the DFT, applying IDFT to the output of the DFT must give back the original data. In other words, if we apply the IDFT to the output array obtained by the corresponding DFT algorithm, without any reorganizations, we obtain the same input array that was originally fed to the DFT, so that no reorganizations are required here. And indeed, it is important to emphasize that a spectral-domain reorganization is usually needed only when we wish to *plot* the resulting output array, because this array is not ordered in the conventional way we use to display our data. But if we need to further process the spectral data obtained by DFT and then apply the IDFT to bring it back into the signal-domain, processing should proceed with the non-reorganized (i.e. DFT-organized) data. In other words, whenever we wish to plot the intermediate

Figure 3.3: Example in which both input and output reorganizations are required for correctly plotting the DFT results. (a) The same continuous function $g(x) = e^{-x}$ step(x) as in Fig. 3.1, represented by a continuous line; this time, however, its discrete counterpart, represented by the black dots, is obtained by sampling $g(x)$ within the symmetric range $-4...4$ (with the same step $\Delta x = 0.25$, so that again $N = 32$). (b) The CFT (continuous Fourier transform) of $g(x)$, $G(u) = 1/(1+[2\pi u]^2) - 2\pi ui/(1+[2\pi u]^2)$ [Bracewell86 p. 418], represented by a continuous line, and its discrete counterpart, represented by black dots, as obtained by applying DFT to the discrete sequence shown in (a), without any corrections. (c) Same as (b), after reorganization of the DFT output. Unlike in Fig. 3.1(c), this output reorganization is not yet sufficient in the present case. (d) Same as (b), but this time only the input array has been reorganized before applying the DFT. (e) Same as (b), but this time with both input and output

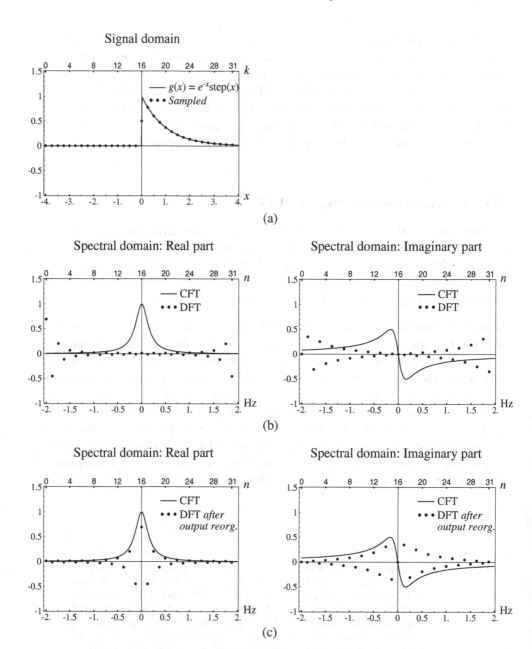

reorganizations. (f) Same as (e), after having applied to the DFT output the required vertical scaling correction, too, as explained in Sec. 4.4; note the close match between the corrected DFT and the CFT. In each of the plots the horizontal axis at the bottom corresponds to the continuous function being plotted, and the

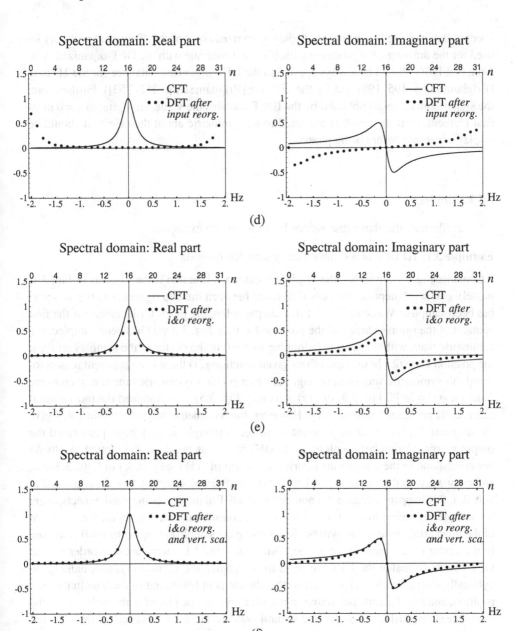

(d)

(e)

(f)

horizontal axis at the top corresponds to the integer index of the discrete data sequence that is represented by the dots. (Note that in (a) the value of the sampled function at the discontinuity point of $g(x)$, $x = 0$, is neither 0 nor 1, but rather the midvalue 0.5; this is explained in Sec. 8.2.)

spectral data we need to prepare a display-organized copy of the array, that will only be used for the drawings, but processing itself should continue with the DFT-organized data. Note that IDFT should be always applied to the DFT-organized data (see for the 1D case [Brigham88 pp. 195–199] and for the 2D case [Brigham88 pp. 252–253]). Furthermore, the signal-domain result obtained by the IDFT contains the value belonging to $x = 0$ in its first element, so if the result is expected to be symmetric about the origin, it should be display-organized before being plotted.

3.5 Examples

Let us illustrate the above discussions by a few simple examples.

Example 3.1: 1D DFT of a symmetrically sampled function:

Consider Fig. 3.3, in which the original continuous function is the same as in Fig. 3.1, namely $g(x) = e^{-x}$ step(x). The only difference between the two figures is in the way $g(x)$ has been sampled: While in Fig. 3.1 the sample belonging to $x = 0$ (the origin) is the first element of the samples array, at the position $k = 0$, in Fig. 3.3 $g(x)$ has been sampled in a symmetric way, with the sample belonging to $x = 0$ in the center of the samples array, at the position $k = N/2$.[7] In the case of our given function $g(x)$ there is no apparent reason for using this symmetric way of sampling, but when $g(x)$ is a symmetric function, such as the function $g(x) = \frac{1}{2}e^{-|x|}$ (Fig. 3.4) or $g(x) = $ rect(x) (Fig. 3.5), this is indeed the most natural and usually preferred sampling way. However, this symmetric way of sampling may give the unaware DFT user an unpleasant surprise: Although he may have performed the output reorganization before plotting the DFT results as required, the plotted results do not correspond to the continuous Fourier transform of $g(x)$ (see part (c) in Figs. 3.3–3.5, and compare with part (c) in Fig. 3.1 where this problem does not occur). As explained in Sec. 3.3, this happens because the non-centered DFT algorithm being used expects to get the sample belonging to $x = 0$ (the origin of the continuous function) in the first element of the input array, whereas in symmetric sampling the sample belonging to $x = 0$ is located in the center of the input array, at the position $k = N/2$.[8] In such cases, in order for the DFT results to match the CFT we need to reorganize the input array, too, namely, to cyclically shift it (with wraparound) so that the element belonging to $x = 0$ be in the first position, and the elements belonging to negative x values be placed at the right end of the sequence. Obviously, this is not the natural way we are used to visualize the sampled signal, and it may look particularly awkward in familiar functions such as $g(x) = $ rect(x). But as we can see in Figs. 3.3–3.5, when applying the DFT to the reorganized input array,

---

[7] Note that the continuous function $g(x)$ in Fig. 3.3 is exactly the same as in Fig. 3.1, and not a shifted version thereof. Otherwise, the continuous Fourier transform here wouldn't remain the same as in Fig. 3.1; rather, it would behave as predicted by the shift theorem (point 6 in Sec. 2.4.1).

[8] Note that this is equivalent to a cyclical shift of $N/2$ elements in the input array, so that the DFT output indeed behaves as predicted by the discrete counterpart of the shift theorem (see point 6 in Sec. 2.4.4 or [Brigham88 p. 107, 114]) — while the continuous Fourier transform remains unchanged (since $g(x)$ has not been shifted).

the DFT results (after the output reorganization) do match the corresponding CFT, up to a certain vertical scaling factor that will be discussed in detail in Sec. 4.4. (In fact, after performing this vertical scaling correction there may still remain some visible discrepancy between the CFT and the DFT results, as we can clearly see for example in Fig. 3.3(f) or in Fig. 3.5(f); this discrepancy is due to the *aliasing* and *leakage* artifacts, inherent to the DFT, that will be discussed later in Chapters 5 and 6.) ■

Example 3.2: 2D DFT of a symmetrically sampled function:

The 2D counterpart of Fig. 3.5 is shown in Fig. 3.6, where the original continuous function is the 2D unit rectangle function, $g(x,y) = \text{rect}(x)\,\text{rect}(y)$. Just as in the 1D case (Fig. 3.5), it is natural and commonplace to sample this continuous function in a symmetric way, with the sample belonging to $(x,y) = (0,0)$ in the center of the 2D sampling array, at the position $(N/2,N/2)$. However, just as in the 1D case, this symmetric way of sampling may give trouble to an unexperienced DFT user: Once again, even if the output reorganization is correctly performed before plotting the DFT results, the resulting plot does not correspond to the continuous Fourier transform of $g(x,y)$, which is the 2D sinc function $G(u,v) = \text{sinc}(u)\,\text{sinc}(v)$. But as shown in Fig. 3.6(e), if we remember to reorganize the input array before applying the 2D DFT (so that the sample corresponding to $(x,y) = (0,0)$ appears in the first position of the 2D array, $(k,l) = (0,0)$, and the 3 negative parts of the rectangle are moved to the 3 other corners of the array, as shown in Fig. 3.2(b)) then the 2D DFT results (of course, after reorganization of the output array) are indeed correct. ■

Example 3.3: Spectral manipulation (filtering) of a signal (see Fig. 3.7):

Suppose that we are given the signal $g(x) = \text{rect}(x/2)$, whose spectrum is $G(u) = 2\text{sinc}(2u)$, and that we wish to low-pass filter it by multiplying its spectrum $G(u)$ by a filter such as $F(u) = \text{rect}(u/2)$. We start by sampling the original signal $g(x)$. Because this function is symmetric about the origin, we sample it within a symmetric range, say, $-r...r$. This input array is ready for being plotted (see the signal domain plot in the top row of Fig. 3.7), but before we pass it to the DFT we first have to DFT-organize it so that the sample belonging to $x = 0$ be indeed in the first position, $k = 0$. We now feed the DFT-organized input array to the DFT, and obtain in the output array a sampled version of the spectrum $G(u) = \text{sinc}(u)$, whose frequency $u = 0$ is located in the first position of the array, $n = 0$. If we wish to plot this intermediate result, we can make a copy of this output array, display-organize it, and pass it to the plotting routine (see the spectral domain plots in the top row of Fig. 3.7); but we keep the original DFT-organized output array for the following processing steps.

We now consider the filter $F(u)$. Because this function is symmetric about the origin, we sample it within a symmetric range, as shown in the second row of Fig. 3.7, and we then DFT-organize it so that the frequency $u = 0$ appears in the first position. Now, we multiply element-by-element the sampled spectrum of $G(u)$ and the sampled filter (both of which contain the frequency $u = 0$ in the first position) to obtain the filtered spectrum. Note that in order for this element-by-element multiplication to be meaningful, both arrays

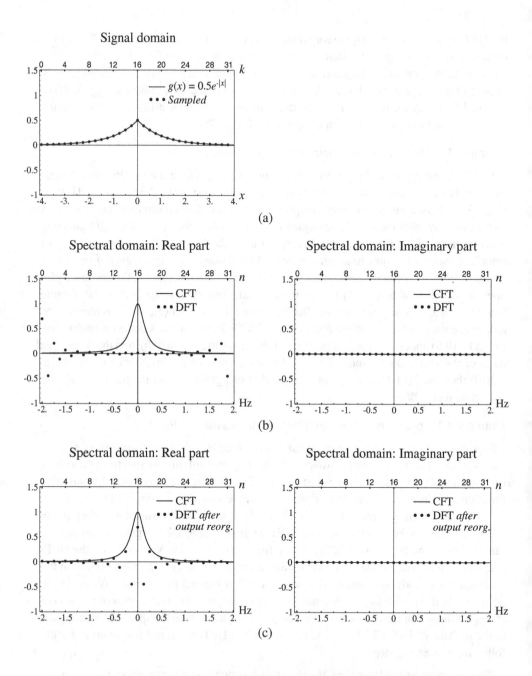

Figure 3.4: Same as Fig. 3.3, but this time with a symmetric continuous function $g(x) = \frac{1}{2}e^{-|x|}$. In such symmetric cases it is indeed natural to sample $g(x)$ in a symmetric way, as shown in (a); but as explained in Example 3.1, this implies the need to reorganize the input array of the

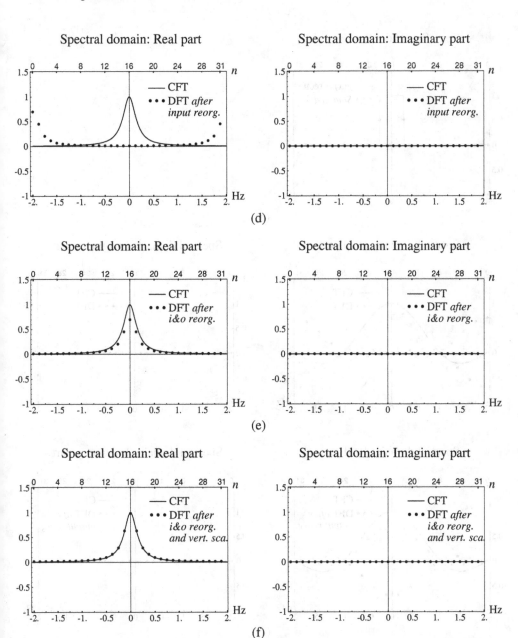

(d)

(e)

(f)

DFT, too. In each of the plots the horizontal axis at the bottom
corresponds to the continuous function being plotted, and the
horizontal axis at the top corresponds to the integer index of the
discrete data sequence that is represented by the dots.

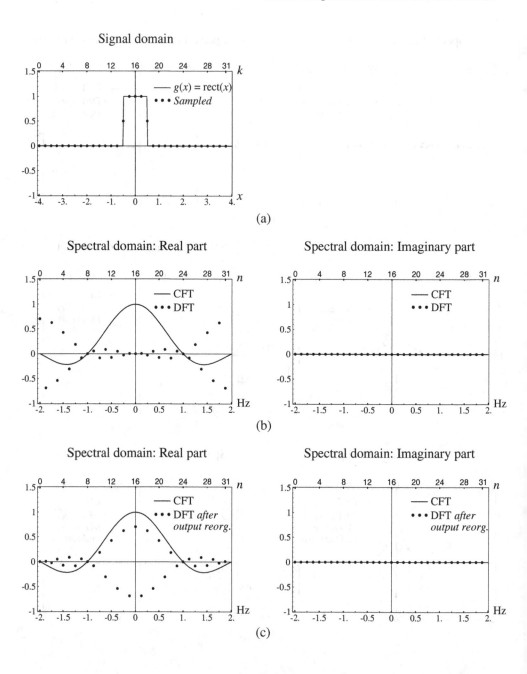

Figure 3.5: Same as Fig. 3.3, but this time using the familiar symmetric continuous
function $g(x) = \text{rect}(x)$. In this case, too, it is natural to sample $g(x)$ in a
symmetric way, as shown in (a); and as explained in Example 3.1, this
implies the need to reorganize the input array of the DFT, too. In each of the
plots the horizontal axis at the bottom corresponds to the continuous

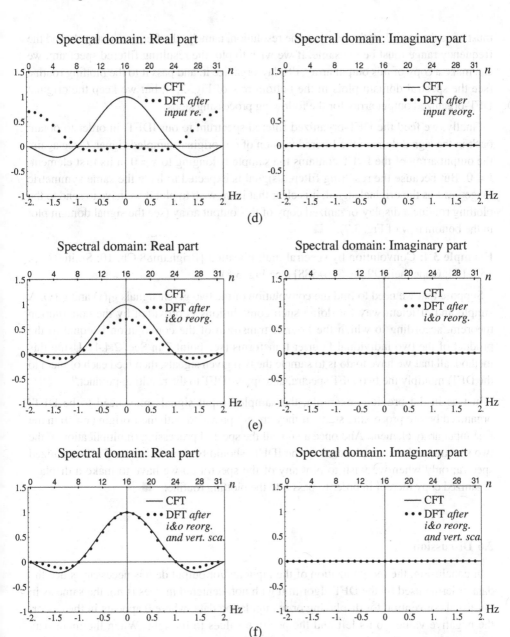

Spectral domain: Real part   Spectral domain: Imaginary part

(d)

Spectral domain: Real part   Spectral domain: Imaginary part

(e)

Spectral domain: Real part   Spectral domain: Imaginary part

(f)

function being plotted, and the horizontal axis at the top corresponds to the integer index of the discrete data sequence that is represented by the dots. (Note that in (a) the value of the sampled function at the two discontinuity points of $g(x)$, $x = -0.5$ and $x = 0.5$, is neither 0 nor 1, but rather the midvalue 0.5; this is explained in Sec. 8.2.)

must have the same length and the same resolution, namely, the frequency step Δu and the frequency range must be the same. If we wish to plot the resulting filtered spectrum, we can make a copy of this output array, display-organize it, and pass it to the plotting routine (see the spectral domain plots in the bottom row of Fig. 3.7); but we keep the original DFT-organized output array for the following processing steps.

Finally, we feed the DFT-organized filtered spectrum to our IDFT, in order to obtain back in the signal domain the filtered version of our original sampled signal. Once again, the output array of the IDFT contains the sample belonging to $x = 0$ in its first element, $k = 0$. But because the resulting filtered signal is expected to have the same symmetric range $-r...r$ as the original signal, it is clear that if we wish to plot it we have to send to the plotting routine a display-organized copy of this output array (see the signal domain plot in the bottom row of Fig. 3.7). ∎

Example 3.4: Convolution by spectral multiplication [Brigham88 Ch. 10; Smith07 pp. 152–154; Oppenheim99 pp. 576–588] (see Fig. 3.8):

Suppose that we need to find the convolution of the two given signals $g_1(x)$ and $g_2(x)$. A simple and efficient way for doing such convolutions is provided by the convolution theorem, according to which the Fourier transform of the convolution is equal to the product of the two individual Fourier transforms (see point 7 in Sec. 2.4.4). Using this method, all that we have to do is to sample the two given signals, then feed each of them to the DFT, multiply the two DFT spectra, and apply IDFT to the resulting product.[9]

Just as in the previous example, the sampled input signals may need to be DFT-organized before processing starts, if they are not provided with their origin ($x = 0$) in the first input array element. And once again, all the spectral processing (multiplication of the two spectra, and then application of the IDFT) should be done using the DFT-organized spectra; only when we wish to plot any of the spectra do we have to make a display-organized copy thereof in order to pass it to the plotting routine. ∎

3.6 Discussion

In conclusion, the reorganization of the input and/or output data is necessary since the data ordering used by the DFT algorithm with non-centered indices is not the same as in our usual convention for displaying or manipulating data, where 0 appears in the center, the negative values to its left and the positive values to its right. When the input data (obtained by sampling) does not correspond to the convention used by the DFT algorithm, in which the value belonging to $x = 0$ appears in the first element of the array, a DFT-organization is required. And when the output data provided by the DFT does not correspond to our displaying conventions, we have to display-organize the data before plotting it. These rules are obviously true for the 2D and MD cases, too.

---

[9] Note, however, that in order that the cyclic convolution obtained by DFT correctly correspond to the original continuous convolution $g_1(x)*g_2(x)$ some precautions may be needed, such as zero padding of the sampled input data. For further details, see Sec. B.7 in Appendix B or the provided references.

Based on this insight, it should be clear by now in which cases a reorganization is required. What happens, however, if we forget to reorganize the data, or if, on the contrary, we perform a superfluous reorganization that is not required? If the wrongly reorganized data is the *output* of the DFT, its plot will simply appear wrongly reorganized (see, for example, part (b) in Fig. 3.1). But if the *input* of the DFT is wrongly reorganized, how will this affect the DFT results? The answer is given, for the 2D case, in references such as [Briggs95 p. 150], [Gonzalez87 p. 77] or [Rosenfeld82 pp. 26–27]: When applying DFT to such a wrongly reorganized input array, each element in the output array is multiplied by $(-1)^{m+n}$, resulting in a checkerboard of positive and negative values as shown, for example, in Fig. 3.6(c).[10] This effect results from the shift theorem (see point 6 in Sec. 2.4.5) for a shift of $N/2$ elements along each dimension. The extension of this result to DFTs of any other dimension is straightforward (see, for example, part (c) in Figs. 3.3–3.5 for the 1D case). Therefore, if we expect to obtain a rather smooth result, but our DFT-based processing yields a violently jumpy result consisting of alternating positive and negative values (which may sometimes look completely dark when densely plotted), one may reasonably suspect that this is related to a reorganization problem in the input of the DFT in question.

Figure 3.6: A 2D generalization of Fig. 3.5, plotted at the same low resolution so that the individual pixels can be clearly seen. The continuous rectangular function $g(x,y) = \text{rect}(x)\,\text{rect}(y)$ is sampled in a symmetric way, but once again this implies the need to reorganize the input array of the DFT, too. (a) The discrete counterpart of the rectangular function $g(x,y)$, as obtained by sampling it within the symmetric range $-4...4$ along both axes (with the same step $\Delta x = \Delta y = 0.25$, so that again $N = 32$). (b) The 2D DFT of the 2D discrete data sequence shown in (a), without any corrections. (c) Same as (b), after reorganization of the DFT output. As explained in Sec. 3.3, this output reorganization is not yet sufficient in the present case. (d) Same as (b), but this time only the input array has been reorganized before applying the DFT. (e) Same as (b), but this time with both input and output reorganizations. (f) Same as (e), after having applied to the DFT output the required vertical scaling correction, too, as explained in Sec. 4.4. In each of the plots the horizontal axis at the bottom and the vertical axis at the left hand side correspond to the continuous function, while the horizontal axis at the top and the vertical axis at the right hand side correspond to the integer indices of the 2D discrete data sequence. But unlike in the 1D case (Fig. 3.5), the original continuous function $g(x,y)$ and its CFT cannot be added here to the same plots for comparison. See Sec. 1.5.2 for further details on the use of 2D density plots and on the possible printing artifacts they may contain, and which should not be confused with true gray level fluctuations. (Note that in (a) the value of the sampled function at the discontinuity points along the rectangle's edges is neither 0 nor 1 but rather the midvalue 0.5, except for the 4 corners where the value is 0.25; see Sec. 8.2.)

---

[10] If the complex-valued spectrum is expressed in terms of magnitude (absolute value) and phase, this effect will only concern the phase, while the magnitude will remain unchanged.

Signal domain

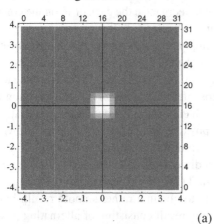

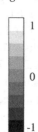

(a)

Spectral domain: Real part   Spectral domain: Imaginary part

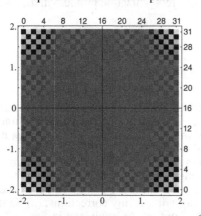   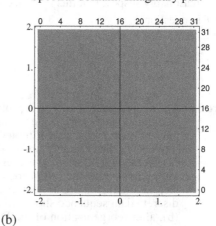

(b)

Spectral domain: Real part   Spectral domain: Imaginary part

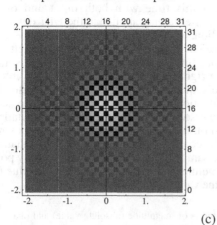   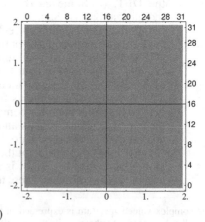

(c)

Spectral domain: Real part   Spectral domain: Imaginary part

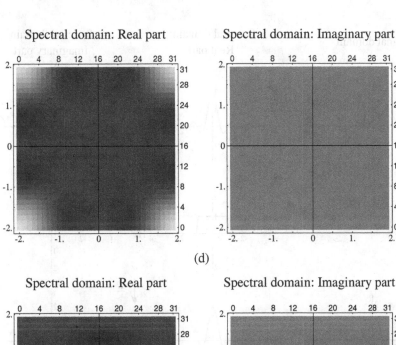
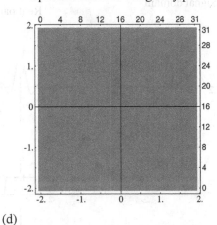

(d)

Spectral domain: Real part   Spectral domain: Imaginary part

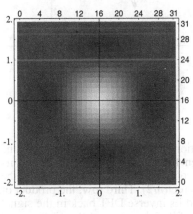
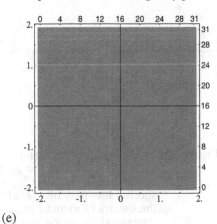

(e)

Spectral domain: Real part   Spectral domain: Imaginary part

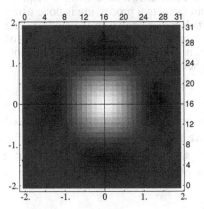
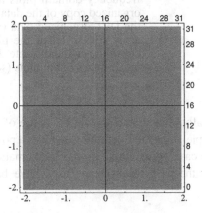

(f)

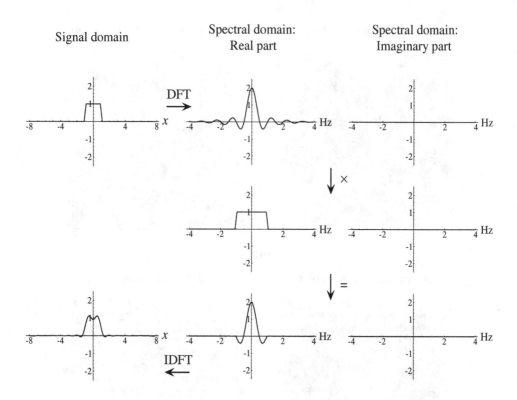

Figure 3.7: Example of spectral manipulation (filtering) by DFT. First row: the
sampled version of the given function $g(x) = \text{rect}(x/2)$ and its
spectrum $G(u) = 2\text{sinc}(2u)$ as obtained by DFT. Second row: the
spectral low-pass filter $F(u) = \text{rect}(u/2)$. Third row: the product of
the spectra $G(u)$ and $F(u)$, and its inverse DFT back in the signal
domain, showing the low-pass filtered version of the given signal
$g(x)$. As explained in Example 3.3, both the signal-domain and
frequency-domain plots have been drawn here from a display-
organized copy of the data, but the actual processing is done using
the DFT-organized data (which is not shown in the figure). The
array size in all of the plots here is $N = 128$. For the sake of
convenience the discrete data is drawn using connected line plots
rather than symbol plots (see Sec. 1.5.1).

Finally, one should always be aware of the precise DFT definition being used. For
example, if the DFT definition includes a positive exponent sign in the direct transform (as
is the case, for example, in the Mathematica® software package [Wolfram96 p. 868])
while the CFT results being used have been obtained from a CFT definition with a

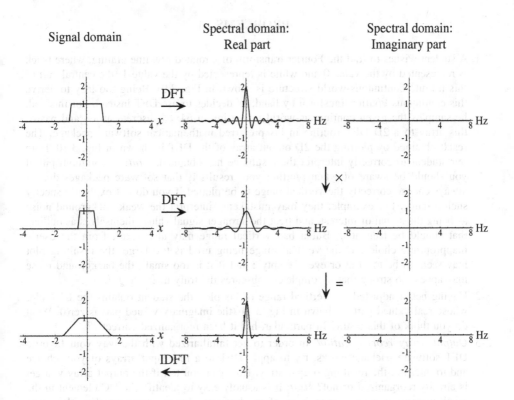

Figure 3.8: Example of convolution by spectral multiplication using DFT. First row: a sampled version of the first given function $g_1(x) = \text{rect}(x/2)$ and its spectrum $G_1(u) = 2\text{sinc}(2u)$ as obtained by DFT. Second row: a sampled version of the second given function $g_2(x) = \text{rect}(x)$ and its spectrum $G_2(u) = \text{sinc}(u)$ as obtained by DFT. Third row: the product of the spectra $G_1(u)$ and $G_2(u)$, and its inverse DFT back in the signal domain, showing the convolution $g_1(x)*g_1(x)$. As explained in Example 3.4, both the signal-domain and frequency-domain plots have been drawn here from a display-organized copy of the data, but the actual processing is done using the DFT-organized data (which is not shown in the figure). The array size in all of the plots here is $N = 128$. For the sake of convenience the discrete data is drawn using connected line plots rather than symbol plots (see Sec. 1.5.1).

negative exponent sign in the direct transform (as in [Bracewell86 p. 6]), then the DFT results we obtain are reversed with respect to the CFT. In such cases the user must remember to invert the resulting DFT output, in addition to all the corrections mentioned so far (see Sec. 8.3 for more details).

PROBLEMS

3-1. A student wishes to find the Fourier transform of a rotated b/w line grating, where black is represented by the value 0 and white is represented by the value 1 (the central part of this infinite continuous-world structure is shown in Fig. 3.9). Being too lazy to derive this continuous Fourier transform by hand, he decides to use DFT instead. For this end, he samples the given continuous-world image into a 64×64 discrete array, and passes this array to a 2D DFT routine in his preferred mathematical software package. The result obtained by plotting the 2D output array of the DFT is shown in Fig. 3.10. Help our student to correctly interpret the results he has obtained. *Hint*: A commom pitfall you should be aware of when plotting your results is that software packages do not always choose correctly the vertical range to be plotted if you do not explicitly specify such a range. For example, they may mistakenly interpret the weak background noise as being the signal of interest, and treat the stronger signal values themselves as outliers that should be truncated. But a bad vertical range may also result from the user's inappropriate choice. If the vertical range being used is too large, the resulting plot may seem to be too flat or even "empty"; and if it is too small, the background noise may appear so strong that it completely obscures the truly interesting data.

3-2. Having better adjusted the vertical range of his plot, the student obtains the DFT plot whose real-valued part is shown in Fig. 3.11 (the imaginary-valued part is zero). What do you think of this result? In particular, has it been reorganized correctly?

3-3. *Output array reorganization.* In order to get familiarized with the way your favorite DFT software package works, try to apply DFT to a few input arrays of your choice and to interpret the resulting output array. How can you tell if the output array you get is already reorganized or not? *Hint*: It is usually easy to identify the DC element in the resulting output array, since it is often (but not always) stronger than the other elements; furthermore, if your input sequence is purely real-valued, the output array must be Hermitian, meaning that its real-valued part must be symmetric to both sides of the DC element. If you see that the DC is located in the first element of the output array, you may guess that the output array is not reorganized. But if the DC is located in the center of the array and surrounded by symmetric elements to both sides (in terms of their real-valued part), you may guess that the output array is already reorganized.

3-4. *Input array reorganization.* Apply your favorite DFT software package to the same input arrays as in the previous problem, but this time after having reorganized them. How does the input reorganization affect the DFT results? When is the input reorganization necessary?

3-5. *Missing or superfluous input array reorganization.* Suppose you forgot to reorganize the input array, or that, on the contrary, you have reorganized it although no such reorganization was required.

(a) What do you expect to see in the DFT output? *Hint*: See Sec. 3.6.

(b) It is easy to design an input signal that remains unchanged after the input reorganization. For example, in the 1D case any signal whose first and second array halves are identical will do. What are the respective conditions in the 2D and MD cases?

(c) In cases where the input signal remains unchanged after the input reorganization, a missing or superfluous input-array reorganization will have no effect on the DFT output. How does this agree with your answer in part (a)?

3-6. *Missing or superfluous input array reorganization.* What do you expect to get in Examples 3.3 and 3.4 if you forget to correctly reorganize the input arrays?

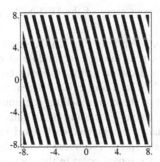

Figure 3.9: A rotated line grating (see Problem 3-1).

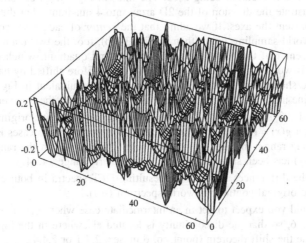

Figure 3.10: The DFT plot obtained by the student in Problem 3-1.

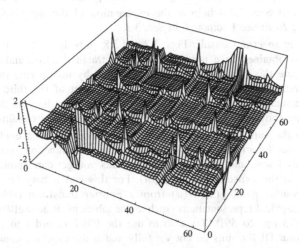

Figure 3.11: The DFT plot obtained by the student in Problem 3-2.

3-7. When do you expect the output array of the DFT to be purely real-valued, and when will it be complex valued?

3-8. Consider Figs. 3.2(b) and 3.6. While in Fig. 3.6 the origin is located in the middle of a pixel and the two axes, too, pass through the middle of pixels, in Fig. 3.2(b) the axes seem to pass *between* pixels. How do you explain this contradiction? Is any of the two figures flawed by an error? *Hint*: There exist different ways to interpret pixels of the discrete world with respect to the corresponding continuous-world image. A pixel is a 2D (or MD) entity having a side dimension of Δx units (in the signal domain) or Δu units (in the spectral domain), but in fact each pixel represents one ideal sampling point of the continuous-world space. The question is, therefore, which point within the pixel corresponds to the true sampled point in the continuous world: is it the center of the pixel? or its bottom left point? or any other point? There exist different possible conventions, and each of them is valid and useful. But confusion may occur if we are not aware of the convention being used. The lines in Fig. 3.2(b) are simply separation lines that illustrate the division of the 2D array into 4 quadrants; but these lines do not intend to represent the axes. If we admit that the *center* of each pixel corresponds to a continuous-world sampling point, then in Fig. 3.2(b), too, the two main axes will pass through the center of the bottom-left pixel of the first quadrant, which corresponds to the continuous-world origin. The true axes will be therefore shifted by half a pixel with respect to the shown separation lines. See also the 1D equivalent in Fig. 3.2(a), where the vertical line in the center is located half a pixel away from the true vertical axis.

3-9. Figs. 3.1 and 3.3 show the DFT results obtained for the same original continuous-world function $g(x) = e^{-x}$ step(x). The difference between the two cases is in the choice of the sampling range: in Fig. 3.1 $g(x)$ has been sampled within the range 0...8, while in Fig. 3.3 $g(x)$ has been sampled within the symmetric range −4...4.

(a) Explain the differences between the resulting DFT spectra in both cases, and their fidelity to the original continuous-world spectrum $G(u)$.

(b) What would you expect to get in an intermediate case where $g(x)$ is sampled within the range −2...6, so that the discontinuity is located elsewhere in the input array? You may try to use the shift theorem (point no. 6 in Sec. 2.4.1 or 2.4.4).

3-10. Can you explain the input and output data reorganizations of the DFT using the figures of Sec. 2.5, which illustrate the graphic development of the DFT from the CFT?

3-11. Can the rules of Sec. 2.4.4 help in the explanation of the input and output data reorganizations? *Hint*: See Footnote 6 in Sec. 3.3.

3-12. *Multiplication of spectra.* Consider Figs. 3.7 and 3.8, in both of which the spectrum in the third row is obtained as a product of the spectra in the first and second rows. Trying to compute this product, a student argues that by multiplying the real-valued parts of the first two rows one gets the real-valued part of the third row, and by multiplying the imaginary-valued parts of the first two rows he gets the imaginary-valued part of the third row. Is this reasoning correct? Under what conditions?

3-13. In order to make sure that you master the data reorganizations explained in this chapter, apply your favorite DFT software package to some functions $g(x)$ whose CFT $G(u)$ you already know, and see if the plotted results you get correspond (up to some scaling factor) to the expected plot of $G(u)$. For this end it may be convenient to choose your Fourier pairs $g(x)$, $G(u)$ from a Fourier transform table which also provides their graphical representations (such as the tables in [Bracewell86, Chapter 21] or [Champeney73 pp. 20...39]). Try also to use the IDFT in order to get $g(x)$ back from $G(u)$. If your DFT results do not yet fully match the expected continuous-world results, try this exercise once again after you have read the following chapters.

Chapter 4

True units along the axes when plotting the DFT

4.1 Introduction

As already mentioned above, the DFT is often used as a tool for the numerical calculation of the Fourier transform (the frequency spectrum) of a given continuous function. For this end, the original continuous function in question must be sampled, and the resulting N-element array (or in the MD case, $N\times...\times N$-element array) is fed as input to the DFT. The resulting N-element (or $N\times...\times N$-element) output array contains the discrete counterpart of the desired frequency spectrum, and after the appropriate reorganization (see Chapter 3) it can be plotted to illustrate graphically the resulting spectrum.

However, there still exists a significant obstacle on our way before we can graphically represent the results obtained by the DFT in a meaningful way: Both the input and output arrays of the DFT consist of a list of values (real or complex numbers), that are indexed by an integer running for example between $0,...,N-1$ (or, in the MD case, by M such indices).[1] But in our graphical plot we would like to mark along the axes the true values in terms of the units that are being used in the original continuous function (meters, seconds, etc.) or in its spectrum (cycles per unit, or Hz). For the input array this is usually not a problem, at least if the sampling of the original continuous function is done by the user himself, so that he knows the correspondence between the input array indices and the true coordinates of the original function. However, in the case of the output array, it is not always evident how to correctly replace in our graphical plot the integer indices of the reorganized output array (say, $0,...,N-1$ along each axis) by the true frequencies. Furthermore, there also exists the question of how to assign the true units along the vertical axis, i.e. what is the connection between the *values* obtained in the output array of the DFT and the true heights in the spectrum of our original function. In the following three sections we will treat each of these questions separately. Some illustrative examples are provided at the end of the chapter, in Sec. 4.5.

4.2 True units for the input array

Although the connection between the true values along the axis of the given function and the index of the DFT's input array is usually known, let us illustrate the situation with a simple 1D example. Suppose we are given the following continuous symmetric function (see Fig. 4.1(a)):

---

[1] Note that in some applications array indices run between $1,...,N$ rather than between $0,...,N-1$. This is the case, for example, in the Mathematica® software package [Wolfram96 p. 41].

$$g(x) = \text{sinc}(x) = \begin{cases} \dfrac{\sin(\pi x)}{\pi x} & x \neq 0 \\ 1 & x = 0 \end{cases} \qquad (4.1)$$

If we wish to feed this function to the DFT, we first need to sample it to create the N-element input array. Since our function $g(x)$ is symmetric, we may want to sample it within the symmetric range $-R/2...R/2$, where R is a chosen real number, so that the indices $k = 0,...,N-1$ of the input array correspond to the sampled values $x = -R/2 ,..., x = R/2$. However, because the input array's length N is usually even (most often a power of 2), it follows that if we assign to $k = 0$ the point $x = -R/2$ and to $k = N-1$ the point $x = R/2$, the center of our symmetric function (the point $x = 0$) will not be sampled. If this is undesirable, the best solution is to assign to $k = 0$ the point $x = -R/2$ and to $k = N/2$ the point $x = 0$;[2] in this case we will have one less sample in the positive end of the x axis, since the maximum positive value being sampled, which is assigned to $k = N-1$, is $x = R/2 - R/N$ rather than $x = R/2$. The index $k = 0,...,N-1$ therefore corresponds to the x values of $-R/2, ... ,R/2 - R/N$, and the step Δx between successive x values (i.e. the sampling interval) is given, thanks to the following simple connection [Gonzalez87 p. 96]:

$$R = N\Delta x \qquad (4.2)$$

by $\Delta x = R/N$. Therefore, the correspondence between the k-th element in the input array and the true x value is given in this example by:

$$x = -R/2 + kR/N \qquad (4.3)$$

A simple verification of this relationship shows, indeed, that for the first array element, $k = 0$, we have $x = -R/2$; for the element $k = N/2$ we have $x = 0$; and for the last element $k = N-1$ we have $x = R/2 - R/N$.

Figure 4.1: True units along the axes when plotting the DFT input and output. (a) The continuous function $g(x) = \text{sinc}(x)$ (represented by a continuous line), and its discrete counterpart (represented by black dots) that is obtained by sampling $g(x)$ within the symmetric range $-4...4$ with a step of $\Delta x = 0.25$, so that $N = 32$. (b) The CFT of $g(x)$, $G(u) = \text{rect}(u)$ [Bracewell86 p. 415] (represented by a continuous line), and its discrete counterpart (represented by black dots) as obtained by applying DFT to the discrete sequence shown in (a), after both input and output reorganizations. (c) Same as (b), after having applied to the DFT output the required vertical scaling correction, too, as explained in Sec. 4.4; note the close match between the corrected DFT and the CFT. In each of the plots the

---

[2] Having the value $x = 0$ at the index $k = N/2$, i.e. as the first element in the second half of the input array, is indeed consistent with our reorganization convention: As we have seen in Chapter 3, the frequency 0 is also located at the same index $n = N/2$ in the output array (after its reorganization). Note that keeping the same conventions for the input and output arrays is desireable due to the reversibility of the Fourier transform (and of the DFT), since our original function $g(x)$ can be also regarded, up to an axis inversion, as the spectrum of the output function $G(u)$ (see point 4 in Sec. 2.4.1).

Signal domain

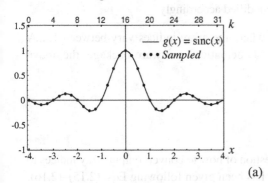

(a)

Spectral domain: Real part   Spectral domain: Imaginary part

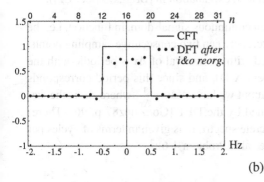

(b)

Spectral domain: Real part   Spectral domain: Imaginary part

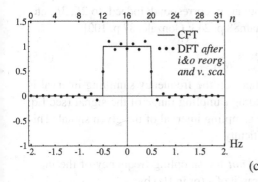

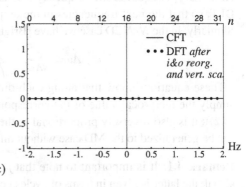

(c)

horizontal axis at the top corresponds to the integer index of the discrete
data sequence that is represented by the dots, and the horizontal axis at the
bottom corresponds to the true units, i.e. the units of the respective
continuous function. (Note that in (c) the value obtained by the DFT at the
discontinuity points of $G(u)$, $u = -0.5$ and $u = 0.5$, is neither 0 nor 1, but
rather the midvalue 0.5; this is explained in Sec. 8.2.)

Note, however, that if we wish to reorganize this input array before feeding it to the DFT (see Sec. 3.3), these indices will have to be modified accordingly.

Finally, it should be noted that in cases where the array indices vary between $1,...,N$ rather than between $0,...,N-1$, like in the Mathematica® software package, the above arithmetics should be adapted accordingly.

4.3 True units for the output array

We now arrive to the more interesting question of how to convert output array indices to true frequencies. A first hint to the answer has been given following Eqs. (2.15)–(2.16). The full answer is as follows (see, for example, the brief discussion in [Bracewell86 p. 358] or [Bracewell95 p. 168], or the more detailed explanation in [Briggs95 Sec. 1.2]):

Let Δx be the sampling interval of the given continuous signal-domain function, i.e. the interval (in terms of seconds, length units, etc.) between two consecutive sampling points. The Fourier transform of a function sampled with an interval of Δx is periodic with the period $\frac{1}{\Delta x}$ (see Sec. 2.5 or Sec. D.1 in Appendix D), and since this period corresponds to N spectral-domain elements in the DFT output we have $N\Delta u = \frac{1}{\Delta x}$, where Δu is the frequency step of the discrete spectrum obtained by the DFT [Gonzalez87 p. 96]. Therefore f_1, the first frequency above 0 in the discrete spectrum, is given in terms of cycles per unit of x, which we denote here by Hz, by (see also Problem 4-6):[3]

$$f_1 = \Delta u = \frac{1}{N\Delta x} \tag{4.4}$$

Note that f_1 is equal to the frequency step Δu of the discrete spectrum obtained by the DFT, so that the next frequencies in the output array correspond, indeed, to $2f_1$, $3f_1$, etc. Similarly, in the $N \times N$ 2D case we have [Brigham88 p. 243; Gonzalez87 p. 100]:

$$\Delta u = \frac{1}{N\Delta x}, \qquad \Delta v = \frac{1}{N\Delta y} \tag{4.5}$$

These equations show that along each dimension the frequency sampling interval is simply the reciprocal value of the corresponding sampling range of the signal (see Eq. (4.2)); it is also inversely proportional to the sampling interval of the given signal. This can be generalized to the MD case without difficulty.

Remark 4.1: It is important to note that f_1 is *not* the sampling frequency of the input signal; the latter is given in terms of cycles per unit of x (or in Hz) by:

$$f_s = \frac{1}{\Delta x} \tag{4.6}$$

[Smith07 pp. 2, 115]. As we can see, it follows that $f_s = Nf_1 = N\Delta u$. ∎

---

[3] Note that $\Delta u = \frac{1}{N\Delta x}$ is often said to be the spectral resolution of the DFT. However, this term should be used with care, since as we will see in Chapter 6 the effective resolution of the DFT is often limited by the leakage effect [Brigham88 p. 180]. On the frequency unit "Hz" and its use here, see in the Glossary.

Let us now show explicitly, based on Eq. (4.4), the connection between the element index in the reorganized output array and the real frequency in Hz. As we have seen in Sec. 3.2, the indices $n = 0,...,N-1$ of the reorganized output array correspond to the frequencies $-(N/2)f_1, ... ,(N/2-1)f_1$ in steps of f_1. Thus, the first element in the reorganized output array, $n = 0$, corresponds to the frequency $-(N/2)f_1 = -\frac{1}{2\Delta x}$; the element $n = N/2$ corresponds to the frequency 0; the element $n = N/2 + 1$ corresponds to the frequency f_1; the element $n = N/2 + 2$ corresponds to the frequency $2f_1$; and the last element, $n = N-1$, corresponds to the frequency $(N/2-1)f_1 = \frac{1}{2\Delta x} - f_1$. The general formula is, therefore:

$$f_{Hz} = -(N/2)f_1 + nf_1$$
$$= (n - N/2)f_1 \qquad (4.7)$$

and using Eq. (4.4) above:

$$f_{Hz} = (n - N/2)\frac{1}{N\Delta x} \qquad (4.8)$$

This relationship allows us, when we graphically plot the output array obtained by the DFT (after its reorganization), to replace the default integer units along the axes (i.e. the array index $n = 0,...,N-1$) by the correponding true frequencies. See, for example, Figs. 3.1(d) or 4.1(c) for the 1D case, and Fig. 3.6(f) for the 2D case.

It is important to stress here that the lowest negative and highest positive frequencies provided by the DFT are (see also Problem 4-6):[4]

$$f_{min} = -(N/2)f_1 = -\frac{1}{2\Delta x} \qquad f_{max} = (N/2-1)f_1 = \frac{1}{2\Delta x} - f_1 \qquad (4.9)$$

or in other words, by virtue of Eq. (4.6):

$$f_{min} = -\tfrac{1}{2}f_s \qquad f_{max} = \tfrac{1}{2}f_s - \Delta u \qquad (4.10)$$

In the MD case this applies, of course, along each of the M axes separately.

Beyond these frequencies the DFT assumes periodicity of the spectrum (see Sec. 2.5 or Sec. D.1 in Appendix D). So what happens if the original continuous function contains frequencies which exceed these limits? We will see the answer in Chapter 5.

Finally, note that the following relationship, that is sometimes found in the literature:

$$f_{Hz} = nf_1 = \frac{n}{N\Delta x} \qquad (4.11)$$

---

[4] In fact, due to the cyclic (periodic) nature of the DFT the element $n = 0$ (after the reorganization of the output array) corresponds to both of the positive and negative frequencies $\pm(N/2)f_1 = \pm\frac{1}{2\Delta x}$ [Press02 p. 510], so that we can say that $f_{max} = -f_{min} = \frac{1}{2\Delta x}$. This is also in line with the well known sampling theory result which says that when sampling $g(x)$ with a sampling interval of Δx, no frequencies beyond $\pm\frac{1}{2\Delta x}$ may appear in the resulting spectrum [Castleman79 pp. 235–236; Briggs95 p. 9; Harris98 p. 713; Smith07 p. 229] (or, in other words, by virtue of Eq. (4.6), that discrete data can only contain frequencies up to half of the sampling frequency f_s [Smith03 pp. 42, 149]). Note that the frequency $\frac{1}{2}f_s = \frac{1}{2\Delta x}$ is often called the *Nyquist frequency*, but we prefer to avoid using this term because of its ambiguity. As explained in [Smith03 pp. 41–42 and 638], it turns out that different authors use this term in four different meanings: the heighest frequency contained in a (band-limited) signal; twice this frequency; the sampling frequency; or half of the sampling frequency.

may be misleading due to the uncertainty about the meaning of its index n: Clearly, it is not the index $0,...,N-1$ of the output array — neither before nor after its reorganization. In fact, this relationship holds for $n = -N/2, ... ,N/2-1$, but not for $n = 0,...,N-1$. This also explains why some authors prefer to use, as we have already noted in Sec. 2.3, a variant of the DFT definition where the indices vary between $-N/2$ and $N/2-1$ (see, for example, [Briggs95 p.67; Bracewell86 p. 362]).

4.3.1 The particular case of periodic functions

We terminate this section with an example illustrating the particular case in which $g(x)$ is periodic. This will help us to better understand the relationships between the signal and spectral domains in the continuous world and their discrete-world counterparts.

Suppose we are given the continuous function $g(x) = \cos(2\pi f x)$, i.e. a cosine with frequency f. Its period is, therefore, $P = 1/f$. Because this is a symmetric function, we may want to sample it within a symmetric range $-R/2...R/2$, as explained in Sec. 4.2 (see Fig. 4.2(a)). The number of periods of $g(x)$ within this range is, of course:

$$q = R/P = Rf \tag{4.12}$$

The same number of periods will also occur within the range $0,...,N-1$ of the sampled version of our function, in the input array of the DFT. This means that in the discrete case the length of each period of the cosine is given, in terms of the array index, by:

$$p = N/q \tag{4.13}$$

This is, indeed, the discrete counterpart of the continuous-world relationship $P = 1/f$.

The Fourier transform of our continuous cosine function consists of two impulses at the frequencies $-f$ and f [Brigham88 pp. 19–20; Bracewell86 p. 412]. By inverting Eq. (4.8):

$$n = N/2 + N\Delta x f_{\text{Hz}} \tag{4.14}$$

Figure 4.2: (a) The continuous periodic function $g(x) = \cos(2\pi f x)$ with frequency $f = 0.5$ (represented by a continuous line), and its discrete counterpart (represented by black dots) that is obtained by sampling $g(x)$ within the symmetric range $-4...4$ with a step of $\Delta x = 0.25$, so that $N = 32$ and $q = 4$. (b) The CFT of $g(x)$, $G(u) = \frac{1}{2}\delta(u-f) + \frac{1}{2}\delta(u+f)$ [Bracewell86 p. 412] (represented by a continuous line), and its discrete counterpart (represented by black dots) as obtained by applying DFT to the discrete sequence shown in (a), after both input and output reorganizations. (c) Same as (b), after having applied to the DFT output the required vertical scaling correction, too, as explained below in Sec. 4.4; the vertical discrepancy between the corrected DFT and the CFT is explained in Sec. A.3 of Appendix A. In each of the plots the horizontal axis

Signal domain

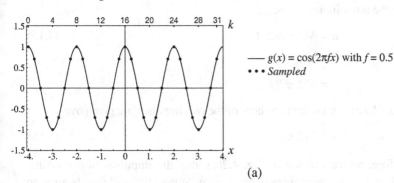

(a)

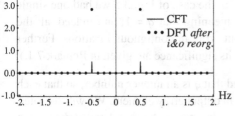

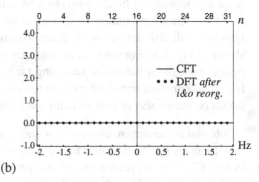

(b)

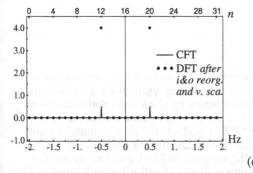

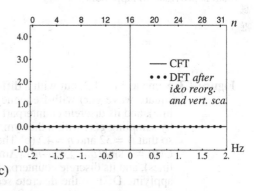

(c)

at the top corresponds to the integer index of the discrete data sequence that is represented by the dots, and the horizontal axis at the bottom corresponds to the true units of the respective continuous function. In terms of the true, continuous-world units the spectral impulses are located at $u = \pm f = \pm 0.5$ Hz, and in terms of the reorganized output array of the DFT they are located, according to Eq. (4.16), at the indices $n = 16 \pm q = 16 \pm 4$.

we see that the frequencies $\pm f$ of these two impulses correspond in the reorganized output array of the DFT to the array indices:

$$n = N/2 \pm N\Delta x f \qquad\qquad (4.15)$$

or by Eq. (4.2):

$$= N/2 \pm Rf$$

This means by Eq. (4.12) that the array indices of the two impulses are given by:

$$n = N/2 \pm q \qquad\qquad (4.16)$$

It follows, therefore, as we can see in Fig. 4.2(c), that the impulses of our cosine function will be located in the reorganized output array of the DFT q elements away to each side of the DC (which is itself located, after the reorganization, at the index $n = N/2$). If we now replace our function $g(x)$ by a periodic rectangular wave having the same period P, whose spectrum contains impulses at all the higher harmonics kf, too, these new impulses will also appear in the discrete output array, spaced by a q-element step, as shown in Fig. 4.3. (For the sake of comparison, in the case of Fig. 3.5 we had one single period throughout the entire sampling range, meaning that $q = 1$; and indeed, all the harmonic impulses appeared there in the output array in contiguous locations. Further details on the number of periods being used and its significance are given in Remark 7.1.)

Note that in the present discussion we assumed that q is an integer number, so that each impulse in the spectrum of $g(x)$ falls exactly on an output array element. We will see later, in Sec. 6.2, what happens when q is not precisely an integer. A more detailed discussion on the roles of p and q in the signal and spectral domains and on the relationship between them is provided in Appendix C.

Figure 4.3: Same as Fig. 4.2, but with a different periodic function. (a) The continuous square wave $g(x)$ with frequency $f = 0.5$ Hz (represented by a continuous line), and its discrete counterpart (represented by black dots) that is obtained by sampling $g(x)$ within the symmetric range $-4...4$ with a step of $\Delta x = 0.25$, so that $N = 32$ and $q = 4.5$ (b) The CFT of $g(x)$, i.e. the infinite impulse train $G(u) = a\ \mathrm{sinc}(bu)\ \Sigma\delta(u{-}nf)$ [Amidror09 p. 22] (represented by continuous lines), and its discrete counterpart (represented by black dots) as obtained by applying DFT to the discrete sequence shown in (a), after both input and output reorganizations. (c) Same as (b), after having applied to the DFT output the required vertical scaling correction, too, as explained in Sec. 4.4; the vertical discrepancy between the corrected DFT and the CFT is explained in Sec. A.3 of Appendix A. In each of the plots the horizontal axis at the top corresponds to the integer index of the discrete data sequence that is

---

[5] Note that at this low resolution the sampled version of our square wave is indistinguishable from a sampled triangular wave. See also Problem 4-10.

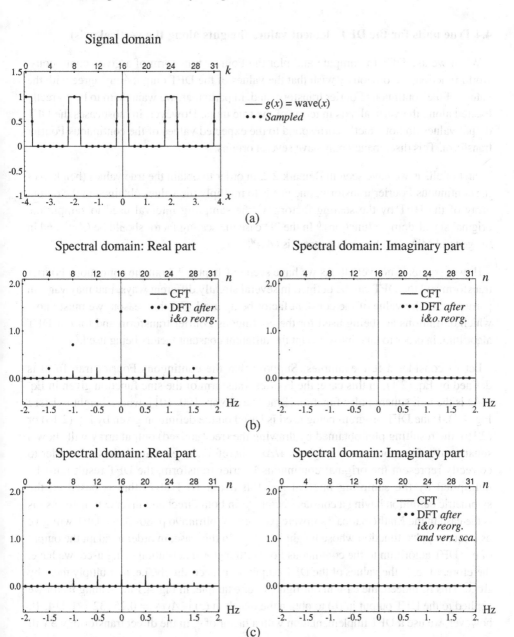

Signal domain

Spectral domain: Real part

Spectral domain: Imaginary part

(a)

(b)

(c)

represented by the dots, and the horizontal axis at the bottom corresponds to the true units, i.e. the units of the respective continuous function. In terms of the true, continuous-world units the spectral impulses are located at $u = \pm kf = \pm 0.5k$ Hz for all integer k, and in terms of the reorganized output array of the DFT they are located, according to Eq. (4.16), at the indices $n = 16 \pm 4k$.

4.4 True units for the DFT element values (heights along the vertical axis)

When we use DFT to compute and plot the Fourier transform of a given continuous-world function, we obviously wish that the values of the DFT output array agree with the values of the continuous Fourier transform and, in particular, we want them to be correctly located along the vertical axis in terms of the true units. However, in most cases the DFT output values do not exactly correspond to the expected values of the continuous Fourier transform. This discrepancy may have several origins.

First of all, as we have seen in Remark 2.2, in order to obtain the true values (heights) of the continuous Fourier transform, one needs to multiply the values obtained in the output array of the DFT by the scaling factor Δx (the sampling interval used to sample the original signal-domain function).[6] In the 2D case the scaling factor should be $(\Delta x)^2$, and in the general MD case the scaling factor is $(\Delta x)^M$.

However, this is not yet all. As we have seen in Chapter 2, both the continuous Fourier transform and the DFT can be defined in several slightly different ways, that may vary, in particular, in the value of the constant factor being used. For this reason, we must know which definitions are being used for the continuous Fourier transform and for our DFT algorithm, in order to take into account the different constant factors being used.[7]

Let us consider a few examples. Suppose that the continuous Fourier transform is defined by Eq. (2.1); in this case, the Fourier transform of the sinc function given in Eq. (4.1) is the unit square pulse function [Bracewell86 pp. 100–101], whose height is 1 (see Fig. 4.1). If the DFT program being used is based on the definition given by Eq. (2.13) or (2.15), the resulting plot obtained by drawing the (reorganized) output array will show a square pulse function whose height is $1/\Delta x$. Therefore, as explained above, in order to correctly represent the original continuous Fourier transform, the DFT result must be multiplied by the sampling interval Δx. But if our DFT algorithm is based on the symmetric definition having a constant factor $\frac{1}{\sqrt{N}}$ in both direct and inverse transforms (as is the case in the Mathematica® software package [Wolfram96 p. 868]), the DFT will give us a square pulse function whose height is $\frac{1}{\Delta x \sqrt{N}}$. In this case, in order to adapt the output of our DFT algorithm to the continuous Fourier transform definition being used, we have, therefore, to scale the values of the DFT output array accordingly, i.e. to multiply them by $\Delta x \sqrt{N}$. This is, indeed, the case in our figures; for example, in Fig. 4.1 the scaling factor we applied to the DFT output in (b) to obtain the match in (c) is $\Delta x \sqrt{N} = 0.25\sqrt{32} \approx 1.414$. If, however, we use a DFT implementation with a factor of $\frac{1}{N}$ in the direct transform, we will

---

[6] One may provide two explanations for this fact: (1) As a result of signal-domain sampling, the frequency domain has been scaled by factor $1/\Delta x$ [Brigham88 pp. 77–79]; therefore, to recover the original spectrum, a scaling by factor Δx is required (see also Sec. 2.5 above). (2) The rectangular rule for the numeric integration of the continuous Fourier transform provides the same sum as the DFT, multiplied by the factor Δx [Brigham88 pp. 100–101]; see also Remark 2.2 in Chapter 2.

[7] Note also that using a DFT definition with reversed signs in the exponential part will result in a reversed list of values in the output [Wolfram96 p. 868]; more precisely, the positive and negative frequencies will be interchanged (since a sign reversal of the exponent can be assimilated with a sign reversal in the index n within the exponent, where due to the cyclic nature of the DFT we have $-n = N - n$ [Smith07 pp. 118–119]). See Sec. 8.3 for more details.

rather have to multiply the DFT output values by the factor $\Delta x N = R$. We will henceforth call the required scaling correction, which takes into account Δx as well as the other scaling factors in question (depending on the continuous and discrete definitions being used), the *standard amplitude scaling correction*.

It is important to note that a similar amplitude scaling correction is also required when using the inverse DFT (IDFT). But because in this case the input array consists of the frequency domain and the output array consists of the signal domain, Δx must be replaced here by Δu (the sample interval in the frequency domain). Note that applying the corrected IDFT to the corrected DFT output must give back the original data. Similarly, applying an uncorrected IDFT to the uncorrected DFT output gives back the original input array, too.

As we can see, the standard amplitude scaling correction is a necessary condition for obtaining the correctly scaled Fourier transform values from the brute results of the DFT, and for correctly placing the true units along the vertical axis. However, even this is not yet all. It turns out that there also exist inherent DFT-related reasons that may cause, in some cases, significant discrepancies between the results obtained by DFT and the true Fourier transform, as clearly seen, for example, in Figs. 4.1(c) and 3.5(f). These artifacts of the DFT (known as aliasing and leakage) will be discussed in the following chapters.

Finally, there exists one more case that deserves particular attention, because it defies the standard amplitude scaling corrections described above even when no DFT artifacts (aliasing or leakage) are present. This case involves all functions whose frequency spectra contain impulses, such as the constant function $g(x) = 1$ whose spectrum is $G(u) = \delta(u)$, i.e. a unit impulse at the origin; the cosine function $g(x) = \cos(2\pi f x)$ whose spectrum is $G(u) = \frac{1}{2}\delta(u{-}f) + \frac{1}{2}\delta(u{+}f)$, i.e. a pair of half-unit impulses[8] at the frequencies $\pm f$; or any other periodic function (or even the sum of an aperiodic function with a periodic function). As shown in Figs. 4.2, 4.3 and Fig. A.2 of Appendix A, it turns out that in such impulsive spectra the strength of the impulses obtained by the DFT is not properly rectified by the standard amplitude scaling correction. This should not really be surprising, since the notion of the amplitude (or height) of an impulse is quite elusive already in the continuous case, so the equivalence with its discrete counterpart is not really obvious. A detailed discussion on this point is provided in Appendix A.

4.5 Examples

Let us illustrate the above discussion by some further examples.

Example 4.1: The same discrete signal representing different continuous functions:

Let us consider Fig. 4.4(a), and compare it with Fig. 4.1(a). The 32-element discrete signals shown in these two figures are strictly identical, as we can clearly see by

---

[8] The reason we use the term "half-unit impulses" rather than "impulses of amplitude $\frac{1}{2}$" is explained in Appendix A.

comparing them element-by-element between $k = 0$ and $k = 31$ (see the horizontal axis representing the index k at the top of the figures). However, while in Fig. 4.1(a) this discrete signal represents the continuous function $g(x) = \mathrm{sinc}(x)$ that has been sampled within the range $-4...4$ with a step of $\Delta x = 0.25$, in Fig. 4.4(a) the same discrete signal corresponds to the continuous function $h(x) = \mathrm{sinc}(2x)$ that has been sampled within the range $-2...2$ with a step of $\Delta x = 0.125$. Because both discrete signals are identical, it is not surprising that their DFTs are identical, too (see the black dots in Figs. 4.1(b) and 4.4(b)). However, the CFTs of the continuous functions $g(x) = \mathrm{sinc}(x)$ and $h(x) = \mathrm{sinc}(2x)$ are obviously different, as we can see, indeed, by comparing the continuous lines in the same two figures: $G(u) = \mathrm{rect}(u)$ while $H(u) = \frac{1}{2}\mathrm{rect}(u/2)$. The difference between the two cases is in the scaling of the frequency axis, as well as in a vertical scaling factor. Both of these are automatically taken care of when we plot the DFT results according to the explanations given in Secs. 4.3 and 4.4 (i.e. when drawing the true frequency axis according to Eq. (4.8), and applying the standard amplitude scaling correction). And indeed, when we follow these steps we obtain in both cases a correct discrete graphic representation of the underlying continuous Fourier transform, as shown in Figs. 4.1(c) and 4.4(c). The different interpretations of the same DFT results in these figures (the different true frequency axis and standard amplitude scaling correction) are simply a consequence of the different sampling range length R and sampling step Δx in Figs. 4.1(a) and 4.4(a).

As we clearly see in this example, when we change the continuous-world interpretation of the same DFT input signal, the DFT output signal remains, of course, unchanged, but its continuous-world interpretation does change accordingly. ■

Example 4.2: The same underlying continuous function as in Example 4.1, using different sampling range and sampling step:

Let us consider again the same continuous function $h(x) = \mathrm{sinc}(2x)$ as in Fig. 4.4(a), but this time we take its 32 discrete samples using the same sampling range $-4...4$ and the same sampling step of $\Delta x = 0.25$ as in Fig. 4.1(a). This is shown in Fig. 4.5. In this case, the range of the true frequency axis and the standard amplitude scaling correction that we obtain according to Secs. 4.3 and 4.4 remain, of course, the same as in Fig. 4.1; but the discrete input signals that are fed to the DFT, and hence the DFT results, are not the same in the two figures (compare Fig. 4.1 and Fig. 4.5). The final graphic result, however, when we plot the DFT according to the explanations given in Secs. 4.3 and 4.4, is correct in both cases, as we can see in Figs. 4.1(c) and 4.5(c). ■

Figure 4.4: (a) The same discrete signal as in Fig. 4.1(a), but this time it represents the continuous function $h(x) = \mathrm{sinc}(2x)$ that has been sampled within the range $-2...2$ with a step of $\Delta x = 0.125$, so that again $N = 32$. (b) The CFT of $h(x)$, $H(u) = \frac{1}{2}\mathrm{rect}(u/2)$ (represented by a continuous line), and its discrete counterpart (represented by black dots) as obtained by applying DFT to the discrete sequence shown in (a), after both input and output

Signal domain

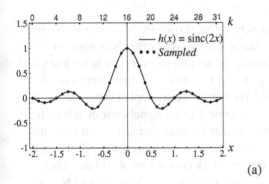

(a)

Spectral domain: Real part   Spectral domain: Imaginary part

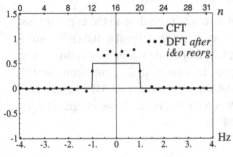 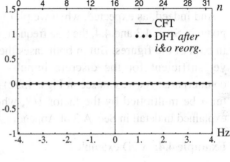

(b)

Spectral domain: Real part   Spectral domain: Imaginary part

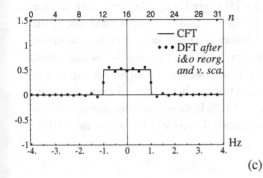

(c)

reorganizations. Note that the DFT output itself is exactly the same as in Fig. 4.1(b), and only the interpretation of the true continuous-world units along the frequency axis is different, as explained in Sec. 4.3. (c) Same as (b), after having applied to the DFT output the required vertical scaling correction, too, as explained in Sec. 4.4; note the close match between the corrected DFT and the CFT.

Example 4.3: The same discrete signal representing different continuous functions — in the periodic case:

Consider Fig. 4.6(a). The 32-element discrete signal shown in this figure is strictly identical to that of Fig. 4.2(a), as we can clearly see by comparing them element-by-element between $k = 0$ and $k = 31$ (see the horizontal axis representing the index k at the top of the figures). However, while in Fig. 4.2(a) this discrete signal represents the continuous function $g(x) = \cos(2\pi f x)$ with $f = \frac{1}{2}$ Hz that has been sampled within the range $-4...4$ with a step of $\Delta x = 0.25$, in Fig. 4.6(a) the same discrete signal corresponds to the continuous function $h(x) = \cos(2\pi f x)$ with $f = 1$ Hz that has been sampled within the range $-2...2$ with a step of $\Delta x = 0.125$. Because both discrete signals are identical, it is not surprising that their DFTs are identical, too (see the black dots in Figs. 4.2(b) and 4.6(b)). However, the continuous Fourier transforms of the continuous functions $g(x)$ and $h(x)$ are obviously different, as we can see, indeed, by comparing the continuous lines in the same two figures: $G(u) = \frac{1}{2}\delta(u-\frac{1}{2}) + \frac{1}{2}\delta(u+\frac{1}{2})$ while $H(u) = \frac{1}{2}\delta(u-1) + \frac{1}{2}\delta(u+1)$.

And indeed, as expected, when we plot the DFT results according to the explanations given in Secs. 4.3 and 4.4, the true frequency units are correctly affixed to the DFT results in both of the figures. But in both cases the standard amplitude scaling correction is not yet sufficient for the discrete impulse heights to match their continuous-world counterparts: As we can see in Figs. 4.2(c) and 4.6(c), in both cases the DFT results still must be multiplied by the factor $1/R$, where $R = N\Delta x$ is the sampling range. This is explained in detail in Sec. A.3 of Appendix A. ∎

Example 4.4: A 2D example:

Consider now Fig. 4.7(a) and compare it with Fig. 3.6(a). The 32×32-element discrete signals shown in these two figures are strictly identical, as we can clearly see by comparing them element-by-element within the 2D range $k = 0,...,31$, $l = 0,...,31$ (see the axes representing the discrete indices k and l at the top and at the right hand side of the figures). However, while in Fig. 3.6(a) this 2D discrete signal represents the continuous function $g(x,y) = \mathrm{rect}(x)\,\mathrm{rect}(y)$ that has been sampled within the range $-4...4$ along both axes with a step of $\Delta x = \Delta y = 0.25$, in Fig. 4.7(a) the same discrete signal corresponds to the continuous function $h(x,y) = \mathrm{rect}(2x)\,\mathrm{rect}(2y)$ that has been sampled along both axes within the range $-2...2$ with a step of $\Delta x = \Delta y = 0.125$. Because both discrete signals are identical, it is not surprising that their DFTs are identical, too (see Figs. 3.6(e) and 4.7(b)).

Figure 4.5: (a) The same continuous signal $h(x) = \mathrm{sinc}(2x)$ as in Fig. 4.4(a), but this time it has been sampled within the range $-4...4$ with a step of $\Delta x = 0.25$, so that again $N = 32$. (b) The CFT of $h(x)$, $H(u) = \frac{1}{2}\mathrm{rect}(u/2)$ (represented by a continuous line), and its discrete counterpart (represented by black dots) as obtained by applying DFT to the

Signal domain

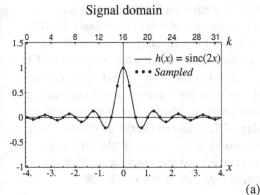

(a)

Spectral domain: Real part

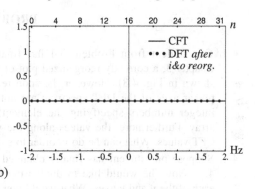

Spectral domain: Imaginary part

(b)

Spectral domain: Real part

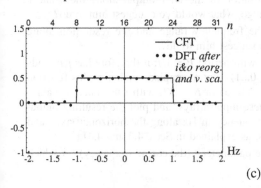

Spectral domain: Imaginary part

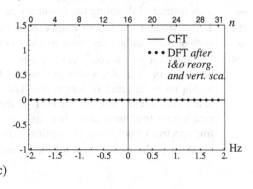

(c)

discrete sequence shown in (a), after both input and output reorganizations. (c) Same as (b), after having applied to the DFT output the required vertical scaling correction, too, as explained in Sec. 4.4; note the close match between the corrected DFT and the CFT. See Example 4.2.

However, the continuous Fourier transforms of the continuous functions $g(x,y)$ and $h(x,y)$ are obviously different: $G(u,v) = \text{sinc}(u)\,\text{sinc}(v)$ while $H(u,v) = \frac{1}{2}\text{sinc}(u/2)\frac{1}{2}\text{sinc}(v/2)$. The difference between the two cases is in the scaling of the frequency axes, as well as in a vertical scaling factor. Both of these are automatically taken care of when we plot the DFT results according to the explanations given in Secs. 4.3 and 4.4 (i.e. when drawing the true frequency axes according to Eq. (4.8), and applying the standard amplitude scaling correction). And indeed, when we follow these steps we obtain in both cases a correct discrete graphic representation of the underlying CFT, as shown in Figs. 3.6(f) and 4.7(c). The different interpretations of the same DFT results in these figures (the different true frequency axes and standard amplitude scaling correction) are simply a consequence of the different sampling range and sampling steps in Figs. 3.6(a) and 4.7(a). ■

PROBLEMS

4-1. Our student from Problem 3-1 has finally obtained, following the explanations of Chapter 3, a correctly reorganized plot of the DFT spectrum (whose real-valued part is shown in Fig. 4.8). However, he soon realizes that the values marked along the two planar axes of the DFT spectrum are not true frequencies in terms of Hz, but rather integer numbers specifying the element's position (index) within the DFT output array. Furthermore, the values along the vertical axis are far away from the expected CFT values. What can he do in order to solve each of these problems?

4-2. Suppose our student has finally obtained the true values along all the axes (see Fig. 4.9). Now, he would like to double the frequency range obtained by the DFT along each of the u and v axes. What would you suggest him to do?

4-3. If our student desired to double the resolution of the DFT output along the u and v axes instead of doubling the frequency range, what would you suggest him to do?

4-4. If our student wanted to double both the frequency range and the resolution along each of the u and v axes, what would you suggest him to do?

4-5. The function $g(x) = \cos(2\pi f x)$ is periodic with period $P = 1/f$; it therefore has f periods per unit (for example, within the interval $0...1$). Plot a sampled version of this function within an input array of length $N = 128$, $N = 64$ or $N = 32$, with true function values along both axes. Compute the DFT of these input arrays, and plot the resulting output arrays with true units along both axes (frequency in Hz along the horizontal axis, and true spectral values along the vertical axis, as explained in Secs. 4.3 and 4.4).

(a) What is the useful range of the cosine parameter f (the cosine frequency) for which

Figure 4.6: (a) The same discrete periodic signal as in Fig. 4.2(a), but this time it represents the continuous periodic function $h(x) = \cos(2\pi f x)$ with frequency $f = 1$ that has been sampled within the range $-2...2$ with a step of $\Delta x = 0.125$, so that again $N = 32$. (b) The CFT of $h(x)$, $H(u) = \frac{1}{2}\delta(u-1) + \frac{1}{2}\delta(u+1)$ (represented by a continuous line), and its discrete counterpart (represented

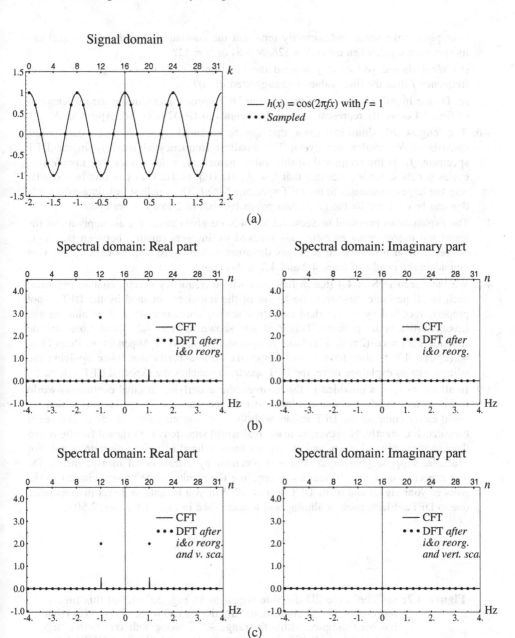

(a)

(b)

(c)

by black dots) as obtained by applying DFT to the discrete sequence shown
in (a), after both input and output reorganizations. (c) Same as (b), after
having applied to the DFT output the required vertical scaling correction,
too, as explained in Sec. 4.4; the vertical discrepancy between the corrected
DFT and the CFT is explained in Sec. A.3 of Appendix A.

your plots make sense and correctly represent the continuous-world function $g(x)$ and its spectrum $G(u)$, when using $N = 128$, $N = 64$ or $N = 32$?

(b) What should you do if you need the function $g(x)$ to have a higher or a lower frequency f than the limit values you suggested in (a)?

(c) Do the impulse heights obtained in your DFT spectra according to the explanations of Sec. 4.4 correctly represent the impulse heights in $G(u)$? *Hint*: See Appendix A.

4-6. The longest full sinusoidal cycle that can be captured by the N given data points consists of N *samples per cycle*. The smallest frequency above zero in the DFT spectrum, f_1, is the reciprocal of this value, namely: $f_1 = 1/N$ *cycles per sample* (i.e. cycles per Δx seconds), meaning that $f_1 = \frac{1}{N\Delta x}$ Hz (Eq. (4.4)). Can you find in a similar way the largest frequency in the DFT spectrum? *Hint*: The smallest full sinusoidal cycle that can be captured by the given data points consists of 2 samples per cycle.

4-7. The explanations provided in Secs. 4.2 and 4.3 are given under the assumption that the input and output array elements are indexed by integers running between $0,...,N-1$. However, in many software packages the array indices must run between $1,...,N$. Can you adapt the results of Secs. 4.2 and 4.3 to this case?

4-8. We have seen in Sec. 4.4 that in functions whose frequency spectra contain impulses, such as all periodic functions, the height of the impulses obtained by the DFT is not properly rectified by the standard amplitude scaling correction (even if no aliasing and leakage artifacts are present). This is clearly shown in Figs. 4.2(c) and 4.6(c) for the function $g(x) = \cos(2\pi f x)$. This issue is discussed in detail in Appendix A. Redo Figs. 4.2(c) and 4.6(c) after having read Appendix A, and verify that when applying the adjustments as explained there, the DFT spectrum matches the expected CFT values.

4-9. In all the examples provided in this chapter, the underlying original continuous-world functions and their respective continuous-world spectra were always known, so that we could easily compare the DFT results with the true continuous-world values and see if they match correctly. However, in many real-world situations the original functions and their spectra are not known, and all that we have at hand is a given discrete signal. For example, suppose your input data was obtained by astronomical measurements. Do you expect to have any difficulties in applying the results of the present chapter to the plots of your signal and of its DFT spectrum? Will you be able to detect discrepancies due to DFT artifacts such as aliasing and leakage, like in Figs. 4.1(c) and 3.5(f)?

Figure 4.7: (a) The same 2D discrete signal as in Fig. 3.6(a), but this time it represents the 2D continuous function $h(x,y) = \text{rect}(2x)\,\text{rect}(2y)$ that has been sampled within the range $-2...2$ along both axes with a step of $\Delta x = \Delta y = 0.125$, so that again $N = 32$. (b) The DFT of the discrete signal shown in (a), after both input and output reorganizations. Note that the DFT output itself is exactly the same as in Fig. 3.6(e), and only the interpretation of the true continuous-world units along both frequency axes is different, as explained in Sec. 4.3. (c) Same as (b), after having applied to the DFT output the required vertical correction, too, as explained in Sec. 4.4. Although the CFTs $G(u,v) = \text{sinc}(u)\,\text{sinc}(v)$ and $H(u,v) = \frac{1}{2}\text{sinc}(u/2)\frac{1}{2}\text{sinc}(v/2)$ are not shown in Figs. 3.6 and 4.7, one can verify that the 2D DFT plot in each of these figures closely matches the respective CFT.

Signal domain

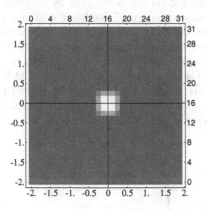

Legend:

(a)

Spectral domain: Real part

Spectral domain: Imaginary part

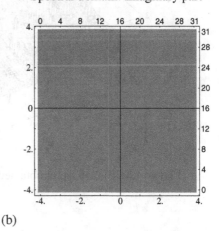

(b)

Spectral domain: Real part

Spectral domain: Imaginary part

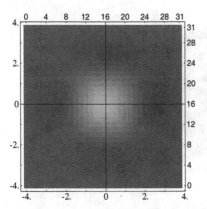

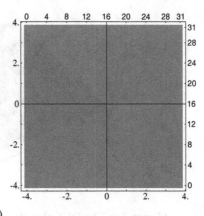

(c)

4-10. Fig. 4.3 shows a square wave $g(x)$ with period $P = 2$ (drawn by a continuous line) and its discrete counterpart (represented by black dots) that is obtained by sampling $g(x)$ within the symmetric range $-4...4$ with a step of $\Delta x = 0.25$, so that $N = 32$ and $q = 4$. However, as mentioned in Footnote 5 there, at this low resolution the sampled version of our square wave is indistinguishable from a sampled triangular wave. Knowing that the CFT of a square wave and the CFT of a triangular wave are not identical, how can you interpret the resulting DFT spectrum? The same question can be also asked for the non-periodic counterpart of $g(x)$ which only contains the central pulse around $x = 0$, and whose spectrum is continuous (non-impulsive). *Hint*: The problems related to insufficient sampling resolution are treated in Chapter 5; see also Problem 5-6.

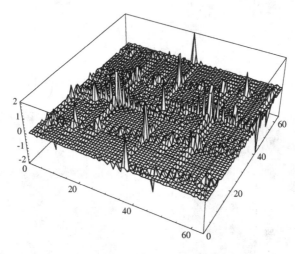

Figure 4.8: The DFT plot obtained by the student in Problem 4-1.

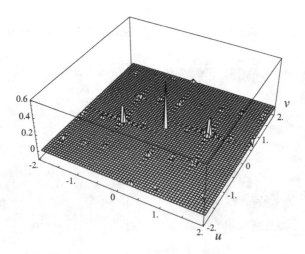

Figure 4.9: The DFT plot obtained by the student in Problem 4-2.

Chapter 5

Issues related to aliasing

5.1 Introduction

When the DFT is used as a numerical approximation to the Fourier transform of an underlying continuous function, some differences can be expected between the two. For example, it is clear that the DFT can only give us the Fourier transform values at the N discrete points for which the computation is done. However, even at these points there may appear in some cases quite significant discrepancies, in spite of the adjustments and scalings discussed above in Chapters 3 and 4. This is not a result of rounding errors in the calculations, which are normally quite insignificant (see Sec. 8.8). The main reason for these discrepancies resides in the discrete and finite nature of the DFT, which implies that the underlying continuous function must be sampled, and that it can only be considered within a finite range. Both of these two operations, *sampling* and *finite-range windowing* (or *truncation*), are inherent to DFT, as it can be seen by comparing Eqs. (2.1)–(2.2) with Eqs. (2.13)–(2.14). Each of these operations involves some loss of data, and introduces its own discrepancy between the values obtained by DFT and the true values of the CFT. These two types of error, both considered as DFT artifacts, are named respectively *aliasing* and *leakage*. In the present chapter we will focus on issues related to aliasing; leakage-related considerations will be discussed in Chapter 6.

5.2 Aliasing in the one dimensional case

Aliasing is a phenomenon which may occur due to the discretization (sampling) of the underlying input function. In fact, aliasing is not only limited to DFT applications, and it may be encountered wherever sampling is used, whether or not the sampled data is then subjected to DFT. Discretizing an analog signal requires that the signal's value be sampled often enough to define the waveform unambiguously. According to the sampling theorem, the sampling frequency must be at least twice the highest frequency present in the signal for its waveform to be defined completely.[1] In other words, there must be at least two samples per cycle for any frequency component (sine or cosine) of the signal. If the sampling points are not taken as densely as required (a situation often called *undersampling*), they will fail to follow the high-frequency fine details of the signal, thus leading to aliasing. In the signal domain this is expressed by the existence of a smoother, lower-frequency signal, known as *alias*, which can be traced through all the sampled

---

[1] There exist in the literature many different variants of the sampling theorem, which are basically all equivalent. This version of the theorem appears, for example, in [Ramirez85 p. 115], [Gaskill78 p. 269] or [Weisstein99 p. 1593].

points and "mimic" or "masquerade" the behaviour of the original signal on its sampled
values (see Fig. 5.1 for a periodic example and Fig. 5.2 for an aperiodic example). Only
based on these too far-apart sampling points, the original signal is undistinguishable from

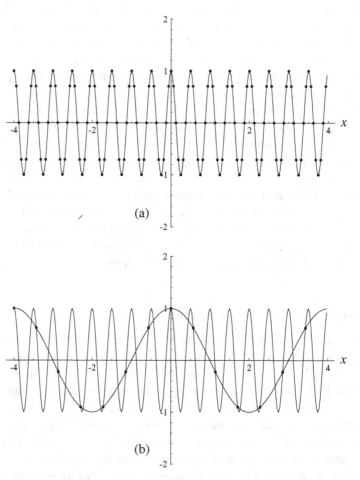

Figure 5.1: Illustration of the signal-domain interpretation of aliasing: a periodic
case. (a) Sampling the continuous periodic signal $g(x) = \cos(2\pi f x)$
having frequency $f = 2$ at a rate higher than twice the maximum signal
frequency gives a correct discrete representation of the original
signal. (b) Sampling the same continuous periodic signal at a rate
lower than twice the maximum signal frequency gives a false, aliased
lower-frequency signal (here: a cosine function with frequency $f_A = \frac{1}{4}$;
see Example 5.4 in Sec. 5.3) that mimics the original signal on its
sampled values. Both signals are drawn with continuous lines, and
their sampled values are indicated by black dots.

its lower-frequency aliased signal, since both of them coincide on all of these sampling points. In other words, aliasing means that high-frequency components in the original signal appear incorrectly as lower frequencies in the sampled result.

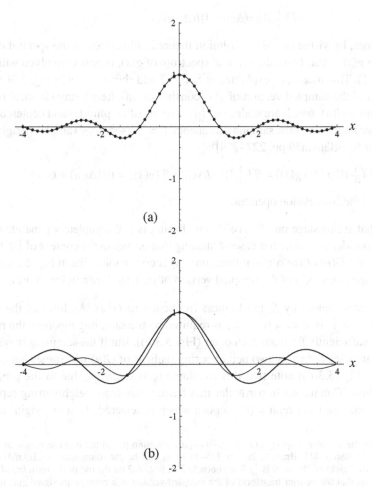

(a)

(b)

Figure 5.2: Same as Fig. 5.1, showing this time an aperiodic case. (a) Sampling a continuous aperiodic signal (here: the function $g(x) = \text{sinc}(x)$) at a rate higher than twice the maximum signal frequency gives a correct discrete representation of the original signal. (b) Sampling the same continuous aperiodic signal at a rate lower than twice the maximum signal frequency gives a false, aliased lower-frequency signal (here: the function $g_A(x) = \frac{3}{2}\text{sinc}(\frac{3}{4}x) - \frac{1}{2}\text{sinc}(\frac{1}{2}x)$; see Example 5.1 in Sec. 5.3) that mimics the original signal on its sampled values. Both signals are drawn with continuous lines, and their sampled values are indicated by black dots.

This is the interpretation of aliasing in terms of the signal domain; but the aliasing phenomenon can be also interpreted from the viewpoint of the spectral domain. Remember that sampling in the signal domain is equivalent to multiplying the original continuous signal $g(x)$ by $\frac{1}{\Delta x}\text{III}(x/\Delta x)$, a periodic unit-height impulse comb (impulse train) having an impulse interval of Δx (see Fig. 5.3).[2] The Fourier transform of this sampling impulse comb is itself an impulse comb with impulse interval of $f_s = \frac{1}{\Delta x}$ (see Remark 4.1 in Sec. 4.3) and impulse height of $\frac{1}{\Delta x}$ [Castleman79 pp. 227–228; Bracewell86 p. 414]:

$$\mathcal{F}[\tfrac{1}{\Delta x}\text{III}(x/\Delta x)] = \text{III}(\Delta x\, u) \tag{5.1}$$

Therefore, by virtue of the convolution theorem, the effect on the spectral domain of sampling $g(x)$ is that $G(u)$, the original spectrum of $g(x)$, is now convolved with impulse comb (5.1). This means, as explained in Sec. 2.5 and shown again in Fig. 5.3(c), that the spectrum of the sampled version of $g(x)$ consists of infinitely many identical replicas of the spectrum $G(u)$, which are scaled by $\frac{1}{\Delta x}$ (in terms of amplitude) and centered about all the integer multiples of the sampling frequency $f_s = \frac{1}{\Delta x}$ (see, for example, [Brigham88 pp. 79–81] or [Castleman79 pp. 227–229]):[3]

$$\mathcal{F}[\tfrac{1}{\Delta x}\text{III}(x/\Delta x)\, g(x)] = \mathcal{F}[\tfrac{1}{\Delta x}\text{III}(x/\Delta x)] * \mathcal{F}[g(x)] = \text{III}(\Delta x\, u) * G(u) \tag{5.2}$$

where $*$ is the convolution operator.

Note that at this stage our discussion on aliasing is still completely general, and we do not yet consider the particular case of aliasing that occurs in the context of DFT. Thus, if the spectrum $G(u)$ of the original function $g(x)$ is continuous, like in Fig. 5.3(a), so is the periodic spectrum (5.2) of the sampled version of $g(x)$, as shown in Fig. 5.3(c).

Let us now denote by f_G the highest frequency in $G(u)$. As long as the sampling frequency $f_s = \frac{1}{\Delta x}$ is at least twice f_G, as required by the sampling theorem, the replicas of $G(u)$ are sufficiently far from each other (Fig. 5.3(c)). But if the sampling frequency f_s is lower than twice f_G,[4] every two neighbouring replicas of $G(u)$ will somewhat overlap, as shown in Fig. 5.3(d); note that this overlapping is *additive*, due to the properties of convolution. This means in particular that frequencies from neighbouring replicas will penetrate into the main replica (the replica which is centered about the origin) and appear

---

[2] Thanks to the impulse property $\delta(x/a) = |a|\delta(x)$ [Bracewell86 p. 76] the impulse height of $\text{III}(x/\Delta x)$ is Δx [Bracewell86 p. 414; Bracewell95 pp. 118–119]; its unit-height counterpart is $\frac{1}{\Delta x}\text{III}(x/\Delta x)$. Similarly, the impulse height of $\text{III}(\Delta x\, u)$ is $\frac{1}{\Delta x}$. See Footnote 3 in Sec. A.2 on the use of the term *impulse height*.

[3] This means that the Fourier transform of the sampled version of a continuous signal $g(x)$ is a periodic function. If no aliasing occurs (see below), each period of this spectrum is equal (within a constant factor Δx) to $G(u)$, the Fourier transform of $g(x)$; see Fig. 5.3(c). What is the spectral meaning of such non-decaying high frequencies that extend to both directions of the spectrum *ad infinitum*? The fact that the spectrum is periodic simply means that the corresponding signal-domain function (our sampled function) is discrete. In fact, our sampled function in the signal domain can be seen as the impulsive inverse Fourier transform of its periodic spectrum. Remember that the Fourier transform $P(u)$ of a periodic function $p(x)$ is an impulsive discrete function (see, for example, [Papoulis68 p. 107], [Brigham88 pp. 77–79] or [Amidror09, Sec. A.2]); the situation in our case is similar, except that the roles of the signal domain and the frequency domain are interchanged (see Chapter 5 in [Brigham88]).

[4] This happens, for example, if $g(x)$ is not *band limited*, i.e. if $f_G = \infty$. Note that the vertical dotted lines going down throughout Fig. 5.3 help to follow the frequencies f_G, $2f_G$ etc. in all rows of the figure.

Signal domain

Spectral domain

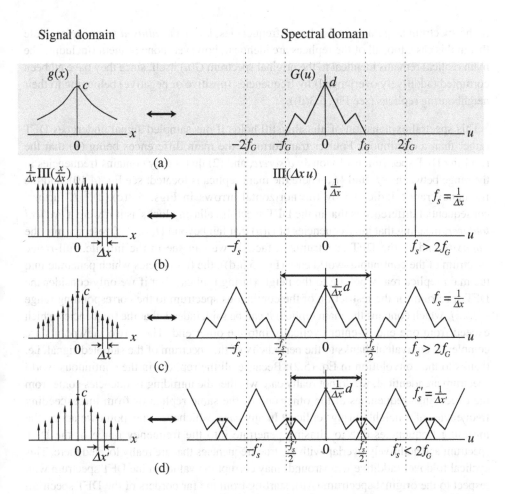

Figure 5.3: Schematic illustration of the Fourier-domain interpretation of aliasing. (a) An original continuous signal $g(x)$ and its continuous-world spectrum $G(u)$. (b) The sampling impulse train and its continuous-world spectrum. (c) The sampled signal is the product of signals (a) and (b), and therefore its continuous-world spectrum is the convolution of their spectra. Note the replicas of the original spectrum, that are located at all integer multiples of the sampling frequency f_s. (d) If the sampling rate is lower than twice the maximum signal frequency f_G (say, f'_s), then every two neighbouring replicas of $G(u)$ overlap (additively), giving false aliased frequencies in the overlapped zones. The aliased parts of the spectrum in (d) are shown by thick dashed lines, which are simply the sum of the original overlapped replicas. The horizontal arrows in the spectra (c) and (d) indicate the part of the infinite continuous-world spectrum that will be represented in the DFT output should we subject the sampled signal to DFT rather than to CFT: As shown in Sec. 4.3, the frequency range of the DFT extends between minus half and plus half of the corresponding sampling frequency. Note that in terms of the DFT frequency range, aliasing can be seen as a *cyclical foldover* where frequencies which exceed beyond one end of the range re-enter at the opposite end. The frequencies f_G, $2f_G$ etc. can be followed through all rows thanks to the vertical dotted lines (note that these lines are not the same as in Figs. 2.1–2.3).

in the spectrum as parasite, false lower frequencies, known as *aliased frequencies*. Note that in this case, too, all of the replicas are identical; however, none of them (including the main replica) remains identical to the original spectrum $G(u)$ itself, since they have all been corrupted (additively overlapped) by frequencies (positive or negative) belonging to their neighbouring replicas (see Fig. 5.3(d)).

This spectral explanation of aliasing still holds if our sampled signal undergoes DFT rather than a continuous Fourier transform — the main differences being (1) that the resulting DFT spectrum is obviously *discrete*, and (2) that it only contains frequencies in the range between $-\frac{1}{2}f_s$ and $\frac{1}{2}f_s$, where the main replica is located; see Eq. (4.10), and the frequency range indicated by the horizontal arrows in Figs. 5.3(c),(d).[5] A further, consequential difference is that in the DFT spectrum aliasing looks as if it were a *cyclical foldover*, meaning that the frequencies of $G(u)$ that fall beyond $\frac{1}{2}f_s$ or $-\frac{1}{2}f_s$ re-enter into the *opposite* end of the DFT spectrum. In fact, as we can see in the infinite, full-range spectrum of the continuous-world case (Fig. 5.3(d)), the frequencies which penetrate into the main replica really belong to the neighbouring replicas; but if we only consider the DFT spectrum (or the restriction of the continuous spectrum to the corresponding range $-\frac{1}{2}f_s...\frac{1}{2}f_s$ which contains the main replica) it can be said, indeed, that the frequencies which exceed from one end re-enter cyclically into the other end. These two viewpoints are completely equivalent thanks to the periodicity of the spectrum of the sampled signal, i.e. thanks to the convolution in Eq. (5.2): Because all the replicas in the continuous-world spectrum are identical, we do not really care whether the intruding frequencies come from the exceeding frequencies at the other end of the same replica or from the exceeding frequencies of a neighbouring replica.[6] In any case, whichever viewpoint we adopt, the intruding frequencies due to aliasing penetrate into the frequency range of the DFT spectrum and additively overlap with the true frequencies that are really located there. This cyclical foldover (additive wraparound) may corrupt the values in the DFT spectrum with respect to the original spectrum $G(u)$, starting from the far borders of the DFT spectrum (i.e. $\pm\frac{1}{2}f_s$) inward; see, for example, Fig. 5.6 in Example 5.1 or Fig. 5.7 in Example 5.2. This is, indeed, the spectral-domain interpretation of the aliasing artifacts in the DFT.

Note that if $G(u)$ (and hence the resulting discrete DFT spectrum) is rather smooth and non-impulsive,[7] as in Fig. 5.3(d), this foldover usually modifies the shape of the DFT spectrum only slightly, and its existence may be quite difficult to detect in practice unless

---

[5] Remember that both the input and output sequences of the DFT should be understood as a single N-element period (or "truncation window") from a periodic sequence which repeatedly extends to both directions *ad infinitum* (even if the original underlying data was not periodic).

[6] Note that in some references foldover due to aliasing is described (and even graphically depicted) as mirroring of the frequencies that exceed beyond $\frac{1}{2}f_s$ (or $-\frac{1}{2}f_s$) backward from the *same* end of the DFT spectrum, i.e. as folding around the spectrum edges backward. While this interpretation makes no substantial difference in the case of real-valued 1D signals, in other 1D cases or in 2D or MD cases it should be avoided (see, for example, the 2D DFT in Fig. 5.13(a)). The correct interpretation of the foldover is that whatever exceeds from one side of the DFT spectrum re-enters in a cyclical way from the opposite side. See also Problem 5-7.

[7] The DFT spectrum is said to be "smooth and non-impulsive", in spite of its obvious discrete nature, if the values of its successive discrete elements vary in a continuous, non-impulsive way.

the true values of the original spectrum $G(u)$ are known in advance. However, when $G(u)$ is a purely impulsive spectrum (i.e. when the original function $g(x)$ is periodic or almost periodic [Amidror09, Sec. B.6]) aliasing may be much easier to detect, since the cyclical foldover introduces false "stray impulses" in the DFT results, where no impulses are present in the correct Fourier transform. This is illustrated, for the 1D case, in Fig. 5.11 (see Example 5.5 in Sec. 5.3 below).

Remark 5.1: There exists a particular case in which no such "stray impulses" appear in the DFT spectrum of a periodic function $g(x)$, even if strong aliasing occurs. This happens if the distance q between the origin and the first harmonic impulse of $g(x)$ in the DFT spectrum (in terms of output-array elements; see Sec. 4.3.1) satisfies $N = pq$ for an integer p (note that q itself does not necessarily need to be integer; see Fig. C.2 in Appendix C). In this case higher harmonic impulses which exceed from one side of the DFT spectrum will fall exactly on top of true impulse locations (harmonic frequencies) in the other side of the spectrum, and their heights will be simply summed up. More details on this subject can be found in Appendix C. ■

Remark 5.2: It is important to note that aliasing can be avoided completely only if the given signal $g(x)$ is band limited; under this condition, aliasing will be absent if the sampling frequency is at least twice the highest frequency present in $g(x)$. But in practice, such situations rarely exist. If $g(x)$ is not band limited (this happens, for example, for any finite-duration function $g(x)$ [Brigham88 p. 103]), the only solution is to reduce aliasing as much as possible, as explained below. ■

Aliasing can be reduced in two different ways. One way consists of sampling the given signal at a high enough rate (i.e. using a sufficiently small sampling interval Δx), so that aliasing becomes negligible. The other possibility is to low-pass filter the signal *prior to* sampling in order to make it band limited (or almost band limited); this allows us to avoid (or at least reduce) aliasing by using a sufficiently high sampling rate [Brigham88 p. 86; Cartwright90 p. 205]. Any combination of these two methods can also be used.

More details on the optimal choice of the sampling interval Δx (and the tradeoffs that are involved in this choice) can be found in Sec. 7.3.

For further reading, there exists a large choice of useful references that treat the aliasing phenomenon from various different points of view. One may consult, for example, [Bracewell86 pp. 197–198], [Brigham88 pp. 172–173], [Ramirez85 pp. 115–123], [Briggs95 pp. 95–98] or [Smith07 Sec. 7.2.14].

5.3 Examples of aliasing in the one dimensional case

Let us illustrate the above discussion by a few 1D examples. The first examples concern aperiodic functions $g(x)$, whose spectrum (CFT) $G(u)$ is continuous, and the remaining examples concern periodic functions, whose spectrum is purely impulsive.

Example 5.1: Aliasing due to undersampling in the case of the function $g(x) = \mathrm{sinc}(ax)$:

Figure 5.4 shows a particular case of the generic Fig. 5.3 in which the function being sampled is $g(x) = \mathrm{sinc}(ax)$, whose CFT is the square pulse $G(u) = \frac{1}{a}\mathrm{rect}(u/a)$ (see part (a) of the figure). As long as the sampling frequency $f_s = \frac{1}{\Delta x}$ is at least twice $f_G = a/2$, the highest frequency in $G(u)$, the signal $g(x)$ is sampled sufficiently often as stipulated by the sampling theorem, and the replicas of $G(u)$ in the spectrum of the sampled signal are far enough from each other to avoid overlapping. This means, indeed, that no aliasing occurs (see part (c) of the figure). But when the sampling frequency is smaller than $2f_G$, say, f'_s, the signal is no longer sufficiently often sampled, and every two neighbouring replicas in the spectral domain somewhat overlap (see part (d) of the figure, and notice there the additive nature of this spectral overlapping). This is the *spectral domain* manifestation of the aliasing. The *signal domain* interpretation of this aliasing is that only based on our too far-apart sampling points, the original signal $g(x)$ is undistinguishable from its alias $g_A(x)$, the signal whose spectrum $G_A(u)$ consists of the central period (marked by an arrow) of the periodic spectrum of Fig. 5.4(d). This signal-domain effect is shown in Fig. 5.2(b) for the simple case in which $a = 1$ and hence $f_G = \frac{1}{2}$, and the sampling frequency is $\frac{3}{4}$.

It is interesting to note that in the present example the explicit expression of the aliased signal $g_A(x)$ can be easily found. For this end, we note that its spectrum $G_A(u)$, i.e. the central period of the periodic spectrum in Fig. 5.4(d), is simply a $2/a$-units high rect function extending between $-\frac{1}{2}f'_s...\frac{1}{2}f'_s$ (where f'_s is the corresponding sampling frequency) minus an $1/a$-units high rect function with a slightly narrower width extending between $-b...b$ where $b = \frac{1}{2}f'_s - (f_G - \frac{1}{2}f'_s) = f'_s - f_G$. By taking the inverse Fourier transform of this difference of rect functions we find that $g_A(x)$ is simply a difference of two sinc functions. For example, in the case shown in Fig. 5.2(b) where $g(x) = \mathrm{sinc}(x)$, $f_G = \frac{1}{2}$ and $f'_s = \frac{3}{4}$, $G_A(u)$ can be seen as a 2-units high rect function extending between $-\frac{3}{8}...\frac{3}{8}$ minus a 1-unit high rect function extending between $-\frac{1}{4}...\frac{1}{4}$ (see Fig. 5.5(b)). It follows therefore that the aliased function that mimics $g(x)$ on the too far-apart sampling points of Fig. 5.2(b) is expressed by $g_A(x) = \frac{3}{2}\mathrm{sinc}(\frac{3}{4}x) - \frac{1}{2}\mathrm{sinc}(\frac{1}{2}x)$. Note that $g_A(x)$ differs from our original signal $g(x) = \mathrm{sinc}(x)$ only in its high frequencies. And indeed, if we carefully compare the spectrum of $g_A(x)$ (Fig. 5.5(b)) with the spectrum of the original signal $g(x)$ (Fig. 5.5(a)), we see that: (1) The maximum frequency in $G_A(u)$ is slightly lower than f_G, the maximum frequency in $G(u)$; and (2) The highest frequencies in $G_A(u)$ are simply a result of the additive overlap with the neighbouring replicas. We see, therefore, that $g_A(x)$ is indeed a lower-frequency alias of $g(x)$: *lower frequency* due to point (1); and *alias* because at our too low sampling frequency $f'_s = \frac{3}{4}$ the sampled versions of $g(x)$ and of $g_A(x)$ have precisely the same spectrum, i.e. the infinite periodic spectrum which is shown in Fig. 5.5(c), so they are undistinguishable from each other.[8]

---

[8] It should be noted for the record that $g_A(x)$ is not the only function that shares with $g(x)$ the same values at the given sampling points; for example, $(g(x) + g_A(x))/2$ will also have the same values at these sampling points, and the same goes for any other weighted mean of $g(x)$ and $g_A(x)$. In fact, there exist infinitely many functions that share the same values at these sampling points. Because their sampled versions are identical, the Fourier transforms of their sampled versions are identical, too, and they are all equal to the infinite periodic spectrum shown in Fig. 5.5(c). However, $g_A(x)$ has the lowest maximum frequency, since it does not exceed half of the sampling frequency f'_s (while $g(x)$ does).

Signal domain   Spectral domain

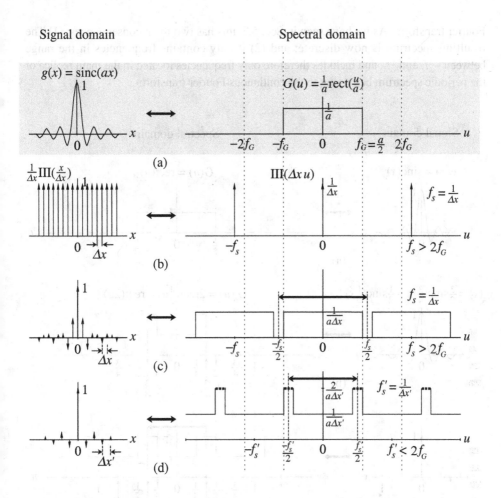

(a)

(b)

(c)

(d)

Figure 5.4: A particular case of Fig. 5.3 showing the signal $g(x) = \text{sinc}(ax)$, whose CFT is the square pulse $G(u) = \frac{1}{a}\text{rect}(u/a)$. The sampled version of $g(x)$ and its CFT are shown in (c). (d) When the sampling rate is lower than twice the maximum signal frequency f_G (say, f'_s), every two neighbouring replicas of $G(u)$ overlap additively on their borders, giving in the overlapped zones false aliased spectral values with respect to the original continuous-world spectrum $G(u)$ shown in (a). The horizontal arrows in the spectra (c) and (d) indicate the frequency range of the DFT, which always extends between minus half and plus half of the corresponding sampling frequency. See also a concrete example in Fig. 5.5.

So far we have only discussed the continuous Fourier transform, so that the spectrum of the sampled version of $g(x)$ (see Figs. 5.4(c),(d)) is periodic (extending throughout the range $-\infty...\infty$) and continuous (being a periodic repetition of the continuous spectrum $G(u)$). Assume now that our sampled signal undergoes DFT rather than a continuous

Fourier transform. As we have seen in Sec. 5.2, this has two main consequences: (1) The resulting spectrum is now discrete; and (2) it only contains frequencies in the range between $-\frac{1}{2}f_s$ and $\frac{1}{2}f_s$, and includes therefore only frequencies located in the main replica of the periodic spectrum belonging to the continuous Fourier transform.

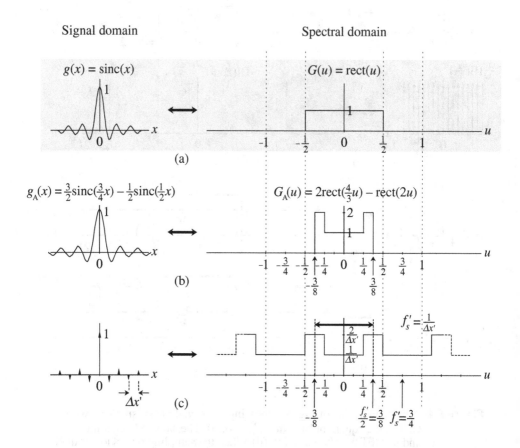

Figure 5.5: The spectral-domain interpretation of the aliasing shown in Fig. 5.2(b). (a) The function $g(x) = \text{sinc}(x)$ and its continuous Fourier transform $G(u) = \text{rect}(u)$. (b) The aliased function $g_A(x) = \frac{3}{2}\text{sinc}(\frac{3}{4}x) - \frac{1}{2}\text{sinc}(\frac{1}{2}x)$ and its continuous Fourier transform $G_A(u) = 2\text{rect}(\frac{4}{3}u) - \text{rect}(2u)$. As shown in Fig. 5.2(b), the sampled versions of $g(x)$ and $g_A(x)$ at the sampling rate of $f'_s = \frac{3}{4}$ are identical; this implies that the spectra of both sampled signals are identical, too, but here we show the reason directly in the spectral domain: When convolving $G(u)$ or $G_A(u)$ with an impulse comb having impulse intervals of $f'_s = \frac{3}{4}$ (and impulse heights of $1/\Delta x' = f'_s$), the periodic spectra we obtain are exactly identical, as shown here in (c). Note that $g_A(x)$ is indeed a *lower-frequency* alias of $g(x)$: Its maximum frequency is exactly $\frac{1}{2}f'_s = \frac{3}{8}$ (see the spectrum $G_A(u)$ in (b)), whereas the maximum frequency of $g(x)$ is $f_G = \frac{1}{2}$ (see the spectrum $G(u)$ in (a)). The horizontal arrow in (c) indicates the frequency range of the DFT.

Figure 5.6 shows the sampled version of the signal $g(x) = \text{sinc}(2x)$ as well as its resulting DFT spectrum (after having performed all the required adjustments as discussed in Chapters 3 and 4), for several gradually decreasing values of the sampling frequency f_s (and of the array length N). As we can see in rows (a)–(c) of the figure, as long as the sampling frequency f_s is higher than twice $f_G = 1$ (the maximum frequency of $g(x)$), the resulting adjusted DFT matches well the expected CFT of $g(x)$, $G(u) = \frac{1}{2}\text{rect}(u/2)$. Row (d) of the figure shows the situation in the limit case when the sampling frequency f_s is exactly twice $f_G = 1$, and rows (e) and (f) show what happens when $f_s < 2f_G$. As we can see, in the last two rows the DFT results no longer match the continuous spectrum $G(u) = \frac{1}{2}\text{rect}(u/2)$, due to the existence of aliasing. Note that because the DFT spectrum only contains frequencies between $-\frac{1}{2}f_s...\frac{1}{2}f_s$, i.e. frequencies in the central replica of the periodic spectrum of Fig. 5.4(d), the overlapping of the neighbouring replicas due to aliasing looks in the DFT spectrum as a cyclical foldover. In other words, frequencies that exceed from one end of the DFT spectrum re-enter in the opposite end and are summed up with the true spectral values belonging to the frequencies that are really located at that part of the spectrum. ∎

Example 5.2: Aliasing due to undersampling in the case of the function $g(x) = \text{sinc}^2(x)$:

This example shows another particular case of the generic Fig. 5.3 in which the function being sampled is $g(x) = \text{sinc}^2(x)$, whose CFT is the triangle $G(u) = \text{tri}(u)$ [Bracewell86 p. 415]. The DFT of the sampled version of this function is shown in Fig. 5.7 for the same gradually decreasing values of the sampling frequency f_s (and of the array length N) as in Fig. 5.6. Once again, as long as the sampling frequency $f_s = \frac{1}{\Delta x}$ is at least twice $f_G = 1$ (the highest frequency in $G(u)$), the signal $g(x)$ is sampled sufficiently often as stipulated by the sampling theorem, and the replicas of $G(u)$ in the spectrum of the sampled signal are far enough from each other to avoid overlapping, so that no aliasing occurs. But when $f_s < 2f_G$, the signal is no longer sufficiently often sampled, and every two neighbouring replicas in the spectral domain somewhat overlap. This gives a cyclical foldover effect in the DFT spectrum (see parts (e)–(f) of the figure, and notice there the additive nature of this spectral overlapping).

Note that in the present example the aliased frequencies have a softer and less drastic impact on the resulting spectrum than in the case shown in Example 5.1. This is, indeed, the situation in most of the real-world cases, where the aliased frequencies only slightly distort the shape of the resulting DFT (starting from both ends of the adjusted DFT spectrum inward, toward the center). The existence of such aliasing may be difficult to detect — unless we know in advance the precise shape of the expected continuous-world spectrum. ∎

Example 5.3: An example whose spectrum is not band limited, and contains both real-valued and imaginary-valued parts (Fig. 5.8):

In the examples discussed so far the functions $g(x)$ were symmetric about the origin, so that their spectra were purely real-valued [Bracewell86 pp. 14–15]. The present example illustrates the situation when both the real-valued and imaginary-valued parts of the

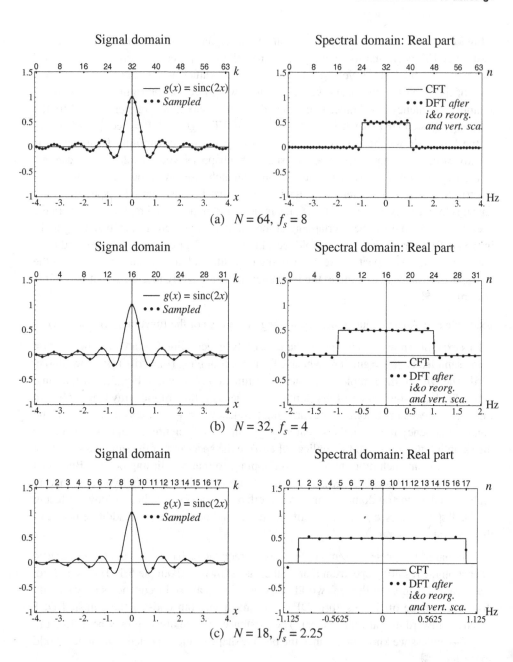

Figure 5.6: The sampled version of the signal $g(x) = \text{sinc}(2x)$ and its resulting DFT (after all the required adjustments as discussed in Chapters 3 and 4). For the sake of comparison, the original continuous function $g(x)$ and its CFT $G(u) = \frac{1}{2}\text{rect}(u/2)$ are also shown, using continuous lines. The successive rows of the figure show the situation for gradually decreasing values of the sampling

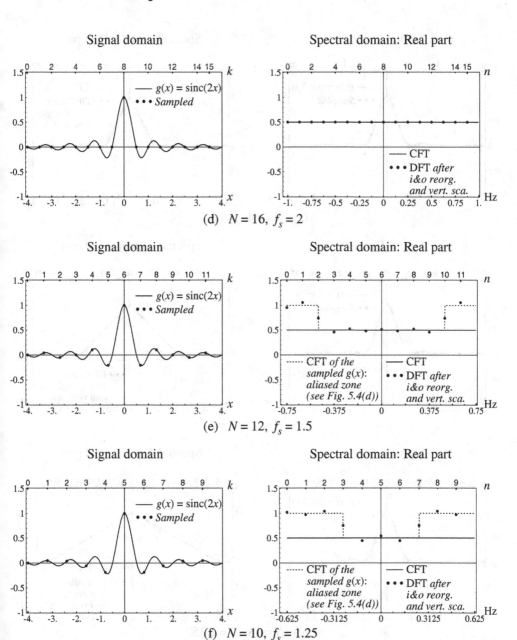

(d) $N = 16$, $f_s = 2$

(e) $N = 12$, $f_s = 1.5$

(f) $N = 10$, $f_s = 1.25$

frequency f_s (and of the array length N). In rows (a)–(c) $f_s > 2f_G$ and no aliasing occurs. In row (d) $f_s = 2f_G$, while in rows (e)–(f) $f_s < 2f_G$ and aliasing appears in the form of a cyclical foldover: Frequencies which exceed from one end of the DFT spectrum re-enter in the opposite end and are additively superposed on top of the existing spectral values there. See Example 5.1.

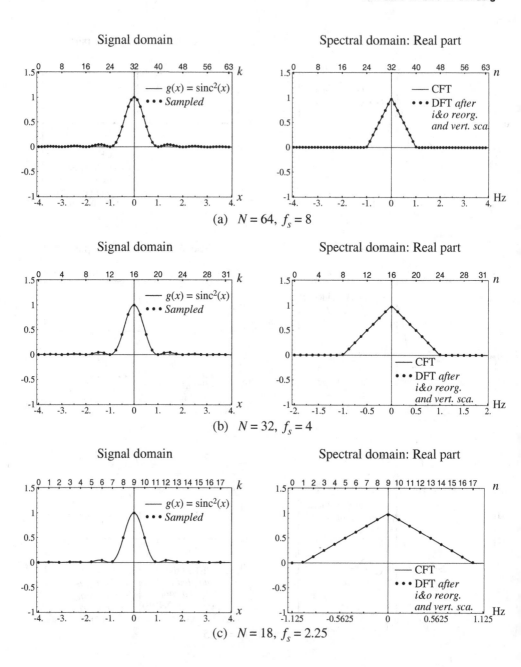

Figure 5.7: Same as Fig. 5.6, for the signal $g(x) = \text{sinc}^2(x)$ whose CFT is $G(u) = \text{tri}(u)$. The successive rows of the figure show the situation for gradually decreasing values of the sampling frequency f_s (and of the array length N). In rows (a)–(c) $f_s > 2f_G$ and no aliasing occurs. In row (d) $f_s = 2f_G$, while in rows (e)–(f)

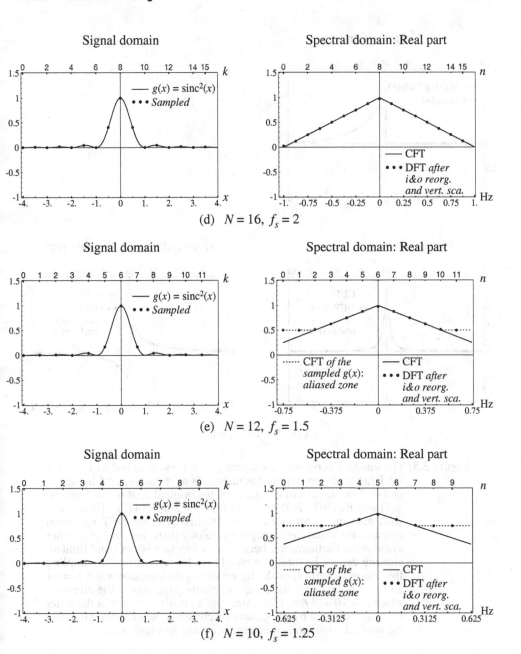

(d) $N = 16$, $f_s = 2$

(e) $N = 12$, $f_s = 1.5$

(f) $N = 10$, $f_s = 1.25$

$f_s < 2f_G$ and aliasing appears in the form of a cyclical foldover:
Frequencies which exceed from one end of the DFT spectrum
re-enter in the opposite end and are additively superposed on top
of the existing spectral values there. See Example 5.2.

Signal domain

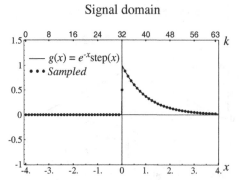

Spectral domain: Real part   Spectral domain: Imaginary part

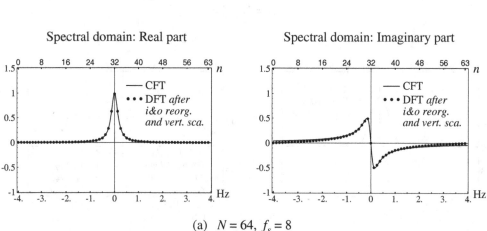

(a) $N = 64$, $f_s = 8$

Figure 5.8: The sampled version of the signal $g(x) = e^{-x}\,\text{step}(x)$ and its resulting DFT (after all the required adjustments as discussed in Chapters 3 and 4). For the sake of comparison, the original continuous function $g(x)$ and its CFT $G(u) = 1/(1+[2\pi u]^2) - 2\pi u i/(1+[2\pi u]^2)$ are also shown, using continuous lines. In this case the DFT spectrum contains both real- and imaginary-valued parts, because $g(x)$ is not symmetric. Furthermore, because this function is not band limited, aliasing cannot be totally avoided here by increasing the sampling frequency f_s. Part (a) of the figure shows the situation when $N = 64$ and $f_s = 8$, and part (b) on the opposite page shows the situation when $N = 10$ and $f_s = 1.25$. Aliasing is clearly stronger in the latter case, as indicated by the greater discrepancy between the DFT and the original continuous spectrum $G(u)$. See Example 5.3.

spectrum are non-zero. Consider the function $g(x) = e^{-x}\,\text{step}(x)$ whose CFT is $G(u) = 1/(1+[2\pi u]^2) - 2\pi u i/(1+[2\pi u]^2)$ [Bracewell86 p. 418]. Note that this function is not band limited (meaning that its maximum frequency is $f_G = \infty$), so that aliasing cannot be avoided here by increasing the sampling frequency f_s. Figure 5.8 shows the sampled version of this function and its adjusted DFT for two different values of the sampling frequency (and of the array length N). As we can see in part (a) of the figure, as long as the sampling

Signal domain

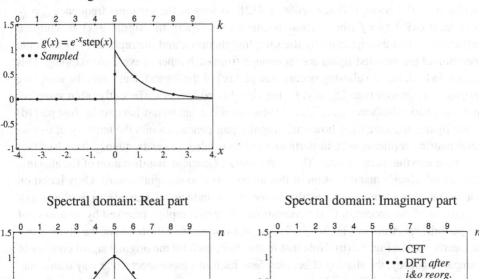

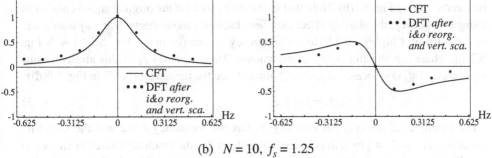

(b) $N = 10$, $f_s = 1.25$

frequency $f_s = \frac{1}{\Delta x}$ is reasonably high, the signal is reasonably well sampled (obviously, except in the sharp peak at the origin), and the replicas of $G(u)$ in the spectrum of the sampled signal do not significantly corrupt each other (although they obviously do overlap). But when the sampling frequency is too low, as shown in part (b) of the figure, the signal is no longer reasonably well sampled, and the overlapping between every two neighbouring replicas in the spectral domain becomes much more significant, giving a stronger cyclical foldover effect in the DFT spectrum. Note that because this function is not band limited, each of the replicas in the spectrum of its sampled version extends *ad infinitum* to both directions. This means that the main replica (the one contained in the DFT spectrum) is additively contaminated not only by its two direct neighbours, but indeed by all of the infinitely many replicas. Nevertheless, the quantitative influence of far-away replicas is marginal here due to the decaying nature of the original spectrum $G(u)$, and most of the aliasing distortion is due to the intrusion of the closest replicas. ∎

Example 5.4: Aliasing due to undersampling in the case of the simple periodic function $g(x) = \cos(2\pi f x)$:

Fig. 5.9 shows a particular case of generic Fig. 5.3 in which the function being sampled is $g(x) = \cos(2\pi f x)$ with frequency f (see part (a) of the figure). The CFT of this function consists of a pair of impulses that are located at a distance f to both sides of the origin: $G(u) = \frac{1}{2}\delta(u-f) + \frac{1}{2}\delta(u+f)$ [Bracewell86 p. 412]. As long as the sampling frequency $f_s = \frac{1}{\Delta x}$ is at least twice $f_G = f$ (the highest frequency in $G(u)$), the signal $g(x)$ is sampled sufficiently often as stipulated by the sampling theorem, and the replicas of $G(u)$ in the spectrum of the sampled signal are far enough from each other to avoid overlapping. This means, indeed, that no aliasing occurs (see part (c) of the figure). But when the sampling frequency is smaller than $2f_G$, say, f'_s, the signal is no longer sufficiently often sampled, and every two neighbouring replicas in the spectral domain somewhat overlap (see part (d) of the figure, and note there how each impulse pair penetrates into the territory of its two neighbouring replicas; note in particular the two false, lower-frequency impulses that penetrate into the main replica). This is the *spectral domain* manifestation of the aliasing. The *signal domain* manifestation of this aliasing is also straightforward: Only based on our too far-apart sampling points, the cosine $g(x)$ is undistinguishable from its alias $g_A(x)$, the cosine whose spectrum $G_A(u)$ consists of the central replica (marked by an arrow) of the periodic spectrum of Fig. 5.9(d). This signal-domain manifestation of the aliasing can be clearly seen in Fig. 5.1(b). Note that in this case, because the original signal consists of a single frequency, the aliasing effect manifests itself in a more spectacular way than usual, in the form of a highly visible lower-frequency cosine (compare Fig. 5.1(b) with Fig. 5.2(b), where the aliasing is less conspicuous). The frequency f_A of this aliased cosine $g_A(x) = \cos(2\pi f_A x)$ is located at f_s (or f'_s) minus f (see the replica around f'_s in Fig. 5.9(d)):

$$f_A = f_s - f \tag{5.3}$$

For example, if the original cosine $g(x)$ has the frequency $f = 2$ and the sampling frequency is $f_s = \frac{1}{\Delta x} = \frac{9}{4}$ (which is indeed lower than $2f$), the resulting aliased cosine $g_A(x)$ has the frequency $f_A = \frac{9}{4} - 2 = \frac{1}{4}$. Note that in the case shown in Fig. 5.1(b), in which the original cosine has the same frequency $f = 2$ but the sampling frequency is only $f_s = \frac{7}{4}$ (which is again lower than $2f$) we obtain $f_A = \frac{7}{4} - 2 = -\frac{1}{4}$, which is obviously equivalent to its positive-sign counterpart $f_A = \frac{1}{4}$. The negative sign of the aliased frequency is obtained here because the sampling frequency $f_s = \frac{7}{4}$ is smaller than the frequency $f = 2$ of the original cosine, meaning that the aliased impulse is located here in the other side of the origin, unlike in Fig. 5.9(d). Note that when the resulting aliased frequency f_A falls close to the spectrum origin the resulting low-frequency aliased cosine may be considered as a *sampling moiré effect*; and indeed, Eq. (5.3) above is precisely the well-known formula of the moiré frequency between two parallel gratings (see for example Eq. (2.11) in [Amidror09 p. 20]). We will return to this subject later, when we discuss the more interesting 2D case (see Examples 5.7 and on in Sec. 5.5, and the discussion in Sec. 5.6).

So far we have only discussed the continuous Fourier transform, so that the spectrum of the sampled cosine contains an infinite number of replicas of the impulse pair $G(u)$ and extends throughout the range $-\infty...\infty$, as shown in Figs. 5.9(c),(d). But if our sampled cosine undergoes DFT rather than a continuous Fourier transform, the resulting DFT spectrum will only contain frequencies in the range $-\frac{1}{2}f_s...\frac{1}{2}f_s$, where the main replica is located (see the frequency range indicated by the horizontal arrow in Figs. 5.9(c),(d)). Figure 5.10 shows, indeed, the DFT of the sampled cosine function with $f = 1$ (after all the

Signal domain   Spectral domain

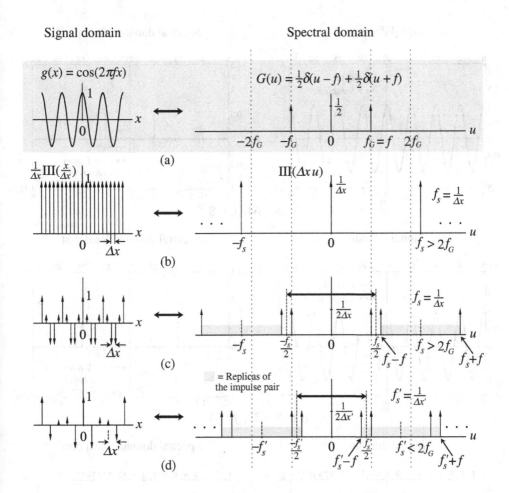

(a)

(b)

(c)

= Replicas of
the impulse pair

(d)

Figure 5.9: A particular case of Fig. 5.3 showing the periodic signal $g(x) = \cos(2\pi fx)$, whose CFT consists of a pair of impulses that are located at a distance f to both sides of the origin: $G(u) = \frac{1}{2}\delta(u-f) + \frac{1}{2}\delta(u+f)$. The sampled version of $g(x)$ and its CFT are shown in (c). (d) When the sampling rate is lower than twice the maximum signal frequency f_G (say, f'_s), each replica of $G(u)$ overlaps additively with the two neighbouring replicas on their borders, meaning that its impulse pair intrudes into the territory of the two neighbouring replicas. In particular, two new false lower-frequency impulses appear now in the main replica; this impulse pair belongs to the new, false aliased cosine that becomes visible in the signal domain. The horizontal arrows in the spectra (c) and (d) indicate the part of the infinite continuous-world spectrum that will be represented in the DFT output should we subject the sampled signal to DFT rather than to CFT; this frequency range extends between minus half and plus half of the corresponding sampling frequency. Note that in terms of the DFT frequency range the aliasing effect can be viewed as a *cyclical foldover*, where frequencies which exceed beyond one end of the range re-enter (additively) at the opposite end.

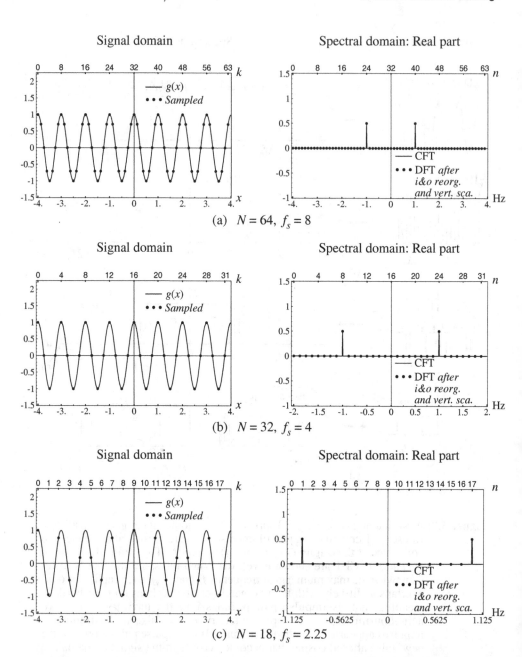

Figure 5.10: Same as Fig. 5.6, for the signal $g(x) = \cos(2\pi f x)$ with frequency $f = 1$ whose CFT is $G(u) = \frac{1}{2}\delta(u-1) + \frac{1}{2}\delta(u+1)$. In rows (a)–(c) $f_s > 2f_G$ and no aliasing occurs. Row (d) shows the limit case where $f_s = 2f_G$; note that because the array length N is even, the DFT spectrum contains one less sample in the positive frequency side (see Footnote 1 in Sec. 3.2), and therefore the

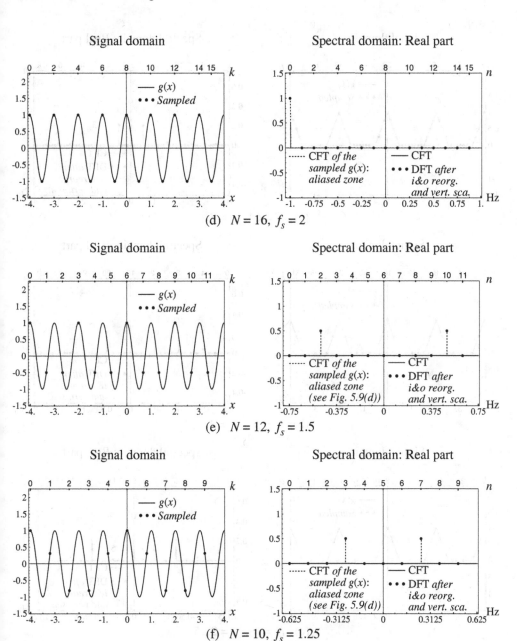

(d) $N = 16$, $f_s = 2$

(e) $N = 12$, $f_s = 1.5$

(f) $N = 10$, $f_s = 1.25$

impulse at the right end of the DFT is folded over and summed up on top of its negative-frequency counterpart at the left end of the DFT. In rows (e)–(f) $f_s < 2f_G$, so each of the two impulses folds over and re-enters from the opposite side of the DFT spectrum, thus simulating a false, lower-frequency cosine function. See Example 5.4.

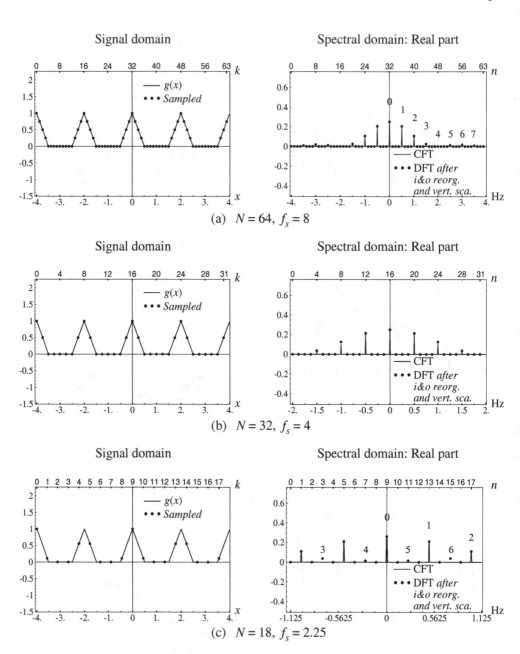

Figure 5.11: Same as Fig. 5.6, for the periodic triangular signal $g(x) = \text{triangwave}(2x)$ with period $P = 2$ (i.e. frequency $f = \frac{1}{2}$). As explained in Example 5.5, the CFT of $g(x)$, $G(u)$, is an impulse train with impulse interval $\Delta u = f = \frac{1}{2}$ whose amplitude is modulated by half the CFT of a single triangular period $\text{tri}(2x)$, namely by $\frac{1}{4}\text{sinc}^2(\frac{1}{2}u)$. Note in (c) how the impulses of the third

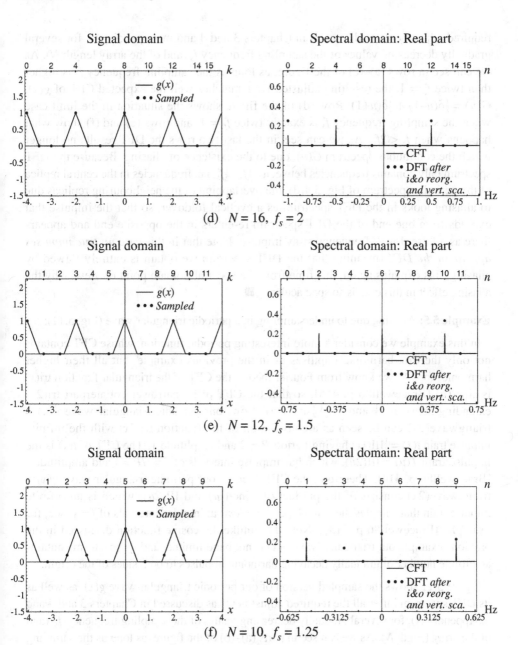

(d) $N = 16$, $f_s = 2$

(e) $N = 12$, $f_s = 1.5$

(f) $N = 10$, $f_s = 1.25$

harmonic and on exceed from one end of the DFT spectrum and re-enter
from the other end in the form of false in-between stray impulses. (This is
not visible in (b), because in this case the folded-over impulses fall exactly
on top of existing impulses, and they only corrupt their amplitudes.) In
(d)–(f) aliasing is already so strong that $G(u)$ is no longer recognizable.

required adjustments as discussed in Chapters 3 and 4 and in Appendix A), for several gradually decreasing values of the sampling frequency f_s (and of the array length N). As we can see in rows (a)–(c) of the figure, as long as the sampling frequency f_s is higher than twice $f_G = 1$, the resulting adjusted DFT matches well the expected CFT of $g(x)$, $G(u) = \frac{1}{2}\delta(u-1) + \frac{1}{2}\delta(u+1)$. Row (d) of the figure shows the situation in the limit case when the sampling frequency f_s is exactly twice $f_G = 1$, and rows (e) and (f) show what happens when $f_s < 2f_G$. As we can see, in the last two rows the DFT results no longer match the continuous spectrum $G(u)$, due to the existence of aliasing. Because the DFT spectrum only contains frequencies between $-\frac{1}{2}f_s...\frac{1}{2}f_s$, i.e. frequencies in the central replica of the periodic spectrum of Fig. 5.4(d), the overlapping of the neighbouring replicas due to aliasing looks in the DFT spectrum as a cyclical foldover, so that the impulse that exceeds from one end of the DFT spectrum re-enters in the opposite end and appears there as a false, lower-frequency stray impulse. Note that in this case *no true impulses appear in the DFT*, meaning that the DFT spectrum we obtain is entirely flawed by aliasing; this explains again, this time from the spectral-domain point of view, why the aliasing effect in this case is so spectacular. ■

Example 5.5: Aliasing due to undersampling in a periodic triangular wave (Fig. 5.11):

In this example we consider a more interesting periodic function, whose CFT contains not only the first harmonic impulses as in the previous example, but all their higher harmonics too. As we know from Fourier theory, the CFT of the triangular function $\mathrm{tri}(x)$ is $\mathrm{sinc}^2(u)$ [Bracewell86 p. 415], so that the CFT of its narrower counterpart $\mathrm{tri}(2x)$ extending between $-\frac{1}{2}$ and $\frac{1}{2}$ is $\frac{1}{2}\mathrm{sinc}^2(\frac{1}{2}u)$. Now, our periodic triangular wave $g(x) = \mathrm{triangwave}(2x)$ can be seen as a convolution of the function $\mathrm{tri}(2x)$ with the infinite impulse train $t(x) = \frac{1}{2}\mathrm{III}(\frac{1}{2}x)$ having period $P = 2$ and amplitude 1. The CFT of $t(x)$ is the impulse train $T(u) = \mathrm{III}(2u)$, which has impulse intervals of $f = 1/P = \frac{1}{2}$ and amplitude $\frac{1}{2}$ [Bracewell86 p. 414]. Therefore, the CFT $G(u)$ of our periodic triangular wave $g(x) = \mathrm{triangwave}(2x)$ consists of the product of $\frac{1}{2}\mathrm{sinc}^2(\frac{1}{2}u)$ and $\mathrm{III}(2u)$, which is an infinite impulse train that samples the "envelope" $\frac{1}{4}\mathrm{sinc}^2(\frac{1}{2}u)$ at impulse intervals of $f = \frac{1}{2}$ (see, for example, [Bracewell86 p. 213]). Note that unlike the cosine function discussed in the previous example our triangular wave $g(x)$ is not band limited, and its spectrum contains an infinite number of gradually decaying harmonic impulses to both sides of the origin.

Figure 5.11 shows the sampled version of our periodic triangular wave $g(x)$, as well as its DFT spectrum (after all the required adjustments as discussed in Chapters 3 and 4 and in Appendix A), for several gradually decreasing values of the sampling frequency f_s (and of the array length N). As we can see in parts (a)–(b) of the figure, as long as the sampling frequency $f_s = \frac{1}{\Delta x}$ is reasonably high, the signal $g(x)$ is reasonably well sampled, and the replicas of $G(u)$ in the spectrum of the sampled signal do not significantly corrupt each other (although they obviously do overlap). But when the sampling frequency is too low, as shown in parts (c)–(f) of the figure, the signal is no longer reasonably well sampled, and the overlapping between every two neighbouring replicas in the spectral domain becomes much more significant, giving a stronger cyclical foldover effect in the DFT spectrum. Note that because the spectrum $G(u)$ is impulsive, the overlap (or cyclical foldover) due to aliasing implies that impulses that exceed from one end of the DFT spectrum re-enter in the opposite end and appear there as false, lower-frequency stray impulses. This can be clearly seen when comparing the DFT spectrum in parts (a) and (c) of Fig. 5.11: While in (a) all the impulses up to the 7th harmonic appear in their correct locations, in (c) only the impulses up to the second harmonic appear in their correct locations, and the impulses of the third harmonic and on fold over and re-enter from the opposite side of the DFT spectrum in the form of new, in-between stray impulses. (Note, however, that in some cases, like in part (b) of the figure, the aliased stray impulses are not visible because they fall precisely on top of true impulses, and they only corrupt their amplitudes; see Remark 5.1 in Sec. 5.2). In parts (d)–(f) of the figure aliasing is already so strong that the impulses of the original spectrum $G(u)$ are no longer recognizable.

Finally, note that because our function is not band limited, each of the replicas in the spectrum of its sampled version extends *ad infinitum* to both directions. This means that the main replica (the one contained in the DFT spectrum) is additively contaminated not only by its two direct neighbours, but by all of the infinitely many replicas. This explains, indeed, the small differences in the impulse amplitudes that we see when comparing the DFT spectra in rows (a)–(f) of Fig. 5.11: These amplitude differences are due to the different ways in which impulses of the infinitely many replicas fold over and fall on top of existing impulses in each of the rows. Nevertheless, the quantitative influence of far-away replicas is marginal because of the decaying nature of the original spectrum $G(u)$, and most of the aliasing distortion is due to the intrusion of the closest replicas. ∎

5.4 Aliasing in two or more dimensions

The above discussion about aliasing can be also extended to the 2D or MD cases. In such cases the sampling in the signal domain and the corresponding convolution in the spectral domain (see Eq. (5.2)) are simply extended to M dimensions:[9]

$$\mathcal{F}[(\tfrac{1}{\Delta x})^M \, \mathrm{III}(\mathbf{x}/\Delta x) \, g(\mathbf{x})] = \mathcal{F}[(\tfrac{1}{\Delta x})^M \, \mathrm{III}(\mathbf{x}/\Delta x)] * \mathcal{F}[g(\mathbf{x})] = \mathrm{III}(\Delta x \mathbf{u}) * G(\mathbf{u}) \qquad (5.4)$$

Figure 5.12 shows schematically how aliasing occurs due to undersampling in the 2D case. The explanation in the MD case, too, is similar.

It should be noted, however, that although Eq. (5.4) is a straightforward generalization of Eq. (5.2), the multidimensional case is much more rich and diversified than the 1D case. For example, the effect of a rotation about the origin in the 2D case on the aliasing (see below) has no equivalent in the 1D case. A discussion on sampling and aliasing in the 2D continuous world can be found, for example, in [Rosenfeld82, Sec. 4.1] or [Bracewell95, Chapter 7], and the connection of such 2D aliasing with the moiré effect is explained in [Amidror09, Sec. 2.13]. More details on the sampling theorem and on aliasing in the MD case can be found in [Petersen62], [Marks91, Sec. 6.2] or [Zayed93, Sec. 3.3].

---

[9] We assume here that the same sampling interval Δx is used along all the M dimensions. If required, an appropriate generalization may allow us to use a different sampling interval along each dimension.

5.5 Examples of aliasing in the multidimensional case

To better illustrate the aliasing phenomenon in the multidimensional DFT, let us consider a few 2D examples with various spectral properties. Note that all our 2D DFT figures are drawn as density plots (see Sec. 1.5.2, as well as Footnote 14 on p. 219).

Example 5.6: Consider the 2D case shown in Fig. 5.13(a). Here, the given continuous-world function $g(x,y)$ is an infinite vertical bar of height 1 and width τ that is centered about the origin and rotated by angle θ; its spectrum $G(u,v)$ is an infinite, continuous straight line-impulse (blade) passing through the origin at the perpendicular direction, whose amplitude is modulated by a sinc function (see Sec. D.10 in Appendix D). The spectrum of this function is not band limited since it contains frequencies going *ad infinitum* to both sides of the origin, and therefore it largely exceeds the maximum frequency that can be represented in the output array of the DFT. As we can see in Fig. 5.13(a), wherever the line-impulse exceeds the borders of the DFT spectrum it simply folds over and re-enters into the DFT spectrum from the opposite side. Note that no parts of the line-impulse are thereby lost, and its folded-over parts simply pursue their course, at the same angle θ, starting from the opposite end of the DFT spectrum. But because in the continuous world the line-impulse extends to both directions *ad infinitum*, it follows that in the discrete world the folded-over blade exceeds the borders of the DFT spectrum again and again, each time re-entering and pursuing its course from the opposite side.[10] This

Figure 5.12: Schematic illustration of aliasing due to undersampling in the 2D case. This figure is a 2D counterpart of the generic 1D case shown in Fig. 5.3. (a) An original continuous band-limited 2D signal $g(x,y)$ and its continuous-world spectrum $G(u,v)$. (b) The sampling impulse nailbed and its continuous-world spectrum. (c) The sampled signal is the product of signals (a) and (b), and therefore its continuous-world spectrum is the convolution of their spectra. Note the replicas of the original spectrum, that are located at all integer linear combinations of the sampling frequencies u_s and v_s. (d) If the sampling rate along any of the two main directions is lower than twice the maximum signal frequency in that direction, u_G or v_G, then every two neighbouring replicas of $G(u,v)$ overlap (additively) along this direction, giving false aliased frequencies in the overlapped zones. The value of the spectrum in the aliased areas in (d) is simply the sum of the values of the overlapped replicas. The horizontal and vertical arrows in the spectra (c) and (d) indicate the part of the infinite continuous-world spectrum that will be represented in the 2D DFT output should we subject the sampled signal to DFT rather than to CFT: The frequency range of the DFT extends along each of the axes between minus half and plus half of the corresponding sampling frequency. See Problem 5-9 for cases where the original signal $g(x,y)$ is not band-limited.

---

[10] In fact, if we consider the infinite 2D continuous-world spectrum explaining the sampling of $g(x,y)$ (see Problem 5-9 and Fig. 5.21), these multiple cycles of foldover are really caused by overlapping of the main replica by farther away replicas of the line impulse. But again, just as in the 1D case (see p. 94), from the point of view of the DFT this is fully equivalent to the folding-over interpretation.

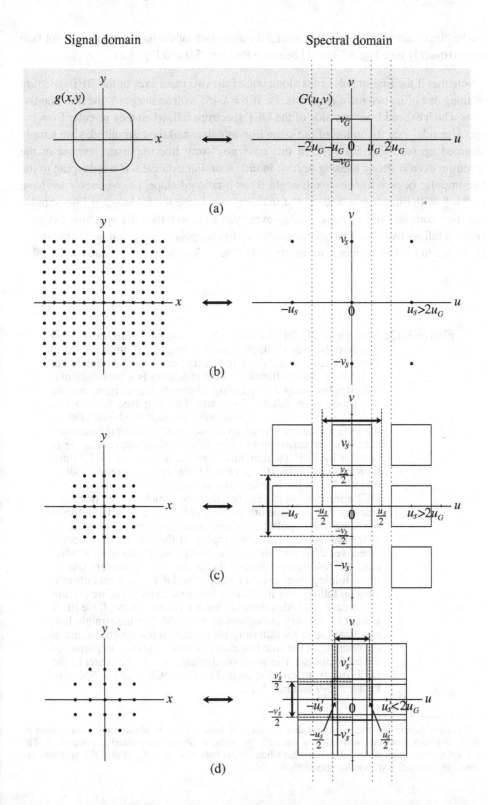

Signal domain   Spectral domain

can be clearly seen in the DFT spectrum if the amplitude of the line-impulse does not fade out too quickly (see Fig. 5.13(a)).[11] See also Problem 5-9 and Fig. 5.21.

Note that if the line-impulse runs along one of the two main axes of the DFT spectrum or along one of its two main diagonals, i.e. if $\theta = k\cdot45°$ with an integer k, the line-impulse parts which exceed from one side of the DFT spectrum fall, when they re-enter from the opposite side, exactly on top of the same line-impulse, and their amplitudes are simply summed up (see Fig. 5.13(c)). In this case no "stray line-impulses" appear in the spectrum even if strong aliasing occurs. In fact, a similar cyclical self-overlapping of the line-impulse occurs whenever the angle θ has a rational slope, i.e. whenever we have $\tan\theta = \frac{n}{m}$ with integer m,n, except that in this general case the folded-over line-impulse may first complete one or more folding-over cycles through the DFT spectrum before it ends up falling over itself (so that some "stray line-impulses" may still be visible in the DFT spectrum). This happend, for example, in Fig. 5.13(a), where $m = 4$ and $n = 1$. ∎

Figure 5.13: Aliasing in 2D DFT: a case whose spectrum consists of a 1D line-impulse (see Example 5.6). The original continuous-world function in the signal domain is a rotated, 1-valued straight bar of width τ. Its continuous-world spectrum is a perpendicular infinite line-impulse passing through the origin, whose amplitude is modulated by a sinc function (see Sec. D.10). Wherever this spectral line-impulse exceeds the boundaries of the DFT spectrum, it folds over and re-enters from the opposite side. The three first rows only differ in their rotation angle θ: (a) $\theta = 14.036°$ (which has a rational slope of 4:1, since $\tan\theta = 1/4$); (b) $\theta = 10°$; (c) $\theta = 45°$ (which has a rational slope of 1:1). When the line-impulse runs along the diagonal of the DFT spectrum, as in case (c), its parts which exceed from one side re-enter from the other side and fall there exactly on top of the same line-impulse; their amplitudes are simply summed up. A similar cyclical self-overlapping of the line-impulse occurs whenever θ has a rational slope, except that in this more general case the folded-over line-impulse may first complete one or more folding-over cycles through the DFT spectrum before it ends up falling over itself. This happens, indeed, in case (a), but not in case (b) which does not have a rational slope. Case (d) is similar to case (a), but differs in the width τ of the straight bar; note that the bar width only influences in the spectral domain the opening of the sinc function that modulates the amplitude of the line impulse. The small oscillations around the center of the DFT spectrum are due to leakage (see Chapter 6). See also Problem 5-9 and Fig. 5.21.

---

[11] This case clearly illustrates the fact that the spectral foldover due to aliasing does not consist of mirroring back from the *same* edge of the DFT spectrum, as already mentioned in Footnote 6. The correct interpretation of the foldover is that whatever exceeds from one side of the DFT spectrum re-enters in a cyclical way from the opposite side.

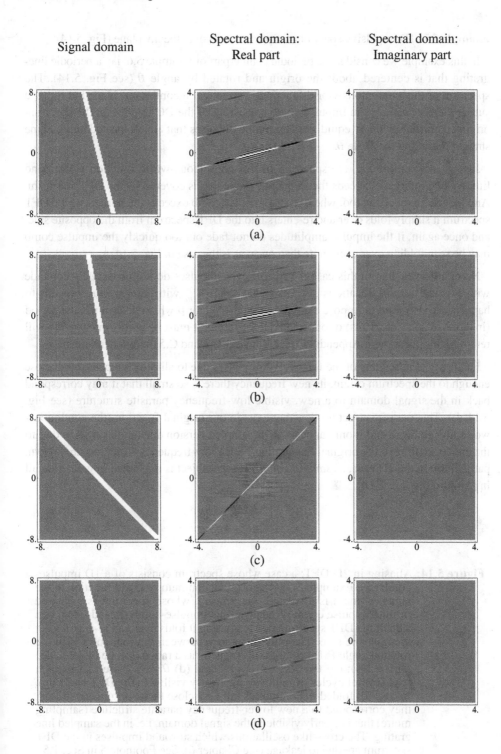

Signal domain   Spectral domain: Real part   Spectral domain: Imaginary part

Example 5.7: 1D impulsive spectrum (impulse comb) over the u,v plane (Fig. 5.14):

In this example we consider the periodic counterpart of Example 5.6, i.e. a periodic line-grating that is centered about the origin and rotated by angle θ (see Fig. 5.14). The spectrum of this continuous-world function is no longer a continuous blade but rather a subset thereof, namely, an impulse comb consisting of the DC impulse at the spectrum origin and infinitely many equidistant harmonic impulses that are all located on the same straight line along the angle θ.

Just as in Example 5.6, the spectrum of this continuous-world function is not band limited, and it largely exceeds the frequency range that is covered by the DFT spectrum. And indeed, in this case, too, wherever the impulse comb exceeds the borders of the DFT spectrum it simply folds over and re-enters into the DFT spectrum from the opposite side; and once again, if the impulse amplitudes do not fade out too quickly the impulse comb may be seen folding over even more than once, as is the case in Fig. 5.14.

Note, however, that in this case the folded-over impulses do not necessarily coincide with existing impulse locations even if we have $\tan\theta = \frac{n}{m}$ with integer m,n. For this to happen, we also need to impose a condition on the vector **p** which determines the period (in terms of array elements) of our rotated line grating: it must be purely integer. We will return to this question in Appendix C (see Remarks C.2 and C.5 there for full details).

Finally, note that if any of the folded-over impulses due to aliasing happens to fall close enough to the spectrum origin, its new frequency there is so small that it may correspond back in the signal domain to a new, visible low-frequency parasite structure (see Fig. 5.14(d)). This parasite structure does not exist in our original function (the continuous-world line grating), and it only appears in its sampled version due to aliasing (i.e. due to the undersampling of the original line grating). Such low-frequency structures are known, particularly in the 2D case, as *sampling moirés*; this subject is explained in greater detail in [Amidror09, Sec. 2.13]. ■

Figure 5.14: Aliasing in 2D DFT: a case whose spectrum consists of a 1D impulse-comb (see Example 5.7). The original continuous-world function in the signal domain is a rotated line grating, whose spectrum is a rotated infinite impulse comb. Whenever this impulse-comb exceeds from one side of the DFT spectrum due to aliasing it folds over and re-enters from the opposite side. The only difference between the four rows is in the rotation angle θ: (a) $\theta = 14.036°$ (which has a rational slope of 4:1, since $\tan\theta = 1/4$); (b) $\theta = 10°$; (c) $\theta = 8°$; and (d) $\theta = 3°$. Note that in each case several cycles of foldover are clearly visible in the DFT spectrum. In case (d) folded-over impulses fall very close to the origin, and indeed, they correspond to a new lower-frequency parasite structure (sampling moiré) that is clearly visible in the signal domain, i.e. in the sampled line grating. The cross-like oscillations which surround impulses in the DFT spectrum are due to leakage (see Chapter 6). See Footnote 5 in Sec. 1.5.

Signal domain

Spectral domain:
Real part

Spectral domain:
Imaginary part

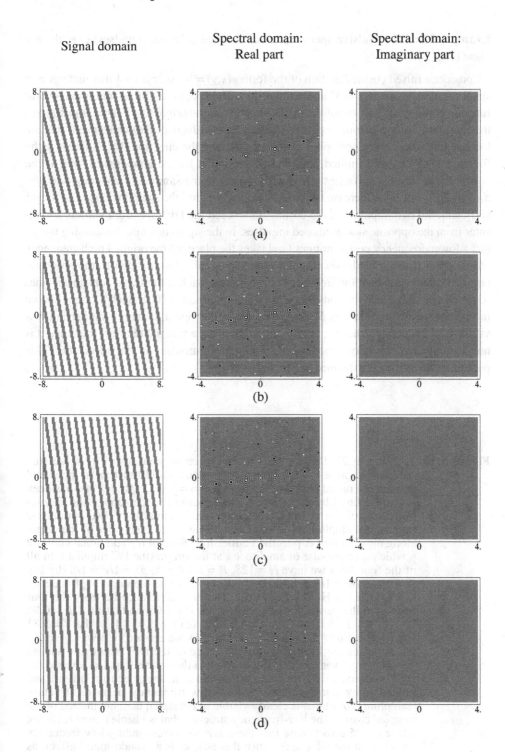

Example 5.8: 1D impulsive spectrum (three isolated collinear impulses) over the u,v plane (Fig. 5.15):

Consider a raised cosine function of the form $g(x,y) = \frac{1}{2}\cos(2\pi fx) + \frac{1}{2}$ that undergoes a small rotation by angle $\theta = 3°$ about the origin. The spectrum of this continuous-world function consists of three impulses: the DC impulse at the origin (the Fourier transform of the constant $\frac{1}{2}$), plus a pair of impulses (the Fourier transform of the cosine itself) that are located at a distance f to both sides of the origin along the direction $\theta = 3°$. Because the function $g(x,y)$ is band limited, its sampled version is indeed alias-free as long as the cosine frequency f is sufficiently low with respect to the sampling frequency (see Fig. 5.15(a)–(c)). But if we increase the cosine frequency f until the sampling rate is no longer sufficient, the two impulses of the cosine will exceed the DFT spectrum borders and re-enter from the opposite side as aliased impulses. In the signal domain, this aliasing means that a lower-frequency cosine mimics (and takes the place of) the original high-frequency cosine. And if the folded-over impulses happen to fall close enough to the spectrum origin, the corresponding low-frequency aliased cosine back in the signal domain becomes so prominent that it may be considered as a sampling moiré effect. Such a case is shown in Fig. 5.15(d). Note that Fig. 5.15(c), too, has in the signal domain a similar but less visible low-frequency structure; but this structure is not a true moiré effect (note that it is not accompanied by corresponding low-frequency impulses in the spectrum). This pseudo-moiré artifact is explained in Sec. 8.6. ∎

Figure 5.15: Aliasing in 2D DFT: a case whose spectrum consists of three isolated collinear impulses. The original continuous-world function in the signal domain is a raised cosine of the form $g(x,y) = \frac{1}{2}\cos(2\pi fx) + \frac{1}{2}$ that has been slightly rotated by angle $\theta = 3°$. Its spectrum is the sum of the Fourier transform of the cosinusoidal term, which consists of two symmetric impulses of amplitude $\frac{1}{4}$ located at a distance f to both sides of the origin along the direction θ, plus the Fourier transform of the additional constant $\frac{1}{2}$, which is an impulse of amplitude $\frac{1}{2}$ at the origin (the DC impulse). In all of the four rows we have $N = 128$, $R = 16$, $F = 8$, $\Delta x = 1/F = 1/8$ (by Eq. (7.4)) and $\Delta u = 1/R = 1/16$ Hz (by Eq. (7.3)), and the sampling frequency is $f_s = 1/\Delta x = 8$ Hz (by Eq. (4.6)). The only difference between the four rows is in the period P of the cosine: (a) $P = 1.05$, so that $f = 1/P = 0.95$ Hz; (b) $P = 1.05 / 2$, so that $f = 1.90$ Hz; (c) $P = 1.05 / 4$, so that $f = 3.81$ Hz; (d) $P = 1.05 / 8$, so that $f = 7.62$ Hz. As we can see, in rows (a)–(c) all the impulses are located within the range of the DFT spectrum, and no aliasing (foldover) occurs. But in row (d) the cosine impulses exceed the DFT borders, fold-over and end up close to the spectrum origin; and indeed, they correspond to a new low-frequency parasite structure (sampling moiré) that is clearly visible in the signal domain, instead of the sampled cosine. The low-frequency structure that is visible in row (c) is not a true moiré effect (note that there are no corresponding low-frequency impulses in the DFT spectrum); this artifact is a pseudo-moiré effect, as explained in Sec. 8.6. See also Footnote 5 in Sec. 1.5.2.

Signal domain

Spectral domain:
Real part

Spectral domain:
Imaginary part

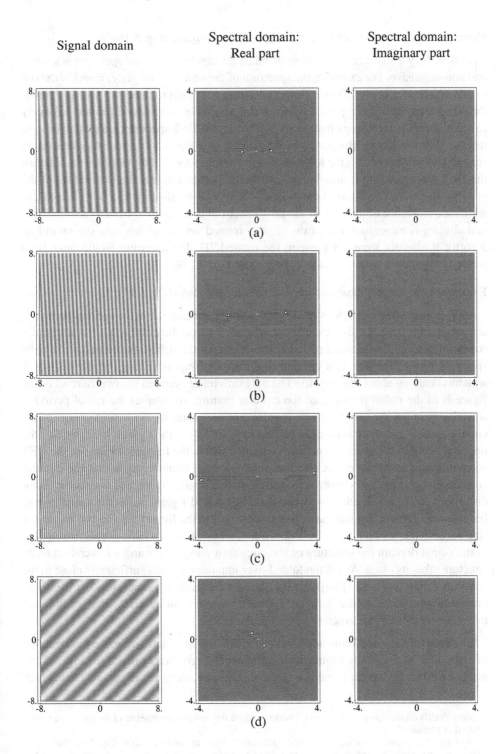

Example 5.9: 2D non-impulsive spectrum over the u,v plane (Fig. 5.16):

Aliasing may also occur in 2D cases in which the continuous-world spectrum is smooth and non-impulsive. For example, the spectrum of the square pulse $g(x,y) = \text{rect}(x/\tau, y/\tau) = \text{rect}(x/\tau)\text{rect}(y/\tau)$ is the 2D sinc function $G(u,v) = \tau^2 \text{sinc}(\tau u)\text{sinc}(\tau v)$ which extends throughout the entire u,v plane. Because this spectrum is not band limited it largely exceeds the frequency range that is covered by the 2D DFT spectrum, so that aliasing in this case is unavoidable. But just as in the 1D case (see Fig. 3.5(f)), since $G(u,v)$ is rather smooth and non-impulsive, the foldover due to aliasing only slightly modifies the shape of the DFT spectrum, and it may be quite difficult to detect unless the true values of the original spectrum $G(u,v)$ are known in advance. This is illustrated by Fig. 5.16(a), in which the existence of aliasing in the DFT spectrum is not easy to detect visually. Note that aliasing is easier to detect in the slightly rotated version of this case shown in Fig. 5.16(b): if aliasing were not present, the rotated 2D sinc spectrum would have been perfectly symmetric with respect to its new rotated axes. ∎

Example 5.10: A curvilinear line-impulse over the u,v plane (Fig. 5.17):

Another interesting example of aliasing involving a line impulse in the spectral domain is shown in Fig. 5.17. In this case the continuous-world function $g(x,y)$ is a circular cosinusoidal grating with radial period T, which extends throughout the entire x,y plane. Its spectrum $G(u,v)$ consists of a circular line-impulse (impulsive ring) of radius $f = 1/T$ which is centered about the origin.[12] The only difference between the twelve rows of the figure is in the radial period T of the circular grating. As long as the radial period is sufficiently large, the sampling rate is sufficient for capturing the main features of the circular grating and there is no aliasing (see rows (a)–(d) in the figure). In such cases the impulsive ring in the spectrum is fully contained within the frequency range of the DFT spectrum, and no foldover occurs. The signal-domain counterpart is that the circular cosinusoidal grating can be easily recognized in the sampled figure, and only some minor distortions can be observed.[13] But when the radial period T gets smaller the sampling rate becomes insufficient. In such cases (see rows (e)–(l) in the figure) the spectral impulsive ring exceeds the borders of the DFT spectrum and folds-over, so it is not surprising that in the signal domain the structure of the circular grating is lost and a lower-frequency structure takes its place. And if the folded-over impulse ring falls sufficiently close to the spectrum origin (see rows (h)–(j) and (l)), the new aliased low-frequencies near the origin manifest themselves in the form of a strong sampling moiré effect that completely overshadows the original structure of the circular grating.

When the folded-over frequencies fall *exactly* on the spectrum origin, they correspond to a *singular moiré*, i.e. a moiré having frequency zero. For example, in Fig. 5.17(i) the moiré effect is singular along the two main axes, i.e. along the directions from which

---

[12] More details on this continuous-world function and on the analytic expression of its spectrum can be found in [Amidror97].

[13] Note that these distortions are not caused by aliasing, since in rows (a)–(d) of Fig. 5.17 there is no aliasing. The minor distortions in the signal domain, particularly visible in row (d), are discussed in Problem 5-18; the minor ripple artifact in the DFT spectra is due to leakage (see Chapter 6).

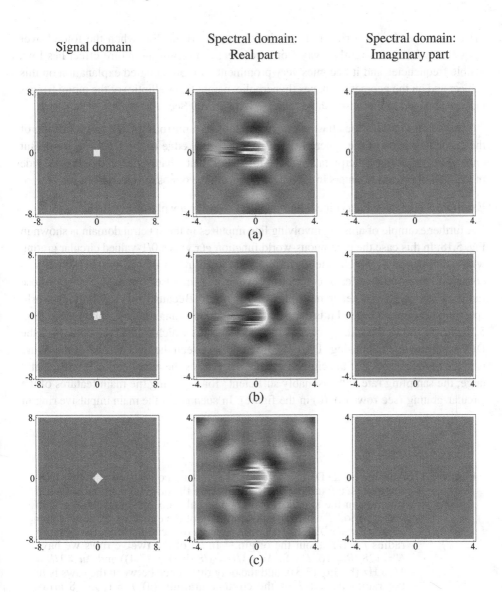

Figure 5.16: Aliasing in 2D DFT: a case whose spectrum is smooth and continuous in all directions (see Example 5.9). (a) The spectrum of a square pulse in the continuous-world is a 2D sinc function. Because this function extends throughout the entire u,v plane it is clear that the DFT spectrum in this case is flawed by aliasing; and yet, aliasing artifacts are not easy to detect visually. (b) Aliasing is easier to detect in a slightly rotated version of this function; note that without aliasing the rotated sinc function would have been perfectly symmetric with respect to its rotated axes. (c) In the 45° rotated version of this function aliasing is again not easy to detect. See also Footnote 5 in Sec. 1.5.2 concerning the gray levels used in the figure.

the folded-over impulse-ring touches the spectrum origin. But when the folded-over frequencies fall just slightly away from the spectrum origin the moiré effect has low, visible frequencies and it becomes very prominent. A more detailed explanation on this subject and on the geometric shapes (hyperbolic, parabolic or elliptic) of the moiré fringes that are obtained in this case can be found in [Amidror09 Sec. 10.7.5].

Note that in Fig. 5.17 the aliasing effect may look as a mirroring-back or a reflection of the exceeding parts of the spectral rings from the *same* edge of the DFT spectrum. But once again, the correct interpretation of the foldover is that whatever exceeds from one side of the DFT spectrum re-enters in a cyclical way from the opposite side. ■

Example 5.11: Several curvilinear line-impulses over the u,v plane (Fig. 5.18):

A further example of aliasing involving line impulses in the spectral domain is shown in Fig. 5.18. In this case the continuous-world function $g(x,y)$ is a 0/1-valued circular grating with radial period T, which extends throughout the entire x,y plane. Its spectrum $G(u,v)$ consists of an infinite series of concentric circular line-impulses (impulsive rings) whose radiuses are positive integer multiples of $f = 1/T$.[14] Because this continuous-world spectrum is not band limited it is clear that aliasing here is unavoidable. And indeed, Fig. 5.18 clearly shows the folding-over of the spectral rings which exceed the borders of the DFT spectrum due to aliasing. The only difference between the twelve rows of the figure is in the radial period T of the circular grating. As long as the radial period is sufficiently large, the sampling rate is "reasonably sufficient" for capturing the main features of the circular grating (see rows (a)–(d) in the figure). In such cases the main impulsive ring in

Figure 5.17: Aliasing in 2D DFT: a case whose spectrum consists of a circular line-impulse (see Example 5.10). In this case the original continuous-world function in the signal domain is a circular cosinusoidal grating with radial period T which extends throughout the entire x,y plane: $g(x,y) = \cos(2\pi\sqrt{x^2+y^2}/T)$. Its spectrum consists of a circular impulsive ring of radius $f = 1/T$ about the origin.[14] In all of the twelve rows we have $N = 128$, $R = 16$, $F = 8$, $\Delta x = 1/F = 1/8$ (by Eq. (7.4)) and $\Delta u = 1/R = 1/16$ Hz (by Eq. (7.3)), and the only difference between the rows is in the radial period T of the circular grating: (a) $T = 1$, i.e. 8 array elements; (b) $T = 0.75$, i.e. 6 array elements; (c) $T = 0.5$, i.e. 4 array elements; (d) $T = 0.3$, i.e. 2.4 array elements; (e) $T = 0.25$, i.e. 2 array elements; (f) $T = 0.2$, i.e. 1.6 array elements; (g) $T = 0.15$, i.e. 1.2 array elements; (h) $T = 0.13$, i.e. 1.04 array elements; (i) $T = 0.125$, i.e. 1 array element; (j) $T = 0.12$, i.e. 0.96 array elements; (k) $T = 0.1$, i.e. 0.8 array elements; (l) $T = 0.09$, i.e. 0.72 array elements. As long as the radial period T is sufficiently large, the sampling rate is sufficient for capturing the main features of the circular grating (see rows (a)–(d)). But when the radial period T gets smaller the sampling rate becomes insufficient, as in rows (e)–(l). See also Problems 5-16 and 5-18.

---

[14] More details on this continuous-world function and on the analytic expression of its spectrum can be found in [Amidror97]. See also Footnote 5 in Sec. 1.5.2 concerning the gray levels used in the figure.

Signal domain   Spectral domain: Spectral domain:
Real part   Imaginary part

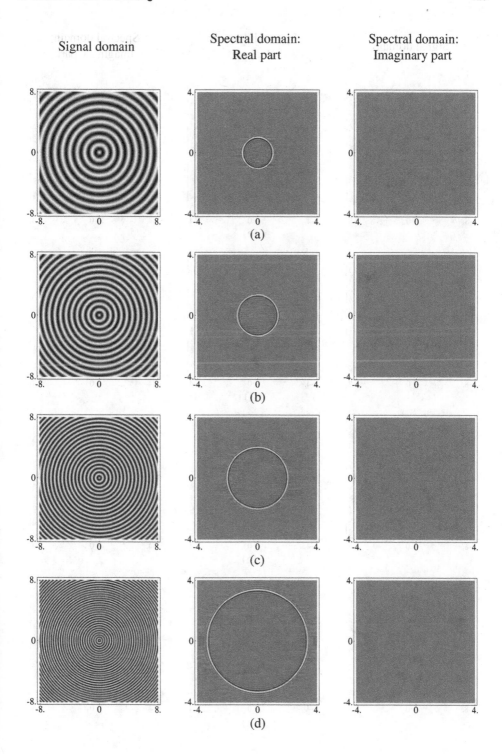

Signal domain
Spectral domain:
Real part
Spectral domain:
Imaginary part

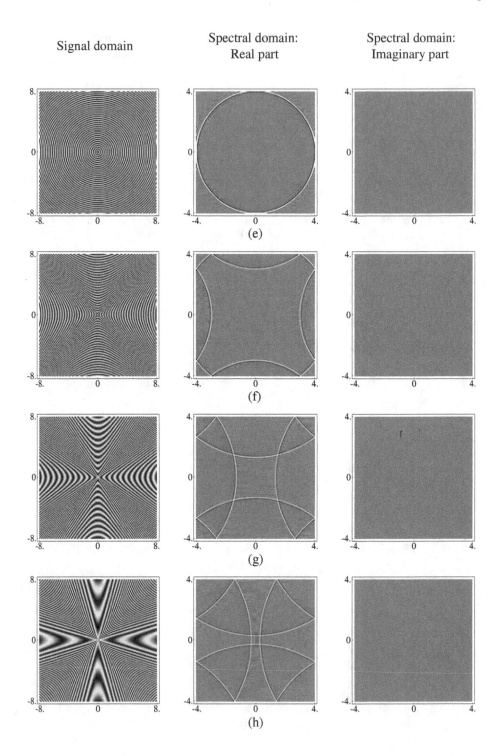

Signal domain | Spectral domain: Real part | Spectral domain: Imaginary part

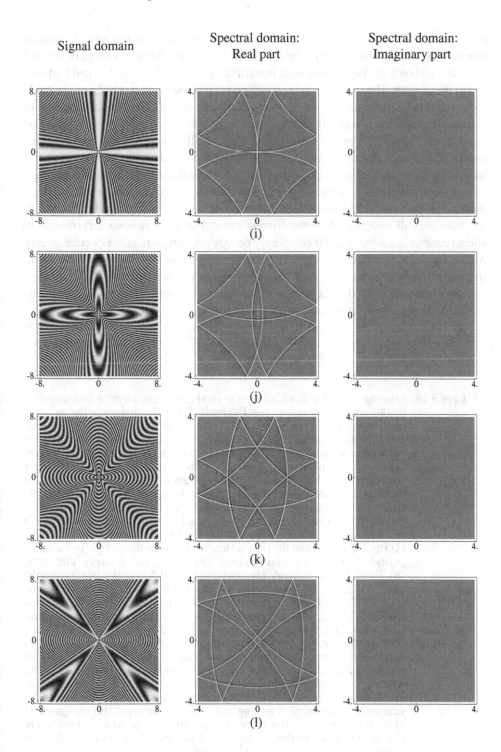

(i)

(j)

(k)

(l)

the spectrum (i.e. the inner, first-harmonic ring, which contains the most significant spectral information about $g(x,y)$) is fully contained within the frequency range of the DFT spectrum, and only the higher-harmonic impulsive rings exceed the borders and fold-over due to the aliasing. The signal-domain counterpart is that the circular grating can be easily recognized in the sampled figure, and only some minor aliasing distortions can be observed (compare rows (a)–(d) with the same rows in Fig. 5.17 where aliasing does not occur at all). But when the radial period T gets smaller the sampling rate becomes insufficient. In such cases (see rows (e)–(l) in the figure) the main spectral impulsive ring itself is folded over, so it is not surprising that in the signal domain the circular grating is so much corrupted that it can no longer be recognized. As we can see, when aliasing is moderate it disturbs only slightly, and the structures of the circular grating and of its spectrum are still recognizable in the discrete world. But when aliasing gets stronger its effect becomes devastating and it completely destroys the structure of the circular grating. And if the main impulse ring itself is folded-over and falls sufficiently close to the spectrum origin (see rows (h)–(j) and (l)), the new aliased low-frequencies near the origin manifest themselves in the form of a strong sampling moiré effect that completely overshadows the original structure of the circular grating.

Figure 5.18: Aliasing in 2D DFT: a case with several concentric circular line-impulses in the spectral domain (see Example 5.11). In this case the original continuous-world function is a 0/1-valued circular grating with the same radial period T as in Fig. 5.17, which extends throughout the entire x,y plane. Its spectrum consists of a DC impulse at the origin plus an infinite series of concentric impulsive rings whose radiuses are positive integer multiples of $f = 1/T$.[14] Because this spectrum is not band limited it is clear that aliasing here is unavoidable. And indeed, we can clearly see in the figure the folding-over of the spectral rings which exceed the borders of the DFT spectrum due to aliasing. In all of the twelve rows we have $N = 128$, $R = 16$, $F = 8$, $\Delta x = 1/F = 1/8$ (by Eq. (7.4)) and $\Delta u = 1/R = 1/16$ Hz (by Eq. (7.3)), and the only difference between the rows is in the radial period T of the circular grating: (a) $T = 1$, i.e. 8 array elements; (b) $T = 0.75$, i.e. 6 array elements; (c) $T = 0.5$, i.e. 4 array elements; (d) $T = 0.3$, i.e. 2.4 array elements; (e) $T = 0.25$, i.e. 2 array elements; (f) $T = 0.2$, i.e. 1.6 array elements; (g) $T = 0.15$, i.e. 1.2 array elements; (h) $T = 0.13$, i.e. 1.04 array elements; (i) $T = 0.125$, i.e. 1 array element; (j) $T = 0.12$, i.e. 0.96 array elements; (k) $T = 0.1$, i.e. 0.8 array elements; (l) $T = 0.09$, i.e. 0.72 array elements. As long as the radial period T is sufficiently large, the sampling rate is "reasonably sufficient" for capturing the main features of the circular grating (see rows (a)–(d)). But when the radial period T gets smaller the sampling rate becomes insufficient, as in rows (e)–(l). As we can see, when aliasing is low the structures of the original grating and of its spectrum are still recognizable in the discrete world. But when the sampling rate is no longer sufficient aliasing becomes so devastating that it completely alters in the discrete world the structures of the circular grating.

Signal domain
Spectral domain:
Real part
Spectral domain:
Imaginary part

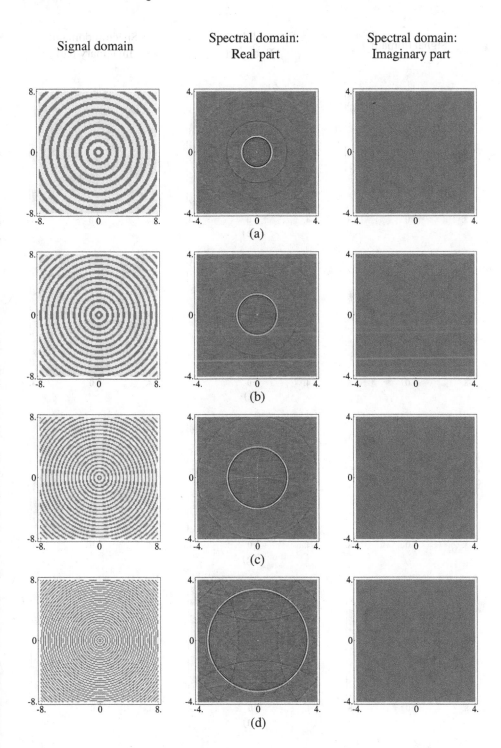

Signal domain   Spectral domain: Spectral domain:
                           Real part   Imaginary part

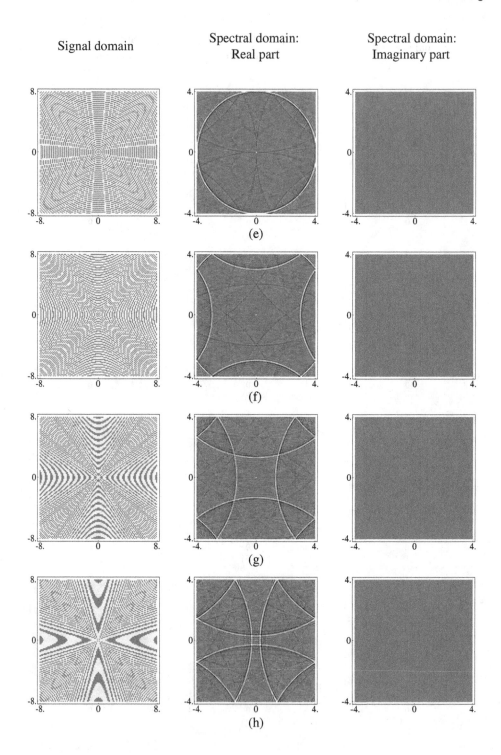

Signal domain	Spectral domain: Real part	Spectral domain: Imaginary part

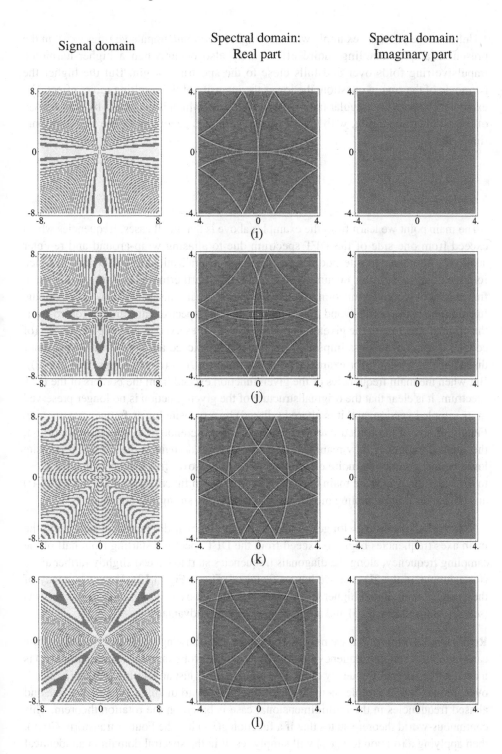

(i)

(j)

(k)

(l)

Unlike in the previous example where no higher-harmonic impulsive rings exist, in the present example sampling moiré effects may also occur when a higher-harmonic impulsive ring folds over and falls close to the spectrum origin. But the higher the harmonic of the ring in question, the lower the intensity of the resulting moiré effect. For example, compare the singular moiré effect of row (c), which is generated by the foldover of the 4-th harmonic ring, with that of row (i), which is generated by the first harmonic ring. ■

5.6 Discussion

The main point we learn from the examples above is that in all cases, frequencies which exceed from one side of the DFT spectrum due to aliasing wrap-around and re-enter additively from the opposite side. This is, indeed, a generalization of the 1D case, in which foldover due to aliasing occurs along the u axis. Furthermore, as long as the main frequencies of the given function (like the first harmonic impulses in the case of a periodic function) do not exceed beyond the borders of the DFT spectrum and are not folded-over, the original structure of the given function is basically preserved in its sampled version (of course, if some of the less important frequencies *do* exceed and fold over, some minor distortions may occur; for example, compare Figs. 5.17(a)–(d) with Figs. 5.18(a)–(d)). But when the main frequencies of the given function exceed from the borders of the DFT spectrum, it is clear that the original structure of the given function is no longer preserved in its sampled version, and it is altered (aliased) by a parasite lower-frequency structure. Finally, if any of the folded-over frequencies falls close enough to the spectrum origin, this aliased frequency may manifest itself in the signal domain in the form of a parasite low-frequency sampling moiré effect. And if the folded-over frequencies which fall close to the spectrum origin are main frequencies of the given function, as in Figs. 5.18(h)–(j) and (*l*), the resulting sampling moiré effect is particularly strong and dominant.

Note that aliasing takes longer to appear along diagonal directions: While along the main axes frequencies begin to exceed from the DFT spectrum starting from half of the sampling frequency, along the diagonals frequencies start to exceed slightly farther away (for example, in the 2D case we have a factor of $\sqrt{2}$; see Fig. 5.12). However, many of these gained non-aliased higher frequencies are infested by strong pseudo-moiré artifacts (see Sec. 8.6 and Fig. 8.15) and are therefore not really advantageous.

Remark 5.3: At this point, we may ask ourselves the following interesting question: How can we tell if a certain frequency in the DFT spectrum (for example, a certain impulse) is indeed a true spectral frequency of the given function, or just an aliasing artifact (a folded-over frequency)? A simple method which often allows to distinguish between true and aliased frequencies in the multidimensional case is based on the rotation theorem. This continuous-world theorem states that if a function $g(x,y)$ has the Fourier transform $G(u,v)$, then applying a rotation to $g(x,y)$ will simply result in the spectral domain in an identical rotation of $G(u,v)$ [Bracewell95 p. 157]. In the discrete world, however, this theorem remains valid only for the true frequencies in the DFT spectrum,[15] but not for aliased frequencies. For example, Figs. 5.14(a)–(d) show what happens in the DFT spectrum to the impulse comb when we apply DFT to $g(x,y)$ after having rotated it by a small angle θ (note that we continue using the same unrotated sampling grid as before the rotation). As we can see, all the true impulses of $G(u,v)$ simply undergo a rotation by angle θ, while the folded-over (aliased) impulses move in a different way. (In fact these impulses, too, undergo a rotation by angle θ, but their center of rotation is not the origin of the DFT spectrum, but rather the center of the replica to which they actually belong, which is obviously located outside the range of the DFT spectrum.) Other examples illustrating the effect of rotation on true and aliased frequencies in the DFT spectrum are shown in Fig. 5.13 and in Fig. 5.16. Note, however, that this method does not work in cases such as Figs. 5.17 and 5.18, which are circularly symmetric and hence invariant under rotations.

A similar test can be also obtained by applying to $g(x,y)$ a stretching transformation, since according to the dilation theorem (see point 2 in Sec. 2.4.2) the Fourier transform of the continuous-world function $g(ax,by)$ is $\frac{1}{|ab|}G(u/a,v/b)$ [Bracewell95 p. 154]; however, in this case shrinking of $g(x,y)$ results in an expansion of $G(u,v)$ and vice versa. Note that this test can be also used in the 1D case, in which the method based on rotations is not applicable since rotations have no 1D equivalent. The same effect can be also obtained by varying the sampling rate that is being used to sample the continuous-world input function (see Sec. B.3 in Appendix B). ∎

Remark 5.4: (MD counterpart of Remark 5.2): Just like in the 1D case, in the multidimensional case, too, aliasing can be avoided completely only if the given signal $g(\mathbf{x})$ is band limited; under this condition, aliasing will be absent if the sampling frequency along each of the axes is at least twice the highest frequency present in $g(\mathbf{x})$. But in practice, such situations rarely occur. If $g(\mathbf{x})$ is not band limited (this happens, for example, for any finite-duration function $g(\mathbf{x})$), the only solution is to reduce aliasing as much as possible, using the same methods as in the 1D case. ∎

Another issue that is often associated with aliasing in the 2D or MD cases concerns the "jaggies" or "staircasing" which appear in the discrete world along slanted or curved lines due to their sampling. The interesting question of how precisely these jaggies are related to aliasing is treated separately in Sec. 8.5.

5.7 Signal-domain aliasing

The use of DFT as an approximation to the CFT involves sampling both in the input and in the output (see Sec. 2.5). Therefore, although aliasing is usually considered as an artifact which occurs due to undersampling in the signal domain, a similar phenomenon

---

[15] With the understanding that after the rotation an impulse may end up in a non-integer location within the DFT spectrum, or even outside the range of the DFT spectrum. The first case will simply result in leakage (see Chapter 6), but in the second case the impulse in question will be folded-over and become an aliased impulse.

may also occur as a result of undersampling in the spectral domain, thanks to the frequency sampling theorem — the frequency-domain dual of the classical sampling theorem [Thornhill83]. This phenomenon, known as *signal-domain aliasing* (or possibly in the 1D case *time-domain aliasing*), may be regarded under two different angles:

The first point of view is based on the inverse DFT. Due to the duality between DFT and the IDFT, it follows that all our previous discussions on the aliasing phenomenon under DFT remain true under IDFT, too, except that the roles of the input and output arrays are reversed. For example, each row in Fig. 5.7 shows in the right-hand column the result obtained by applying DFT to the discrete sequence shown in the left-hand column. But the same figure can be also interpreted the other way around, where the left-hand column is the result of applying IDFT to the discrete sequence shown in the right-hand column. Now, just as in the DFT, the sampling which occurs in the input of the IDFT (the sampling of the continuous-world spectrum $G(u)$) causes a periodic replication of the output of the IDFT; but this time, the replication occurs in the signal domain. Therefore, if the input array of the IDFT does not sample the continuous-world spectrum $G(u)$ densely enough (which is not the case in Fig. 5.7), the replicas of $g(x)$ back in the signal domain will overlap each other (additively) and cause signal-domain aliasing. Stated in other words, if a continuous-world frequency function $G(u)$ is sampled with sampling interval Δu, and if $g(x)$, the inverse CFT of $G(u)$, does not effectively die out by the distance (or time) $R = 1/\Delta u$ (see Eq. (7.3)), then the IDFT of the sampled version of $G(u)$ will be signal-aliased [Thornhill83 p. 234].

The second possible point of view for considering signal-domain aliasing does not explicitly involve the use of the IDFT. As we have seen in Sec. 2.5, the DFT operation can be obtained from the continuous-world Fourier transform by applying three successive steps: Sampling of the signal domain (which results in periodic replications in the spectral domain); truncating the signal domain to a finite range of length R (which results in convolving the spectral domain by the function $R \operatorname{sinc}(Ru)$); and finally sampling of the spectral domain (which results in periodic replications in the signal domain). And indeed, just as the sampling operation in the *signal domain* may result in aliasing if it is not performed densely enough, as required by the sampling theorem, so the sampling operation in the *frequency domain* may result in aliasing if it is not performed densely enough, as required this time by the frequency sampling theorem. This last aliasing affects

Figure 5.19: When the original continuous-world function $g(x)$ exceeds beyond the borders of the input array, we can, by taking into account the signal-domain aliasing (cyclical foldover in the input array of the DFT), reduce the mismatch between the DFT spectrum and the original continuous-world spectrum $G(u)$. (a) When the input function $g(x) = \operatorname{sinc}(x)$ is truncated at the borders of the DFT input array without taking care of the foldover, a mismatch occurs between the DFT results and $G(u)$ (this

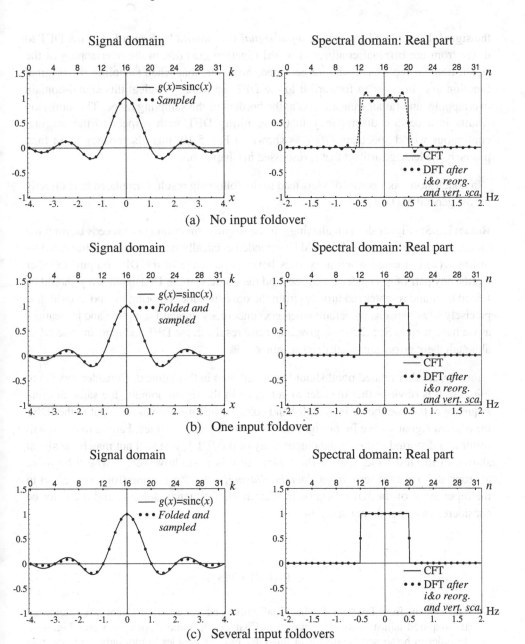

(a) No input foldover

(b) One input foldover

(c) Several input foldovers

discrepancy is precisely the leakage artifact that is discussed in Chapter 6; see Fig. 6.5). (b) Allowing a single additive foldover to each side of the input array already improves the DFT results. (c) Allowing several foldover cycles to each side further reduces the discrepancy. Allowing an infinite number of additive foldovers will result in a perfect match between the DFT results and $G(u)$.

the signal domain, and causes the *input signal* that should be forwarded to the DFT to differ from the original continuous-world function $g(x)$ due to the overlapping of the signal-domain replicas. It should be noted, however, that when sampling the original function $g(x)$ in order to forward it to the DFT we usually neglect this signal-domain overlapping, and rather truncate $g(x)$ at the borders of the sampling range. This omission results in a certain discrepancy in the resulting DFT with respect to the original continuous-world spectrum $G(u)$, as shown in Fig. 5.19; this discrepancy is, in fact, precisely the leakage artifact that is discussed in Chapter 6.

In conclusion, both points of view lead to the following result, formulated here directly for the general MD case:

Remark 5.5: (Signal-domain aliasing): If the original function $g(\mathbf{x})$ exceeds beyond the sampling range of the DFT, it should be extended cyclically rather than being truncated — just as $G(\mathbf{u})$ is treated when it exceeds beyond the range of the DFT output. In other words, any part of $g(\mathbf{x})$ that exceeds beyond the borders of the DFT input array should be folded over and re-enter (additively) from the opposite side. Ignoring this point (which is precisely what happens by default when $g(\mathbf{x})$ undergoes standard sampling and truncation, as we have seen in Sec. 2.5) will give imprecise results in the DFT and in its inverse DFT, although these errors are usually rather minor. ■

Finally, another related point should be mentioned in this context. Consider row (e) of Fig. 5.6. It is obvious that in order to get, back in the signal domain, the same discrete sequence as the one shown in the left-hand column, we must provide as input to the IDFT the *aliased* signal shown in the right-hand column by dashed lines. Failing to do so will result in a distorted signal in the output array of the IDFT, that will not match the signal shown in the left-hand side of the figure. In this case, however, it would be more appropriate to use the term *input-domain aliasing*, since the aliasing (foldover) effect in the input array of the IDFT occurs, in fact, in the frequency domain, and it cannot be considered as signal-domain aliasing.

PROBLEMS

5-1. Our student from Problem 4-1 has finally obtained a correct DFT plot, after doing all the required adjustments as discussed in Chapters 3, 4 and Appendix A (see Fig. 4.9). In order to better see the details of this DFT plot, he decides to plot only the lower part of the vertical scale, i.e. up to the value 0.1 (see Fig. 5.20). This allows him to realize that his DFT plot significantly differs from the expected CFT results. For example, instead of getting a single comb of impulses located along a rotated straight line passing through the origin, the DFT spectrum contains impulses that are dispersed throughout the entire spectral plane.

(a) Can you find the analytic expression of the CFT? *Hint*: See, for example, [Amidror09 pp. 21–25]. You may assume that the grating's period is 1, the width of each black line is 40% of the period, and the rotation angle is $\theta = 15°$.

(b) How do you explain the results obtained in the DFT spectrum? Can you give the precise interpretation of each of the impulses in the DFT spectrum?

(c) Can you guess why the impulses in the DFT spectrum (after all the required adjustments) are not sharp as in the CFT, and look rather smeared and leaked-out? *Hint*: See Chapter 6.

5-2. What would you expect to see in the DFT spectrum of Fig. 5.20 if the rotation angle of the original line grating were slightly increased or decreased?

5-3. What would you expect to see in the DFT spectrum of Fig. 5.20 if the period of the original line grating were slightly increased or decreased?

5-4. Looking at Figs. 5.1 and 5.2, a student asks the following question: When a continuous function $g(x)$ is being sampled, there always exist many other continuous-world functions that can be traced through the resulting sample points, and thus mimic the behaviour of $g(x)$ on its sampled values and be considered as aliases. This is true even if the sampling frequency satisfies the requirements of the sampling theorem, like in Figs. 5.1(a) and 5.2(a). This means that aliased signals exist in all cases, even when the requirements of the sampling theorem are satisfied. So why do we say that aliasing does not occur when these requirements are satisfied?

(a) How would you answer the student's question?

(b) Suppose that $g_1(x)$ is a continuous function that passes through the same sampling values as $g(x)$. Is it necessarily a lower-frequency alias of $g(x)$? *Hint*: See Example 5.1 and Footnote 8 there.

(c) What can you say about the DFT spectra of $g(x)$ and $g_1(x)$?

5-5. What would you expect to see if in the successive rows of Fig. 5.6 we gradually reduced the sampling frequency f_s but left the array length N unchanged?

5-6. After having studied the present chapter, consider again Problem 4-10. Can you provide a detailed explanation like in the examples given in Sec. 5.3?

5-7. *Foldover*. The interpretation of the foldover effect due to aliasing is that "whatever exceeds from one side of the DFT spectrum re-enters in a cyclical way from the opposite side". But as mentioned in Footnote 6 in Sec. 5.2, some references describe

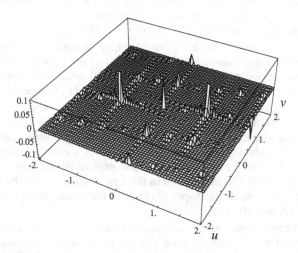

Figure 5.20: The DFT plot obtained by the student in Problem 5-1.

the foldover effect (and even graphically depict it) as a mirroring of the frequencies that exceed beyond $\frac{1}{2}f_s$ (or $-\frac{1}{2}f_s$) backward from the *same* end of the DFT spectrum, i.e. as folding around the spectrum edges backward (see for example the figures in [Press02 p. 507], [Ramirez85 p. 121] or [Ramirez74 p. 102]). Explain why this interpretation is equivalent to ours in the case of 1D real-valued signals. What happens in the case of complex or purely imaginary 1D signals? Why should this interpretation be avoided in 2D or MD cases? *Hint*: See, for example, the 2D DFT in Fig. 5.13(a).

5-8. For any 1D real-valued signal, the leftmost element of the reorganized DFT plot, which corresponds to the lowest negative frequency provided by the DFT (and symmetrically also to the highest positive frequency; see Footnote 4 in Sec. 4.3) must be purely real valued.

(a) How do you explain this fact? *Hint*: See Sec. 8.4 and point (1) in Sec. D.2.

(b) This means that for any 1D real-valued signal having a continuous CFT, the imaginary-valued part of the DFT spectrum must gradually go down to zero toward both ends of the DFT spectrum, even if the CFT itself behaves differently (see, for example, Fig. 5.8(b)). This imposes an obvious discrepancy between the CFT and the corresponding DFT. Can this discrepancy be also explained as a result of aliasing? In particular, can it occur when no aliasing is present?

5-9. Fig. 5.12 schematically shows what happens in the continuous-world spectrum when sampling a *band-limited* function $g(x,y)$ whose maximum frequencies to both directions are u_G and v_G, respectively. Row (c) of this figure shows the sampled signal and its continuous-world spectrum when the sampling frequencies u_s and v_s are higher than $2u_G$ and $2v_G$, respectively, so that the replicas of $G(u,v)$ in the spectrum do not overlap and no aliasing occurs. Row (d) of the figure shows the situation when the sampling frequencies u_s and v_s are slightly lower than $2u_G$ and $2v_G$, and the replicas of $G(u,v)$ in the spectrum slightly overlap with their direct neighbours. While clearly illustrating the spectral overlapping-effect of aliasing for a band-limited function $g(x,y)$, this figure is no longer useful when $g(x,y)$ is not band limited and its maximum frequencies are $u_G = \infty$ and $v_G = \infty$, like in the case of the infinite rotated bar $g(x,y)$ discussed in Example 5.6. Draw a schematic figure that clearly shows what happens in the continuous-world spectrum when this function is sampled along both directions with sampling rates of u_s and v_s, showing in particular the overlapping replicas of the infinite line impulse $G(u,v)$. Compare your continuous-world spectrum with the DFT spectrum of this case shown in Fig. 5.13. *Hint*: See Fig. 5.21.

5-10. Suppose that you redo Fig. 5.17, but this time instead of keeping the sampling resolution constant and reducing the radial period of the circular cosine function, you keep the radial period of the function constant and reduce the sampling resolution. What would you expect to get in the signal domain and in the spectral domain?

5-11. Sometimes, the aliasing phenomenon due to insufficient sampling resolution gives unexpected, spectacular or even "artistic" results. For instance, Fig. 5.17 shows the evolution of the shapes produced by aliasing both in the signal and in the spectral domains, as we gradually reduce the radial period of a circular cosine function (and hence increase the effect of aliasing). Can you explain the geometric shapes obtained in these figures? *Hint*: The explanation in this case is quite easy in the spectral domain. If you wish to understand the shapes obtained in the signal domain, you may consult, for example, [Amidror09 Sec. 10.7.5].

5-12. Fig. 5.22 shows some other unexpected results of aliasing due to insufficient sampling rates. Can you explain what you see in the signal and spectral domains in each case? *Hint*: Compare with the non-aliased versions shown in Fig. 5.23. As for the

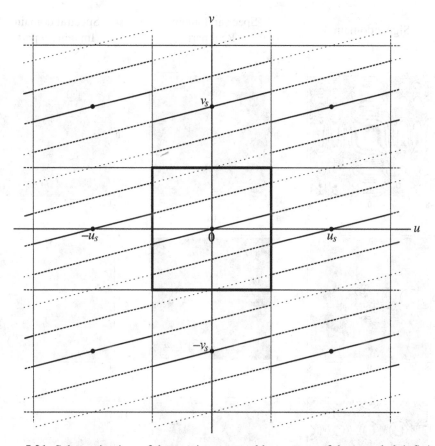

Figure 5.21: Schematic view of the continuous-world spectrum of the sampled, infinite rotated bar discussed in Example 5.6. The original continuous bar $g(x,y)$ was sampled to both directions with the sampling frequencies of u_s and v_s, respectively, and therefore the spectrum of $g(x,y)$, the infinite sinc-modulated line impulse $G(u,v)$, is replicated here about each integer multiple (mu_s, nv_s), $m,n \in \mathbb{Z}$. Each of the squares represents the central part of a replica of $G(u,v)$; note, however, that each replica contains an infinite line impulse, and extends *ad infinitum* to all directions, overlapping (additively) *all* the other replicas on its way. The highlighted square in the center, belonging to the main replica, indicates the part of the continuous-world spectrum that will be included in the 2D DFT output should we subject our sampled signal to DFT rather than to CFT. The frequency range of the DFT extends along each of the axes between minus half and plus half of the corresponding sampling frequency. The DFT counterpart of this continuous-world spectrum is shown in Fig. 5.13(a).

formation of circular or hyperbolic moiré shapes in the sampled signal in Fig. 5.22(a), see, for example, analog discussions in Sec. 10.7.4 of [Amidror09] and Example H.4 in Appendix H of [Amidror07]. As for the foldover effect in the DFT spectrum in Fig. 5.22(b), see a similar case in Fig. 5.17. And as for the formation of a 2D periodic structure in the sampled signal in Fig. 5.22(c), and hence the formation of an

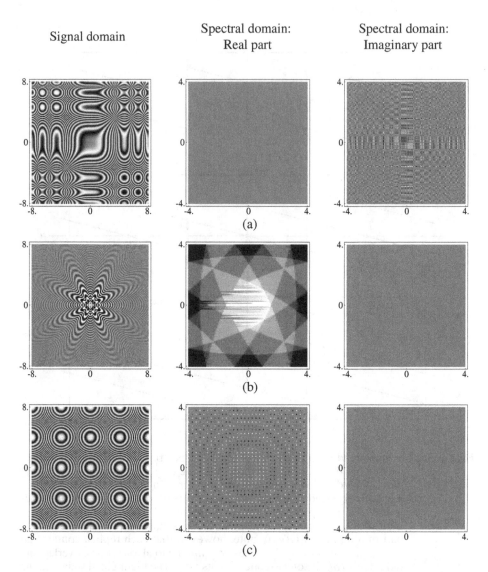

Signal domain   Spectral domain: Spectral domain:
                       Real part   Imaginary part

(a)

(b)

(c)

Figure 5.22: Sometimes aliasing due to an insufficient sampling rate may give quite unexpected shapes. Each row shows the sampled version of a given continuous 2D function $g(x,y)$ and the resulting DFT. In all of the 3 cases we have used $N = 128$, a sampling range of $R = 16$, and a sampling interval of $\Delta x = 1/8$. The functions being sampled are, respectively: (a) $g(x,y) = \sin(2\pi f [x^3 - y^3]/8)$ with $f = 1$. (b) $g(x,y) = \frac{\pi}{4} d^2 \text{somb}(d\sqrt{x^2+y^2}) = d J_1(\pi d \sqrt{x^2+y^2})/(2\sqrt{x^2+y^2})$ with $d = 38.5$. (c) $g(x,y) = \cos(2\pi f [x^2 + y^2])$ with $f = 1$; note in this last case the unexpected periodicity of the signal domain, and the resulting impulsive nature of the DFT spectrum. Non-aliased counterparts of these cases, which better reflect the continuous-world functions and their expected CFTs, are shown in Fig. 5.23. See Problem 5-12.

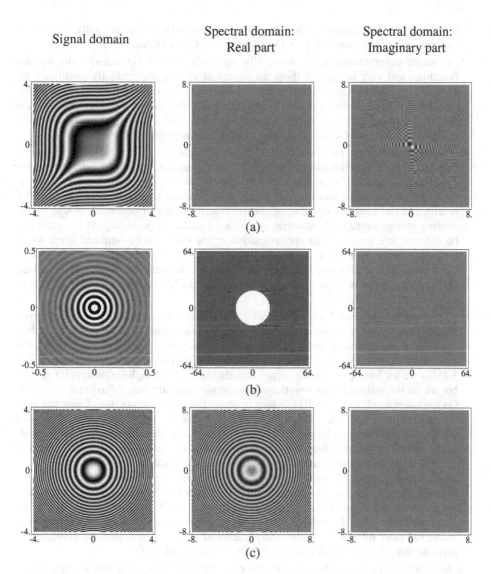

| Signal domain | Spectral domain: Real part | Spectral domain: Imaginary part |

(a)

(b)

(c)

Figure 5.23: A more faithful counterpart of Fig. 5.22, which is obtained by using higher sampling rates. In this figure we still use $N = 128$, like in Fig. 5.22, but in rows (a) and (c) we use the sampling range $R = 8$ and the sampling interval $\Delta x = 1/16$, and in row (b) we use the sampling range $R = 1$ and the sampling interval $\Delta x = 1/128$. The functions being sampled are the same as in Fig. 5.22: (a) $g(x,y) = \sin(2\pi f [x^3 - y^3]/8)$ with $f = 1$. (b) $g(x,y) = \frac{\pi}{4}d^2 \mathrm{somb}(d\sqrt{x^2+y^2}) = d\,J_1(\pi d\sqrt{x^2+y^2})/(2\sqrt{x^2+y^2})$ with $d = 38.5$ (compare also with the second row of Fig. 6.8, after a domain reversal). (c) $g(x,y) = \cos(2\pi f [x^2 + y^2])$ with $f = 1$. Unlike in Fig. 5.22, in the present figure the sampling frequencies being used are sufficiently high for capturing all (or almost all) the frequencies present in the signal domain. Hence the DFT spectrum reflects correctly the expected CFTs and no aliasing is visible (except in row (c), where weak aliasing still exists).

impulsive DFT spectrum (although both $g(x,y)$ and $G(u,v)$ are originally continuous and aperiodic, as shown in Fig. 5.23(c)), see Sec. 10.8 in [Amidror09].

5-13. Generate other examples involving aliasing (both 1D and 2D cases) using various functions, and vary in each of them the degree of aliasing by gradually modifying the sampling resolution. Observe the effects of the gradually varying aliasing, both in the signal and in the spectral domains. Beside being fun, this should develop your intuition as for the many different forms that aliasing may take, and allow you to easily recognize aliasing artifacts as such, and to avoid any possible misinterpretations.

5-14. *Interpretation of the DFT results.* In a paper published in 1970 [Cooley70, Figures 1 and 2] the DFT approximation to the CFT of the continuous-world function $g(x) = e^{-x}$ step(x) is plotted in a different way than in our Fig. 3.1 (see Chapter 3). It is explained in the paper that the significant deviation one sees there between the DFT results and the continuous curve of the CFT at the higher frequencies is due to the aliasing present in the DFT, and that by taking a smaller sampling step Δx this error can be reduced and a reasonable approximation to the CFT can be obtained for a wider frequency range. A student claims that the significant deviation observed in the plots there is not due to aliasing, but rather due to a misinterpretation of the unreorganized output data. What is your opinion? *Hint*: You may also compare with Fig. 9.1 in [Brigham88 pp. 168–169].

5-15. *Can both the original high frequency and its low-frequency alias be perceived simultaneously in a sampled signal?* We have seen that if aliasing occurs when sampling a continuous signal, a new lower-frequency signal appears that mimics the true high-frequency signal (and actually takes its place). For example, looking at Fig. 5.15(d) we no longer see in the signal domain the original high-frequency 2D cosine, but we do see instead a new low-frequency cosinusoidal structure (alias) that mimics it. On the other hand, in Fig. 5.1(b), which explains this phenomenon, we clearly see both the original high-frequency structure and the new low-frequency structure that mimics it through the same sampling points. How do you explain this difference? *Hint*: In real-world situations we are given the sampled signal alone, and unlike in Fig. 5.1(b), the original continuous-world signal is not shown in the background. And indeed, if we delete all the continuous lines from Fig. 5.1(b) and leave there only the black dots (the sampled signal), our eye will interpret these dots as a low-frequency signal, and it will not perceive the original high-frequency. The same result is also true in the case of aperiodic functions, like in Fig. 5.2(b), but it is less obvious there since our eye (or brain) has more difficulties in the interpretation of aperiodic structures (as compared to periodic structures in which a simple pattern repeats regularly).

5-16. Interestingly, figures such as 5.17(f) offer an opportunity to see within a single figure both aliased and non-aliased versions of the same sampled signal: Observe in the spectral domain the angular sectors in which the impulsive ring already exceeds the DFT borders and the sectors where it does not, and compare *in the signal domain* the structures of these sectors. Does this contradict our findings in the previous problem?

5-17. Suppose you are given a random or pseudo-random figure, such as an image consisting of randomly positioned dots. What would you expect to see in the signal and in the spectral domains as you gradually reduce the sampling resolution of this image?

5-18. Figure 5.17(d) contains some visible distortions in its signal domain, although it is free of aliasing (unlike Fig. 5.18(d)). Student A claims that these are printing artifacts due to an interference with the halftone screen that is used for printing this gray level figure (see Sec. 8.7). Student B, however, claims that these distortions are sub-Nyquist artifacts (see Sec. 8.6). What is your opinion? How can you verify it?

Chapter 6

Issues related to leakage

6.1 Introduction

After having discussed in Chapter 5 the aliasing artifact, we arrive now to the second major source of discrepancy between the DFT results and the continuous Fourier transform, the leakage artifact. Once again we start with the 1D case, and only then we proceed to the more general multidimensional case.

6.2 Leakage in the one dimensional case

Unlike the continuous Fourier transform, the DFT can only take into account a finite range of the original, underlying function (compare, for example, Eqs. (2.1)–(2.2) with Eqs. (2.13)–(2.14)). This fact, however, is not as innocent as it may first seem, and it can cause in the DFT output another type of discrepancy, known as *leakage*.

Leakage is a frequency-smearing artifact that may occur in the DFT spectrum due to the finite-range truncation of the underlying signal. This truncation operation can be viewed as a multiplication of the underlying function $g(x)$ by a rectangular "truncation window" $w(x)$ whose length equals the sampling range length $R = N\Delta x$ (see Eq. (4.2)), namely, $w(x) = \text{rect}(x/R)$.[1] From the viewpoint of the signal domain, this truncation may cause a discontinuity between the last and the first elements of the input array (because data is considered by the DFT as cyclical), and thus introduce into the DFT spectrum new high frequencies that do not exist in the original signal (see, for example, [Brigham88 pp. 102–103] or [Cartwright90 pp. 212–213], and Figs. 6.3(b)–(d) below). However, as is often the case, this phenomenon can be more easily explained from the viewpoint of the spectral domain, still in the continuous world. By virtue of the convolution theorem, the signal-domain multiplication of $g(x)$ with $w(x) = \text{rect}(x/R)$ causes in the frequency domain a convolution of the original Fourier transform $G(u)$ with the Fourier transform of the truncation window $w(x)$, which is given by (see, for example, [Brigham88 pp. 13–14]):[2]

$$W(u) = \mathcal{F}[\text{rect}(x/R)] = R\text{sinc}(Ru) \tag{6.1}$$

---

[1] This truncation can be done either before or after sampling; see Sec. D.8 in Appendix D.

[2] It is assumed here, for the sake of simplicity, that the range R of the truncation window is symmetric about the origin, namely, that the original continuous function $g(x)$ is sampled within a symmetric range $-R/2...R/2$. This allows us to assume that the Fourier transform of the truncation window is purely real valued, as in Eq. (6.1). If the rectangular truncation window is located in a different position, its spectrum is no longer purely real valued (see, for example, [Brigham88 pp. 15–16]); but according to the shift theorem the magnitude (absolute value) of its spectrum remains unchanged, and only the phase of its spectrum is affected by a linear increment. However, as mentioned in [Brigham88 p. 100], had we considered the complex spectrum instead of Eq. (6.1), similar results would have been obtained.

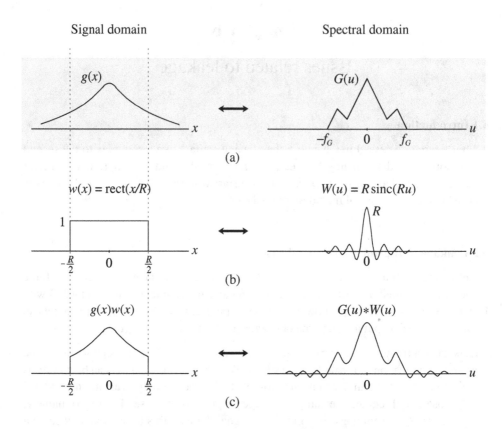

Signal domain   Spectral domain

(a)

(b)

(c)

Figure 6.1: Schematic illustration showing the effect of truncation on the spectrum
of a signal $g(x)$ in the continuous world. (a) An original continuous
signal $g(x)$ and its continuous-world spectrum $G(u)$. (b) The truncating
function $w(x) = \mathrm{rect}(x/R)$ and its continuous Fourier transform $W(u) = R\,\mathrm{sinc}(R\,u)$. (c) The truncated version of $g(x)$ extending between $-R/2$
and $R/2$ is a product of $g(x)$ and $w(x)$; therefore its spectrum is a
convolution of $G(u)$ and $W(u)$. By comparing the spectra of the
original signal in (a) and of its truncated version in (c) we see that the
Fourier-domain effect of the truncation consists of smearing or
smoothing out the original spectrum, and adding a ripple effect (i.e.
decaying oscillations)

As shown in Fig. 6.1, this convolution may affect the spectrum in two ways with respect
to the original spectrum $G(u)$: it may *smear* or *smooth out* the spectrum due to the main
lobe of the sinc function, and it may add a *ripple* effect (i.e. decaying oscillations) due to
the sidelobes of the sinc function (see, for example, [Brigham88 pp. 90–91 and 98–103]).
These effects occur in the DFT spectrum, too, since the DFT operates on the truncated
version of $g(x)$, and thus yields discrete samples of $G(u)*W(u)$ rather than discrete
samples of $G(u)$. This is, indeed, the origin of the leakage artifact in the DFT output.

Signal domain   Spectral domain

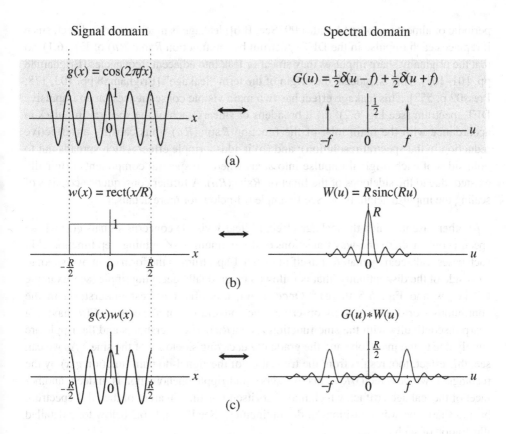

(a)

(b)

(c)

Figure 6.2: A particular case of the generic Fig. 6.1, showing the effect of
truncation on the spectrum of the function $g(x) = \cos(2\pi f x)$ in the
continuous world. (a) The function $g(x)$ and its continuous Fourier
transform $G(u)$, which consists of a pair of half-unit impulses at the
frequencies $\pm f$. (b) The truncating function $w(x) = \text{rect}(x/R)$ and its
continuous Fourier transform $W(u) = R\,\text{sinc}(Ru)$. (c) The truncated
version of $g(x)$ extending between $-R/2$ and $R/2$ is a product of $g(x)$
and $w(x)$; therefore its spectrum is a convolution of $G(u)$ and $W(u)$.
By comparing the spectra of the original function in (a) and of its
truncated version in (c) we see that the Fourier-domain effect of the
truncation consists of vertical scaling by R and smearing each of
the two impulses into a narrow sinc function that is centered about
the original impulse location, and whose oscillations gradually
decay to both directions. See also Problem 6-23.

Note that just as in the case of aliasing, if the original spectrum $G(u)$ (and hence the
resulting discrete DFT spectrum, too) is rather smooth and non-impulsive, leakage does
not have a spectacular effect on the shape of the DFT spectrum, and its existence may be
quite difficult to detect unless the true values of the original spectrum $G(u)$ are known in
advance. However, if the original spectrum $G(u)$ is impulsive (i.e. if the given signal $g(x)$ is

periodic or almost periodic [Amidror09, Sec. B.6]) leakage is much easier to detect, since it replaces each impulse in the DFT spectrum by the function $R\,\text{sinc}(Ru)$ of Eq. (6.1), so that the originally sharp impulses may smear or leak into adjacent frequencies [Brigham88 pp. 101–103]. This is, indeed, the origin of the term "leakage" [Brigham88 pp. 103, 178; Press02 p. 557]. This leakage effect has two main visible consequences on an impulsive DFT spectrum (see Fig. 6.2): (a) it broadens or smears each of the spectral impulses in accordance with the main lobe of the function $R\,\text{sinc}(Ru)$ (thus causing an effective reduction in the spectral resolution); and (b) it adds a ripple effect which spreads out to both sides of each original impulse into areas where no spectral components originally existed, due to the sidelobes of the function $R\,\text{sinc}(Ru)$. A further consequence consists of scaling the impulse height by R. See Example 6.1 below for more details.

Another case in which the leakage effect is clearly visible concerns signals $g(x)$ whose spectra contain sharp one-way transitions or discontinuities resembling step functions.[3] In such cases leakage may manifest itself in the DFT spectrum in the form of an overshoot to each side of the discontinuity, that is followed by gradually decaying ripple (see Example 6.2 below and Fig. 6.5 there).[4] Once again, this effect is best understood in the continuous-world spectrum, by observing the convolution of a spectrum $G(u)$ having a sharp discontinuity with the sinc function $R\,\text{sinc}(Ru)$: The overshoots and the ripple are simply due to the main lobe and the gradually decaying sidelobes of the sinc.[5] As we can see, this effect, too, results from the truncation of the signal-domain function $g(x)$ by the rectangular window $\text{rect}(x/R)$. This overshoot-and-ripple phenomenon is, indeed, another facet of the leakage artifact, which manifests itself this time in areas of the DFT spectrum having sharp one-way transitions or discontinuities.[6] See Example 6.2 below for a detailed illustration of such a case.

---

[3] This includes also a discontinuity between the last value of the DFT spectrum and its first value, due to the cyclical nature of the DFT.

[4] This phenomenon is essentially equivalent to (or more precisely, it is a Fourier dual of) the Gibbs phenomenon [Smith07 p. 108, Footnote 7]. The Gibbs phenomenon occurs when we reconstruct a *continuous* periodic function $p(x)$ with sharp transitions, such as a rectangular wave, from a truncated part of its Fourier series representation that does not include the higher frequencies [Bracewell86 pp. 209–211]. This gives an overshoot and ripple effect to each side of any discontinuity in $p(x)$. And indeed, our overshooting effect here can be seen as a discrete counterpart of the Gibbs phenomenon, where both the signal domain and the spectral domain are discrete. An illustrated demonstration of this phenomenon in the DFT context can be found, for example, in [Smith03 pp. 218–219]. The similarity of our case with the Gibbs phenomenon becomes clearer if we interchange the roles of the signal domain and of the spectral domain, so that the truncation affects the high frequencies in the spectral domain and the overshoots and ripple occur in the signal domain, as in the classical Gibbs phenomenon. Note that the explanation of the classical Gibbs phenomenon in the continuous world is also based on truncation with a rectangular window in one domain, that gives convolution with a sinc function in the other domain (see, for example, [Bracewell86 pp. 209–211]).

[5] In fact, as one can see by making convolution trials between a step function and a gradually truncated sinc function, the main lobe of the sinc function only rounds the corners of the step function. The overshoots are obtained when we add to the main lobe of the sinc function the first sidelobe to each of its sides; and the decaying ripple effect beyond the overshoots is obtained when we add also the higher-order sidelobes of the sinc function (see Sec. D.7 in Appendix D and Fig. D.6 there). A similar discussion can be found in [Oppenheim99 p. 473] and Fig. 7.23 therein.

[6] As explained in [Bracewell86 p. 209], the function $R\,\text{sinc}(Ru)$ has unit area, so in places where $G(u)$ is slowly varying the result of the convolution will be in close agreement with $G(u)$.

Remark 6.1: Just like aliasing, leakage is inherent to DFT, and in practice it can rarely be avoided completely. And yet, there exist two particular cases in which leakage does not occur, in spite of the finite-range truncation of the underlying signal [Brigham88 pp. 98–101 and 103–104]: (a) when the given function $g(x)$ is periodic, and the range length R of the truncation window (i.e. the sampling range of $g(x)$) is an exact integer multiple of the period of $g(x)$; or (b) when $g(x)$ is a finite-duration function, and it is fully included within the sampling range R so that it is not affected by the rectangular window rect(x/R). Note that case (a) is the only periodic case where each of the impulses in the DFT spectrum is perfectly sharp, meaning that (i) it falls precisely on an element of the output array, and (ii) its width is exactly one frequency sample (a single element of the output array); see Fig. 6.3 and compare the different rows there. And indeed, as we will see in Remark 6.3 below, this is precisely the dual formulation of case (a) in terms of the spectral domain.[7,8] In all cases other than (a) and (b) leakage is unavoidable, and the only solution is, therefore, to reduce leakage as much as possible. ■

Just as in the case of aliasing (see Remark 5.2), leakage, too, can be reduced either by a judicious choice of the DFT parameters, or by applying an appropriate treatment to the input signal *prior to* the application of the DFT; but unfortunately, the remedies here are not the same as in the case of aliasing reduction. The first way to reduce leakage consists of increasing the length of the rectangular truncation window (i.e. the sampling range length R), so that its spectrum (Eq. (6.1)) becomes a narrower sinc function and better approaches an impulse. The more closely this sinc function approximates an impulse, the less smearing and ripple is introduced by the convolution with it [Brigham88 p. 90].[9] The other possible technique for reducing leakage (that can also be used when the range length R cannot be increased) consists of truncating the input signal more softly, i.e. multiplying it *before* the application of the DFT with a windowing function that is smoother than the sharp-edged rectangular window rect(x/R) [Brigham88 p. 180]. Such windowing functions have the advantage that their Fourier transforms have smaller sidelobes than $R\,\mathrm{sinc}(Ru)$, the Fourier transform of rect(x/R). A list of such windowing functions (also called *tapering functions*, *weighting functions* or *apodisation functions*) can be found in the literature, along with a full description of their respective properties, advantages and drawbacks (see, for example, [Harris78], [Nuttall81], [Brigham88, Sec. 9.2] or [Press02 pp. 558–563]). In general the use of such windowing functions reduces the sidelobes, and

---

[7] To see this, note that the fact that the period P of $g(x)$ satisfies $R = qP$ (i.e. $P = R/q$) for a certain integer q means, in terms of the frequency domain, that the frequency f of $g(x)$ satisfies $f = 1/P = q/R = q\Delta u$ (using Eq. (7.3)), i.e. that f is an exact integer multiple of the frequency step Δu, and falls exactly on an output array element of the DFT. And vice versa, if the period P of $g(x)$ does not exactly satisfy $R = qP$ for any integer q (so that there may exist a cyclical discontinuity between the beginning and the end of the input array), it means in terms of the frequency domain that the frequency f of $g(x)$ does not fall exactly on an output array element of the DFT. We will return to this point in greater detail in Sec. 6.5; more information on the roles of q and p (the discrete counterpart of P) is provided in Appendix C.

[8] On the dual formulation of case (b) in terms of the spectral domain, see Sec. D.4 in Appendix D.

[9] In fact, we will see in Sec. 6.4 that the width of the main lobe of $R\,\mathrm{sinc}(Ru)$ is always two DFT output array elements, and the width of each of its sidelobes is always one array element. However, by increasing the range length R we decrease the frequency step Δu (see Eq. (7.3)), so that although the lobe widths remain unchanged in terms of array elements, in terms of true frequencies in the continuous-world spectrum their bandwidths become narrower.

hence allows to detect weaker spectral features that were hidden by the sidelobes; this, however, at the expense of broadening the main lobe and therefore decreasing the *spectral resolution*, i.e. the capacity to distinguish or resolve closely spaced peaks in the spectrum [Brigham88 p. 180]. This tradeoff between sidelobe level and main lobe bandwidth means that there is no single, universal best solution, and in each application one needs to find the most suitable windowing function depending on the case.[10]

Remark 6.2: By combining Remark 5.2 and Remark 6.1 we see that the only case where both aliasing and leakage are completely absent happens when the given signal $g(x)$ is both band limited and periodic, the sampling rate being used is at least twice the largest frequency component of $g(x)$, and the range length R of the truncation window (i.e. the sampling range of $g(x)$) is an exact integer multiple of the period of $g(x)$. This is, indeed, the only case where the (reorganized) DFT and the continuous Fourier transform perfectly coincide on the N discrete points of the DFT spectrum (within a scaling constant, as already explained above in Sec. 4.3) [Brigham88 p. 101]. Note that case (b) of Remark 6.1 is excluded here, since a finite-duration function $g(x)$ cannot be band limited [Brigham88 p. 103], so that case (b) cannot be aliasing-free. ■

It is interesting to note also that fighting against leakage requires increasing the sampling range length R of the given signal $g(x)$, while fighting against aliasing requires, as we have seen in Remark 5.2, a denser sampling of the signal (the use of a higher sampling rate, i.e. a smaller sampling interval Δx). Each of these goals requires, therefore, its own increase in the number N of samples; but if for some reason N cannot be freely increased, one has to choose between reducing aliasing or reducing leakage. In other words, because for a given N there is a tradeoff between the sampling resolution of $g(x)$ and its sampling range, there is also a tradeoff between aliasing and leakage.

More details on the optimal choice of the sampling range length R (and the tradeoffs that are involved in this choice) can be found in Sec. 7.4.

Interested readers can find many references that treat the leakage phenomenon from various different points of view. For example, one may mention [Smith03 pp. 174–176], [Smith07 pp. 105–108], [Ramirez85 pp. 102–109], [Brigham88 pp. 98–107, 176–179], [Cartwright90 pp. 206–208] and [Briggs95 pp. 98–99].

6.3 Examples of leakage in the one dimensional case

Let us illustrate the above discussion by two 1D examples, one featuring a periodic function $g(x)$ (whose spectrum $G(u)$ is therefore purely impulsive), and the other featuring an aperiodic function $g(x)$ (whose spectrum $G(u)$ is continuous).

---

[10] Note that in some references such as [Weisstein99 pp. 55–56] this tapering process is called *apodization*, literally meaning "foot removal". The "foot" refers here to the trailing sequence of sidelobes; the origin of this term is more obvious in the 2D case (see Sec. 6.6 below).

Example 6.1: Leakage due to truncation in the periodic function $g(x) = \cos(2\pi f x)$:

Fig. 6.2 shows a particular case of the generic Fig. 6.1 in which the original continuous function is $g(x) = \cos(2\pi f x)$, whose CFT consists of a pair of half-unit impulses at the frequencies $\pm f$ (see part (a) of the figure): $G(u) = \frac{1}{2}\delta(u{-}f) + \frac{1}{2}\delta(u{+}f)$ [Brigham88 pp. 19–20]. The truncation of $g(x)$ into the finite range $-R/2...R/2$ can be seen as a multiplication of $g(x)$ by the truncating window $w(x) = \text{rect}(x/R)$, a square pulse of length R and height 1 which extends between $-R/2$ and $R/2$. The truncating window $w(x)$ and its CFT $W(u) = R\,\text{sinc}(Ru)$ are shown in part (b) of the figure, and the resulting truncated function $g(x)w(x)$ and its spectrum are shown in part (c) of the figure. By virtue of the convolution theorem, the spectrum (CFT) of the truncated function $g(x)w(x)$ is the convolution of the spectra $G(u)$ and $W(u)$, which is given by (see [Bracewell86 p. 416]):

$$G(u)*W(u) = \tfrac{1}{2}R\,\text{sinc}(R(u-f)) + \tfrac{1}{2}R\,\text{sinc}(R(u+f)) \qquad (6.2)$$

As we can see in part (c) of the figure, the effect of the signal-domain truncation on the spectral domain consists of vertically scaling each of the two impulses by R, and smearing it into a narrow sinc function that is centered about the original impulse location, and whose oscillations gradually decay to both directions.

So far we have only discussed the continuous Fourier transform, so that the spectrum of the truncated function $g(x)w(x)$ is continuous (see Fig. 6.2(c)). Assume now that instead of considering $g(x)w(x)$ and its CFT, we sample $g(x)w(x)$ and subject it to DFT (note that the truncating function $w(x)$ corresponds here to the finite range of $g(x)$ that is taken into account by the DFT). As we have seen in Sec. 5.2, the use of DFT has two main consequences: (1) The resulting spectrum is now discrete; and (2) it only contains frequencies in the range between $-\frac{1}{2}f_s$ and $\frac{1}{2}f_s$ (where f_s is the sampling frequency), and includes therefore only frequencies located in the main replica of the periodic spectrum of the continuous Fourier transform.

Figure 6.3: Illustration of the leakage artifact in the case of the periodic signal $g(x) = \cos(2\pi f x)$. The figure shows the sampled version of the signal $g(x)$, after its truncation into the range $-4...4$ that is included in the input array, and the resulting DFT (after all the required adjustments as discussed in Chapters 3 and 4 and in Appendix A). For the sake of comparison, the original continuous function $g(x)$ and its CFT $G(u) = \frac{1}{2}\delta(u{-}f) + \frac{1}{2}\delta(u{+}f)$ are also shown, drawn by continuous lines. The successive rows of the figure show the situation for a few gradually increasing values of the cosine frequency f, varying between $f = 0.5$ in row (a) and $f = 0.75$ in row (f). Note that the sampling frequency $f_s = 4$ and the array length $N = 32$ remain unchanged in all cases (they have been chosen sufficiently high so as to guarantee that no aliasing artifacts occur here). Similarly, the signal domain range length $R = 8$, the signal domain step $\Delta x = \frac{1}{4}$ and the spectral domain step $\Delta u = \frac{1}{8}$ remain unchanged, too. In rows (a), (e) and (f) the cosine frequency f is

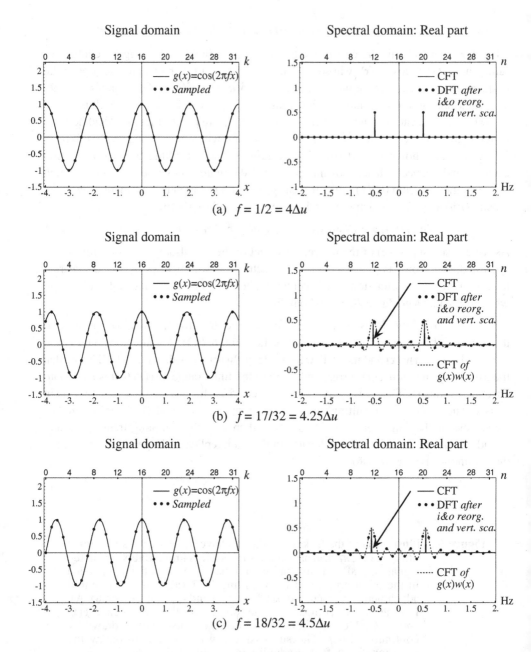

(a) $f = 1/2 = 4\Delta u$

(b) $f = 17/32 = 4.25\Delta u$

(c) $f = 18/32 = 4.5\Delta u$

exactly an integer multiple of the frequency step of the DFT spectrum, Δu, so that the two impulses are represented in the DFT results correctly. But in rows (b)–(d) the cosine frequency f falls in-between elements of the DFT spectrum, resulting in a significant leakage effect: each of the two impulses is smeared between its two neighbouring output array elements, and also shows gradually

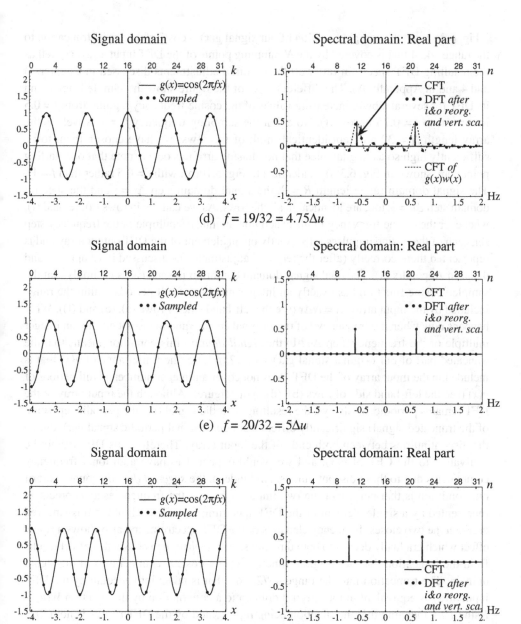

(d) $f = 19/32 = 4.75\Delta u$

(e) $f = 20/32 = 5\Delta u$

(f) $f = 3/4 = 6\Delta u$

decaying oscillations to both directions. The dashed lines in the spectra of rows (b)–(d) correspond to the CFT of $g(x)w(x)$, i.e. to the continuous function $G(u)*W(u)$ (Eq. (6.2)). Note the perfect agreement between these dashed lines and the black dots that represent the DFT results (after the adjustments according to Chapters 3 and 4 and Appendix A). See Example 6.1.

Fig. 6.3 shows the sampled version of our signal $g(x) = \cos(2\pi fx)$ after its truncation to the range $-4...4$ that is covered by the N sampling points of the DFT input array, as well as the resulting DFT spectrum (after all the required adjustments discussed in Chapters 3 and 4 and in Appendix A). The different rows of the figure show the sampled signal and its DFT for several slowly increasing values of the cosine frequency f, going from $f = 0.5$ in row (a) to $f = 0.75$ in row (f). Note that the sampling frequency $f_s = 4$ as well as the array length $N = 32$ are kept identical in all of the rows; we have chosen them to be sufficiently high so as to guarantee that no aliasing artifacts occur (note that our starting point here, shown in Fig. 6.3(a), is identical to Fig. 5.10(b), with $f = 0.5$ rather than $f = 1$). The signal domain range length $R = 8$, the signal domain step $\Delta x = \frac{1}{4}$ and the spectral domain step $\Delta u = \frac{1}{8}$, too, are identical in all the rows. As we can see in rows (a), (e) and (f), whenever the cosine frequency f falls exactly on an integer multiple of the frequency step Δu, each of the cosine impulses falls exactly on an element of the DFT output array and is represented there correctly (after the required adjustments as discussed in Chapters 3 and 4 and in Appendix A). The dual signal-domain condition in such cases is simply that the sampled cosine function has exactly an integer number of full periods within the range included by the input array, $R = N\Delta x$ (see the left-hand side of rows (a), (e) and (f)). What happens now when the frequency f of our original cosine function is not exactly an integer multiple of the frequency step Δu? In the *signal domain* this condition means that the truncated part of our original signal $g(x) = \cos(2\pi fx)$, i.e. the finite range of $g(x)$ that is included in the input array of the DFT, does not cover an integer number of full periods of $g(x)$ (see the left-hand side of rows (b)–(d) in our figure). Although the input array of the DFT remains periodic, as always, the resulting periodic signal (i.e. the periodic repetition of the truncated signal) significantly differs from our original periodic signal $g(x)$, due to the discontinuities between both ends of the input array. Therefore, its DFT cannot be equivalent to the CFT of $g(x)$, and we would expect it to have additional frequency components due to these discontinuities. And indeed, the *spectral domain* counterpart of our condition is that neither of the two impulses having the frequencies $\pm f$ is precisely represented by a single element of the DFT spectrum; instead, each of them is smeared between the two closest frequency elements of the DFT spectrum, and also shows a ripple effect which gradually decays to both directions. As we have seen above, this effect results from the truncation that our original function $g(x) = \cos(2\pi fx)$ undergoes in the signal domain: The truncation into the range $-R/2...R/2$ that is taken into account by the DFT input array is equivalent in the spectral domain to a convolution by the function $W(u) = R\,\mathrm{sinc}(Ru)$, meaning that the function being represented by the DFT results is not $G(u)$, but rather the convolution $G(u)*W(u)$ as given in Eq. (6.2) above. And indeed, looking at the DFT spectra in rows (b)–(d) of the figure we note the perfect agreement between the DFT results represented by the black dots and the continuous function $G(u)*W(u)$ which is plotted there by dashed lines. This is, indeed, the explanation of the leakage artifact in the DFT spectrum of our given function $g(x) = \cos(2\pi fx)$.

So why such a smearing effect does not occur in rows (a), (e) and (f), where the cosine frequency f is exactly an integer multiple of the frequency step Δu (and the input array of the DFT contains exactly an integer number of full periods of $g(x) = \cos(2\pi fx)$)? The answer is that $G(u)$ is indeed convolved with $W(u) = R\text{sinc}(Ru)$ in all of the cases, because the same truncation is operated in all of them. However, when the cosine frequency f is exactly an integer multiple of the frequency step Δu the functions $W(u\pm f) = R\text{sinc}(R(u\pm f))$ have their peaks exactly at an integer multiple of Δu, where the true impulse is located, and all the other points of the DFT spectrum, i.e. the other integer multiples of Δu, fall precisely on a zero point of these sinc functions. Clearly, in such cases the convolution of the spectral impulses with the narrow sinc function $W(u) = R\text{sinc}(Ru)$ has no visible effect on the DFT output (except for an impulse height scaling by R, which can be corrected as explained in Appendix A). This is clearly shown in Fig. 6.4. More details on the precise effects of the truncation and on impulses that fall in-between elements of the DFT spectrum are given in Secs. 6.4 and 6.5 below (see also references such as [Brigham88 pp. 98–103]). ∎

Example 6.2: Leakage due to truncation in the aperiodic function $g(x) = \text{sinc}(2ax)$:

Let us now consider the aperiodic function $g(x) = \text{sinc}(2ax)$, whose CFT is $G(u) = \frac{1}{2a}\text{rect}(\frac{u}{2a})$, i.e. a square pulse of height $\frac{1}{2a}$ extending between $-a$ and a. Once again, the truncation of $g(x)$ into a finite range $-R/2...R/2$ can be seen as a multiplication of $g(x)$ by the truncating window $w(x) = \text{rect}(x/R)$, a square pulse of length R and height 1 which extends between $-R/2$ and $R/2$. By virtue of the convolution theorem, the spectrum (CFT) of the truncated function $g(x)w(x)$ is the convolution of the spectra $G(u)$ and $W(u) = R\text{sinc}(Ru)$, which is given by (see Sec. D.5 in Appendix D):

$$G(u)*W(u) = \frac{1}{2a\pi}\left[\text{Si}(\pi R(u+a)) - \text{Si}(\pi R(u-a))\right] \qquad (6.3)$$

where $\text{Si}(u)$ is the sine integral function. This convolution with $W(u)$ means that the effect of the signal-domain truncation on the spectral domain consists of smoothing out the original square pulse $G(u)$; moreover, the nature of the sine integral function implies the addition of an overshoot to each side of the two sharp transitions of $G(u)$, followed by gradually decreasing oscillations to both directions.

So far we have only discussed the continuous Fourier transform, so that the spectrum of the truncated function $g(x)w(x)$ is continuous. Assume now that instead of considering $g(x)w(x)$ and its CFT, we sample $g(x)w(x)$ and subject it to DFT (note that the truncating function $w(x)$ corresponds here to the finite range of $g(x)$ that is taken into account by the DFT). Fig. 6.5 shows the sampled version of our signal $g(x) = \text{sinc}(2ax)$ after its truncation to the range $-4...4$ that is covered by the N sampling points of the DFT input array, as well as the resulting DFT spectrum (after all the required adjustments discussed in Chapters 3 and 4). The different rows of the figure show the sampled signal and its DFT for several slowly increasing values of the parameter a, going from $a = 0.5$ in row (a) to $a = 0.75$ in row (f). Note that as in the previous example, the sampling frequency $f_s = 4$ as well as the array length $N = 32$ are kept identical in all of the rows; we have chosen them to be sufficiently high so as to guarantee that no aliasing artifacts occur (note that our starting point with $a = 0.5$, shown in Fig. 6.5(a), is identical to Fig. 4.1 in Chapter 4). Similarly, the signal domain range length $R = 8$, the signal domain step $\Delta x = \frac{1}{4}$ and the spectral domain step $\Delta u = \frac{1}{8}$, too, remain identical in all the rows.

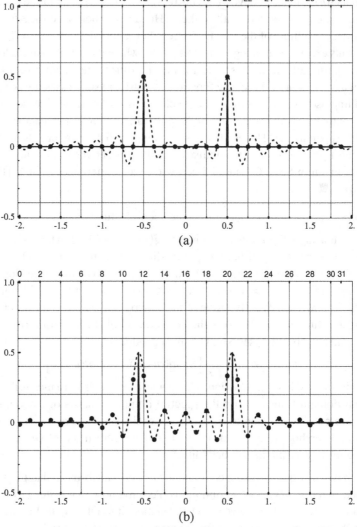

Figure 6.4: A magnified view of the DFT spectra in rows (a) and (c) of Fig. 6.3 (after all
the required adjustments). In each case the continuous black line represents
the original continuous spectrum $G(u)$, consisting of two half-unit impulses,
and the dashed line represents the continuous function $G(u)*W(u)$ of Eq.
(6.2); note the perfect agreement between the dashed line and the DFT
results represented by the black dots. (a) When the cosine frequency is
exactly a multiple of the spectral domain step Δu (here: $f = 4\Delta u$), each of the
two spectral impulses falls exactly on an element of the DFT spectrum, and
so does the corresponding peak of $G(u)*W(u)$; moreover, all the zero points
of $G(u)*W(u)$ fall exactly on the successive elements of the DFT spectrum
(i.e. on all the other integer multiples of Δu), so that $W(u)$ does not affect the
DFT results (up to vertical scaling). (b) When the cosine frequency is not
exactly a multiple of Δu (here: $f = 4.5\Delta u$), each spectral impulse falls in-
between frequency elements of the DFT spectrum, and "leaks" into its two
closest neighbours, and into gradually decaying oscillations to both sides.

Let us now compare in Fig. 6.5 the DFT results with the original spectrum $G(u)$, which is plotted with a continuous line. As we can see, the DFT smoothes out each of the sharp transitions of $G(u)$, and also adds an overshoot to each side of the transition, followed by gradually decaying oscillations to both directions, much like the Gibbs phenomenon (see Footnote 4 in Sec. 6.2). This effect results, as we have seen above, from the truncation that our original function $g(x) = \text{sinc}(2ax)$ undergoes in the signal domain: In order to be forwarded to the DFT, $g(x)$ must be first sampled and truncated into the range $-R/2...R/2$ that is taken into account by the DFT input array. This truncation is equivalent in the spectral domain to a convolution by the function $W(u) = R\,\text{sinc}(Ru)$, meaning that the function being represented by the DFT results is not $G(u)$, but rather the convolution $G(u)*W(u)$ given in Eq. (6.3) above. And indeed, looking at the DFT spectrum in each row of the figure we note the perfect agreement between the DFT results represented by the black dots and the continuous function $G(u)*W(u)$ which is plotted there by dashed lines. This is, indeed, the explanation of the leakage artifact in the DFT spectrum of our given function $g(x) = \text{sinc}(2ax)$.

Note that unlike in Fig. 6.3, in the present case all the rows of the figure suffer from a significant leakage effect, both when the pulse borders $\pm a$ fall exactly on an integer multiple of the frequency step Δu and when they fall in-between elements of the DFT spectrum. From the spectral domain point of view, the reason is that unlike the function $G(u)*W(u)$ of the previous example (Eq. (6.2)), the function $G(u)*W(u)$ in the present case (Eq. (6.3)) does not have zeroes at the integer multiples of Δu, so that even when the sharp transitions of $G(u)$ are centered exactly on a frequency that exists in the DFT output we still get decaying oscillations throughout the spectrum. From the signal-domain point of view, the explanation is based on the fact that our given signal $g(x) = \text{sinc}(2ax)$ is aperiodic: Since the input array of the DFT is periodic, as always, the resulting signal (i.e. the periodic repetition of the truncated signal) significantly differs from our original aperiodic signal $g(x)$. Therefore, we cannot expect its DFT to be equivalent to the CFT of $g(x)$, and this, independently of whether or not the pulse borders $\pm a$ fall exactly on an integer multiple of the frequency step Δu. ■

Figure 6.5: Illustration of the leakage artifact in the case of the aperiodic signal $g(x) = \text{sinc}(2ax)$. The figure shows the sampled version of the signal $g(x)$, after its truncation into the range $-4...4$ that is included in the input array, and the resulting DFT (after all the required adjustments as discussed in Chapters 3 and 4). For the sake of comparison, the original continuous function $g(x)$ and its CFT $G(u)$ $= \frac{1}{2a}\text{rect}(\frac{u}{2a})$ (a square pulse of height $\frac{1}{2a}$ extending between $-a$ and a) are also shown, drawn by continuous lines. The successive rows of the figure show the situation for a few gradually increasing values of the parameter a, varying between $a = 0.5$ in row (a) and $a = 0.75$ in row (f). Note that the sampling frequency $f_s = 4$ and the array length $N = 32$ remain unchanged in all cases (they have been chosen

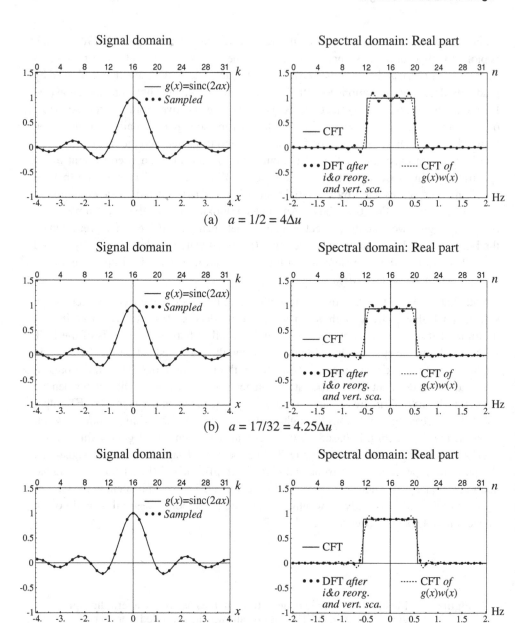

(a) $a = 1/2 = 4\Delta u$

(b) $a = 17/32 = 4.25\Delta u$

(c) $a = 18/32 = 4.5\Delta u$

sufficiently high so as to guarantee that no aliasing artifacts occur here). Similarly, the signal domain range length $R = 8$, the signal domain step $\Delta x = \frac{1}{4}$ and the spectral domain step $\Delta u = \frac{1}{8}$ remain unchanged, too. The dashed lines in the spectra

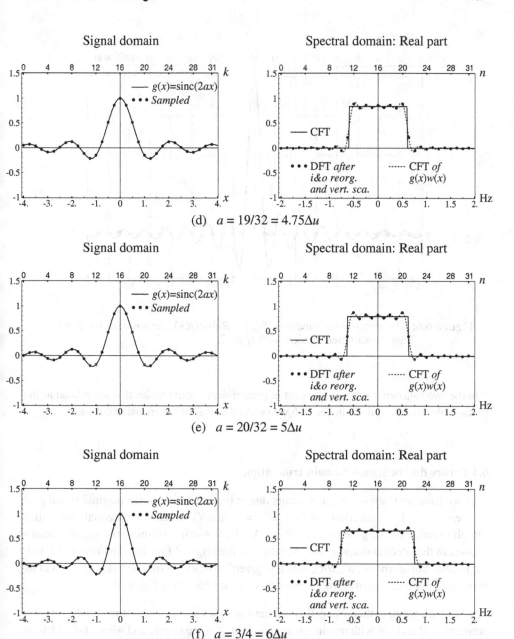

(d) $a = 19/32 = 4.75\Delta u$

(e) $a = 20/32 = 5\Delta u$

(f) $a = 3/4 = 6\Delta u$

correspond to the CFT of $g(x)w(x)$, i.e. to the continuous function $G(u)*W(u)$ (Eq. (6.3)). Note the perfect agreement between these dashed lines and the black dots that represent the DFT results (after the adjustments according to Chapters 3 and 4). See Example 6.2.

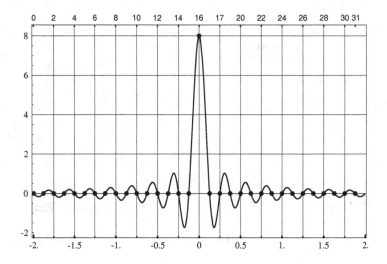

Figure 6.6: The continuous function $W(u) = R\,\mathrm{sinc}(Ru)$, drawn here for $R = 8$, and its sampled version with $N = 32$.

In the two following sections we treat in greater detail some topics that are related to the leakage phenomenon; then, in Sec. 6.6 we proceed to the multidimensional case.

6.4 Errors due to signal-domain truncation

As we have seen above, leakage occurs due to the truncation of the original signal $g(x)$ by the rectangular truncation window $w(x) = \mathrm{rect}(x/R)$ whose length equals the entire sampling range of the given signal, $R = N\Delta x$. This multiplication in the signal domain implies in the spectral domain a convolution of the original Fourier transform $G(u)$ with the Fourier transform of $\mathrm{rect}(x/R)$, which is given by $W(u) = R\,\mathrm{sinc}(Ru)$ (Eq. (6.1)). Let us now study more closely the properties of this sinc function (see Fig. 6.6).

First of all, we note that $R\,\mathrm{sinc}(Ru)$ is a narrow sinc function, whose main lobe extends between $-1/R$ and $1/R$ with maximum amplitude of R at the origin, and whose k-th sidelobe extends between k/R and $(k+1)/R$ for $k = 1, 2, \ldots$ (or symmetrically between k/R and $(k-1)/R$ for $k = -1, -2, \ldots$ in the negative side). The main lobe and the sidelobes are delimited by the zero-crossing points of this function, which occur at the points k/R for all integer values of k except for 0 (the origin). We see, therefore, that the width of the main lobe of this function (between the two zero-crossing points $-1/R$ and $1/R$) is exactly $2/R$. What does this mean in terms of the DFT output array elements? Knowing that the entire frequency range length F of the DFT spectrum extends over N elements, it follows (e.g. using the rule of three [Harris98 p. 15]) that a frequency span of $2/R$ extends over:

$$w = \frac{2N}{RF} \quad \text{output array elements}$$

and using the relation $RF = N$ (see Eq. (7.2)) we obtain:

$$w = 2 \quad \text{output array elements} \tag{6.4}$$

Similarly, because the width of each sidelobe of the function $R\,\text{sinc}(Ru)$ (i.e. the distance between its two bordering zero-crossing points) is $1/R$, half of the width of the main lobe, it follows that the width of each sidelobe corresponds exactly to one output array element.

We see, therefore, that the main lobe and the sidelobes of the function $R\,\text{sinc}(Ru)$, which are the source of the leakage effect in the DFT spectrum, always have the width of 2 or 1 array elements, respectively, independently of the choice of the DFT parameters R, N, Δx, etc. And indeed, this is clearly visible in detailed figures illustrating the leakage of spectral impulses, like Fig. 6.4. This means that the smearing of a spectral impulse due to the main lobe of the function $R\,\text{sinc}(Ru)$ with which the impulse is being convolved may extend to two array elements.[11] Similarly, each of the sidelobes being added to either side of the spectral impulse extends over exactly one array element. Nevertheless, the global effect of leakage extends over a multitude of consecutive sidelobes with gradually decreasing amplitudes. This gives the typical impression of a leaking or melting-down impulse with a decaying "tail" that trails away to each side of the smeared impulse. Note that the individual one-element sidelobes of each trailing tail will have alternating positive and negative amplitudes. But if we plot the magnitude (i.e. the absolute value) of this DFT spectrum (or its squared version, the power spectrum), we will get a continuously decaying positive-valued tail to each side of the smeared impulse (compare Figs. 6.14 and 6.16).

It should be noted, however, that this analysis of the lobe widths due to the leakage phenomenon is basically valid for cases with impulsive spectra, where each spectral impulse will be individually smeared by $R\,\text{sinc}(Ru)$ (compare Figs. 6.4 and 6.6). When the original spectrum $G(u)$ is not impulsive, its convolution with $R\,\text{sinc}(Ru)$ will still give a smearing effect with visible sidelobes, but the widths of these lobes may be different. As we have seen in Example 6.2 and in Fig. 6.5, this is the case when $G(u)$ is a rectangular function; in this case the convolution of $G(u)$ with $R\,\text{sinc}(Ru)$ gives a new function (see Eq. (6.3)) whose lobe widths are no longer exact multiples of output array elements. Although the typical behaviour of leakage still remains the same (smearing and ripple), the precise lobe-width analysis given above is not necessarily preserved.

Thus far, we have described here in greater detail the leakage artifact of the DFT. What happens, however, when the signal-domain function $g(x)$ is truncated by a rectangular window whose length is less than N elements (i.e. less than the entire span R of the sampling range), so that all the remaining input array elements are set to zero? Suppose, for example, that we use a truncation window $w_1(x)$ whose length is $N/4$ (or

---

[11] Note that this smearing causes an effective 2-fold reduction in the spectral resolution, although the theoretical spectral resolution of the DFT spectrum, given by the frequency step $\Delta u = \frac{1}{N\Delta x}$ of Eq. (4.4) remains unchanged [Cartwright90 pp. 207–208].

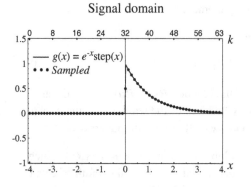

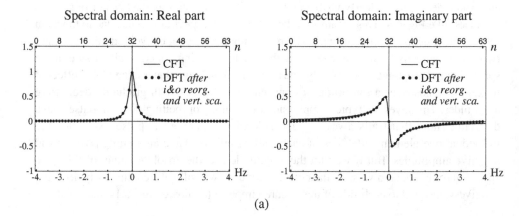

(a)

Figure 6.7: The effect of excessive signal-domain truncation on the DFT spectrum.
(a) The original continuous function $g(x) = e^{-x}\,\text{step}(x)$ and its CFT
$G(u) = 1/(1+[2\pi u]^2) - 2\pi u i/(1+[2\pi u]^2)$ [Bracewell86 p. 418], plotted
by continuous lines, and their DFT counterparts, plotted by black dots.

equivalently $R/4$, assuming that the sampling interval Δx remains unchanged; see Eq.
(4.2)). In this case the convolution in the spectral domain will be done with the sinc
function $W_1(u) = R/4\,\text{sinc}(uR/4)$, whose lobes are 4 times wider than the lobes of the sinc
function $R\,\text{sinc}(Ru)$ that generates the leakage phenomenon. This will cause in the DFT
spectrum a 4-fold wider smearing: Each impulse will be smeared by the main lobe of the
new sinc function into 8 output array elements about its true location, and moreover, each
individual sidelobe will be 4 elements wide, thus causing a visible, wavy ripple effect. Such
a ripple effect in the DFT spectrum due to an excessive signal-domain truncation may also
occur when $g(x)$ is not impulsive (see, for example, Fig. 6.7), but then the lobe widths may
be somewhat different.[12]

_____

[12] Because the truncation of the given function $g(x)$ causes a sharp discontinuity in the signal domain,
the addition of a large spectral range of high frequencies in its DFT spectrum is not really surprising
[Smith07 p. 108].

Signal domain

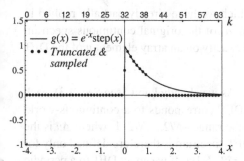

Spectral domain: Real part   Spectral domain: Imaginary part

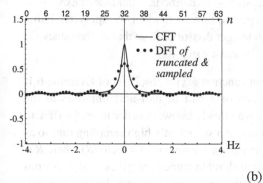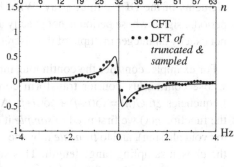

(b)

(b) The same original function $g(x)$ and its CFT $G(u)$, plotted by
continuous lines, but this time the black dots represent the truncated
function $g(x)w_1(x)$ and its DFT.

It should be stressed, however, that although the explanation of this phenomenon is
similar to that of leakage, there exists a fundamental difference between the two: The
N-element windowing effect that causes leakage is inherent to DFT, and it cannot be
avoided. But the windowing effect due to a truncation of the given N-element signal *within*
its sampling range (i.e. using a truncation window of less than N elements) can be avoided
— and in fact, such a truncation should never be done unless its consequences in terms of
the DFT spectrum are intended and well understood.[13]

---

[13] Note that this truncation in the signal domain is in fact the Fourier dual of the low-pass filtering
of $g(x)$ by truncating its DFT spectrum and taking the inverse DFT of the result. As shown in Fig. 3.7
in Chapter 3, low-pass filtering gives, indeed, a "leaked out" (i.e. low-pass filtered) version of $g(x)$; the
difference between the two cases is, however, that the roles of the signal and spectral domains are
interchanged.

6.5 Spectral impulses that fall between output array elements

A related question may be asked now, to which we have already partially replied in Example 6.1: What happens to an impulse $\delta(u-a)$ of the original continuous spectrum $G(u)$ when it does not fall in the discrete world exactly on an array element n of the DFT output, but rather *between* two elements?

Let us formulate this question more precisely. As we already know (see Sec. 4.3), each of the N elements in the output array of the DFT corresponds to a continuous-world frequency of $n\Delta u$ Hz for a certain integer n in the range $-N/2,...,N/2-1$, where Δu is the frequency step of the discrete spectrum. The DFT spectrum is therefore defined only for these discrete frequencies, namely, $n\Delta u = n\frac{1}{N\Delta x} = nf_s/N$ Hz. If we apply DFT to a periodic signal $g(x)$ whose period P is exactly the sampling range length $R = N\Delta x$ or some integer divisor of R, meaning that $P = R/q$ for a certain integer q, $1 \leq q < N/2$, then the frequency of $g(x)$, $f = 1/P = q/R = q\Delta u$ (see Eq. (7.3)), as well as its harmonics, fall indeed exactly on elements of the DFT output array. What happens, however, when we apply DFT to a periodic signal whose period is not exactly an integer divisor of R, so that its frequency f is not exactly an integer multiple of the discrete frequency step Δu?

For example, consider the continuous input function $g(x) = \cos(2\pi f x)$ of Example 6.1, whose continuous Fourier transform consists of a pair of half-unit impulses at the frequencies $\pm f$: $G(u) = \frac{1}{2}\delta(u-f) + \frac{1}{2}\delta(u+f)$. As we already know, in order to apply DFT to the function $g(x)$ we first need to *sample* it (say, at a sufficiently high sampling rate so as to avoid aliasing), and to *truncate* it by the rectangular window $w(x) = \text{rect}(x/R)$ where R is the chosen sampling range length. This signal-domain truncation gives in the spectral domain a convolution of the original two-impulse spectrum $G(u)$ with the sinc function $W(u) = R\text{sinc}(Ru)$, which is the Fourier transform of $w(x)$. In other words, the DFT of the original function consists of samples of $G(u)*W(u)$ rather than samples of $G(u)$. We have seen above that the sinc function $R\text{sinc}(Ru)$ consists of a narrow main lobe extending between the two zero-crossing points $-1/R$ and $1/R$, and of sidelobes which extend to both directions between all the following zero-crossing points k/R and $(k+1)/R$ for $k = 1, 2, ...$ (or in the negative side between k/R and $(k-1)/R$ for $k = -1, -2, ...$). In terms of output array elements, the main lobe of this sinc function extends between the integer points -1 and 1, and its sidelobes extend to both directions between all the integer pairs k and $k+1$ for $k = 1, 2, ...$ (or, respectively, between k and $k-1$ for $k = -1, -2, ...$).

Now, if the frequency f of our given cosine function is exactly an integer multiple of the frequency step $\Delta u = \frac{1}{N\Delta x} = 1/R$ (see Eqs. (4.4) and (4.2)), so that the two impulses of $G(u)$ fall exactly on an output array element, the convolution of $G(u)$ with $R\text{sinc}(Ru)$, which places a replica of $R\text{sinc}(Ru)$ about each of the two impulses, will cause no errors in the DFT spectrum with respect to $G(u)$ (except for multiplying the impulse amplitude by R; see Sec. A.3 in Appendix A). The reason is that all the other output array elements will fall exactly on the zero-crossing points of the replicas of $R\text{sinc}(Ru)$ (see Fig. 6.4(a)). The resulting DFT spectrum will therefore contain exactly two sharp impulses, as shown in this figure.

If, however, the frequency f of our given cosine function is *not* an integer multiple of $\Delta u = \frac{1}{N\Delta x} = 1/R$, so that the two impulses of $G(u)$ fall *between* output array elements, the convolution of each of these impulses with $R\text{sinc}(Ru)$ will cause in the DFT spectrum a leakage effect, as shown in Fig. 6.4(b): Since the two-element wide main lobe of the sinc is now located *between* two output array elements, both of these array elements will fall inside the main lobe of the sinc function and have positive values. And moreover, *each* of the other output array elements to the left and to the right will fall inside one of the one-element wide sidelobes of the sinc replica, and hence they will have alternating negative and positive non-zero values. The resulting DFT spectrum will therefore have additional non-zero values at frequencies other than $\pm f$: It will include two smeared (two-element wide) impulses located at the output array elements nearest to the true cosine frequencies f and $-f$, each of them having long "tails" which trail away to both directions throughout the entire range of the DFT spectrum, as shown in Fig. 6.4(b). Note that the effect of leakage is quantifiable; a mathematical expression giving the exact leakage value generated by the original frequency f at each of the output array elements can be found, for example, in [Smith07 pp. 105–106] or in [Cartwright90 p. 206]. Furthermore, it is shown in [Cartwright90 p. 208] that the power (i.e. the square of the amplitude) of the original frequency f is the sum of the powers of the output array elements among which it is split (this is not true, however, for the amplitudes). Leakage can be seen, therefore, as taking power from the frequency component f existing in the original continuous signal $g(x)$, and transferring it to other frequency components of the DFT output array, which do not exist in the original continuous signal [Ramirez85 p. 107].

Leakage, when it occurs, affects in the general case both the real and the imaginary parts of the DFT spectrum. Also, when converting the DFT spectrum to the polar form, leakage is carried through the conversion to affect both magnitude and phase [Ramirez85 p. 107].

It is important to stress, however, that in spite of this significant discrepancy with respect to the original spectrum $G(u)$, the DFT spectrum is not wrong in itself — it is exactly what it should be for the given discrete signal [Ramirez85 p. 107]; simply, it contains an inherent leakage effect that did not exist in the spectrum $G(u)$ in the continuous case. In other words, from the viewpoint of the continuous case leakage is an "error"; but from the viewpoint of the N samples provided to the DFT the result is correct.

Although we used in this discussion the specific example of a cosine function, the same leakage effect occurs in the DFT spectrum for any spectral impulse that falls between output array elements, and more generally, for any frequency component f (impulsive or not) that does not fall exactly on an output array element. Note, in particular, that if a general periodic signal $g(x)$ is acquired so that the sampling range R contains an integer number of periods, then each of the harmonics, too, has an integer number of periods in R. Thus, each of the spectral components of $g(x)$ will fall exactly on an output array element of the DFT, rather than between elements. But if $g(x)$ is an almost-periodic function (see, for example, Appendix B in [Amidror09]) which is made up of discrete frequencies that are not harmonically related (such as f and $\sqrt{2}f$), some frequency components may exhibit leakage while other don't; such signals are often found in vibration studies where the signal is the result of several vibration sources [Ramirez85 p. 109].

As we can see, the answer to the question in the beginning of this section can be formulated as follows:

Remark 6.3: The fact that the frequency f of a given periodic function $g(x)$ falls exactly on an output array element is equivalent, in terms of the signal domain, to the fact that the period P of the function $g(x)$ precisely satisfies $R = qP$ for a certain integer q, so that $g(x)$ satisfies the leakage-free condition (a) of Remark 6.1. Indeed, if the frequency f of $g(x)$ is an integer multiple of Δu, we have $f = q\Delta u = q/R$ (using Eq. (7.3) below), so that the period $P = 1/f$ of $g(x)$ satisfies $P = R/q$ for a certain integer q; this is precisely condition (a) of Remark 6.1. And vice versa, if the frequency f of $g(x)$ does not fall exactly on an output array element, this means in terms of the signal domain that the period P of $g(x)$ does not exactly satisfy $R = qP$ for any integer q, meaning that leakage will occur in the DFT spectrum. More details on the role of q can be found in Appendix C. ∎

We conclude, therefore, with the following result (see also Remark C.3 in Appendix C):

Remark 6.4: A periodic function $g(x)$ is perfectly cyclical (seamlessly wraparound) within the input array *if and only if* its impulses in the DFT spectrum fall exactly on output array elements (which means, in turn, that no leakage occurs in the DFT spectrum). ∎

We see, as a consequence, that if $g(x)$ is a periodic function which suffers from leakage in the DFT spectrum since q is non-integer, then there are two equivalent ways to rectify the problem: either by resampling $g(x)$ within a new range $R = qP$ where q is integer; or by modifying N to be an integer multiple of p (the period of $g(x)$ in terms of input array elements), so that by virtue of $pq = N$ (Eq. (C.2)) q becomes integer.[14]

In real life, where signals are usually much more complex than simple cosines and contain many more components of different frequencies, it is extremely unlikely that a smooth match occurs between the beginning and the end of the sampled signal, or that all the frequency components fall precisely on output array elements of the DFT. Spectral leakage will therefore almost certainly affect the spectrum of any real-world signal.

Remark 6.5: (*Signal-domain* impulses that fall between array elements): Due to the duality between the signal and frequency domains under DFT, the input array can be seen as the inverse DFT of the output array. Therefore, our discussion above applies also to impulses in the *input* array of the DFT. Suppose we are given a continuous-world signal $g(x)$ that contains one or more impulses, and that we want to find its spectral representation using DFT. For this end, we first need to sample $g(x)$ in order to obtain the input array that will be forwarded to the DFT. Obviously, the best would be to sample $g(x)$ such that all its impulses fall precisely on a discrete element of the input array. But what happens when this is not possible, for example, if the impulse locations in $g(x)$ are not

---

[14] This is not always possible since many DFT algorithms require that the array length N be a power of 2, or at least an even number.

harmonically related (say, $x = 1$ and $x = \sqrt{2}$)? In such cases one (or more) of the impulses of $g(x)$ will actually fall *between* elements of the input array. Due to the duality between DFT and the inverse DFT, it follows that such an in-between impulse should be obtained in the input array by sampling the narrow signal-domain sinc function $F\,\text{sinc}(Fx)$ after centering it on the true impulse location and scaling its amplitude. And if $g(x)$ contains several in-between impulses, their respective signal-domain sinc functions should be summed up additively, just like in Fig. 6.4 (but this time in the signal domain). This may be considered as *signal-domain leakage*. Other possible solutions such as shifting the in-between impulse to the closest input array location, or even worse, ignoring it altogether (which is the default solution when $g(x)$ undergoes a standard "brute force" sampling) will give imprecise results in the DFT as well as in the inverse DFT of these results. We will return to the subject of signal-domain leakage in greater detail in Sec. 6.7 below. ∎

Table 6.1 provides a synoptic review of the main properties of the aliasing and leakage artifacts of the DFT.

6.6 Leakage in two or more dimensions

The above discussion about leakage can be extended to the multidimensional case, too. A short discussion on the 2D case can be found, for example, in [Brigham88 pp. 250–253]. In the 2D case the default truncation window is $w(x,y) = \text{rect}(x/R,y/R) = \text{rect}(x/R)\text{rect}(y/R)$, whose spectrum is the 2D sinc function $W(u,v) = R^2\text{sinc}(Ru)\text{sinc}(Rv)$. This function has a narrow 2D main lobe centered at the origin, and a narrow "tail" of decaying sidelobes trailing off *ad infinitum* to both sides of the main lobe along each of the u and v axes; its sidelobes elsewhere in the u,v plane, away from both axes, are rather negligible (see Fig. 6.8(b)). This is, indeed, a straightforward generalization of the 1D sinc function $R\text{sinc}(Ru)$ of Eq. (6.1) (Fig. 6.6). However, as is often the case, here too the situation in the MD case is richer and more diversified than in the 1D case: While in the 1D case the various alternative window functions that can be used instead of the default truncation window $\text{rect}(x/R)$ basically differ only in their gradual slopes on the borders of the window, in 2D or MD cases window functions may also differ in their spatial *geometric shape*. For example, in the 2D case one may replace the default rectangular truncation window $w(x,y) = \text{rect}(x/R,y/R)$, where R is the length of the sampling range along each dimension, by a circular truncation window with diameter R. This truncation window only differs from $w(x,y)$ in its 2D geometric shape, but it still has a sharp 0/1 transition along its borders (see Fig. 6.8(c)):

$$w_C(x,y) = \text{rect}(\sqrt{x^2 + y^2}/R)$$

The Fourier transform of this function, shown in Fig. 6.8(d), is expressed in terms of the "sombrero" function, or equivalently in terms of the Bessel function J_1 [Gaskill78 pp. 72–73, 329; Bracewell95 pp. 150–151]:

$$W_C(u,v) = \tfrac{\pi}{4}R^2\text{somb}(R\sqrt{u^2 + v^2}) = RJ_1(\pi R\sqrt{u^2 + v^2})/(2\sqrt{u^2 + v^2})$$

where:

$$\mathrm{somb}(r) = \begin{cases} 2J_1(\pi r)/\pi r & r \neq 0 \\ 1 & r = 0 \end{cases}$$

The function $W_c(u,v)$ has the advantage over $W(u,v)$ of being circular, with no protruding tails along the axes. But although $W(u,v)$ and $W_c(u,v)$ are the most obvious choices, there exist in the 2D or MD cases infinitely many other sharped-edged truncation functions, that only differ in their geometric shape (triangular, hexagonal, etc.). And then, just as in the 1D case, one may also devise based on each of these sharp-edged functions many other variants with smoother transitions between 0 and 1.

Artifact: Properties:	Aliasing	Leakage
Effect of the artifact in the signal domain:	A lower frequency signal mimics the original signal on its sampled values	Windowing, mismatch between last and first values
Effect of the artifact in the spectral domain:	All the signal frequencies beyond half of the sampling frequency undergo a cyclical foldover back into the lower frequency range	Smearing, ripple
Cause of the artifact:	Undersampling of the given signal	Truncation of the given signal
Particular cases where the artifact does not occur:	See Remark 5.2	See Remark 6.1
Adequate choice of DFT parameters for reducing the artifact:	Smaller Δx ($\Rightarrow$ Larger F) (For fixed N: $\Rightarrow$ Smaller R)	Larger R ($\Rightarrow$ Smaller Δu) (For fixed N: $\Rightarrow$ Smaller F)
Appropriate treatment to be applied to the given signal (*prior to* DFT) for reducing the artifact:	Low pass filtering; see also Remark 6.10 in Sec. 6.7	Smoother windowing (tapering); see also Remark 6.10 in Sec. 6.7

Table 6.1: Synoptic overview of the main properties of aliasing and leakage.

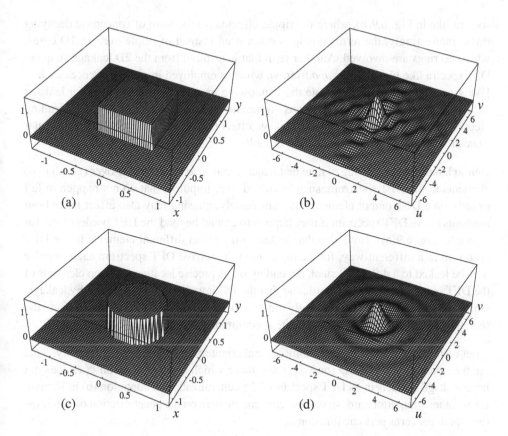

Figure 6.8: (a) The default 2D truncation window, $w(x,y) = \text{rect}(x/R, y/R)$, drawn here for $R = 1$. (b) Its continuous Fourier transform is the narrow 2D sinc function $W(u,v) = R^2 \text{sinc}(Ru)\text{sinc}(Rv)$, the 2D counterpart of the function $R \text{sinc}(Ru)$ of Fig. 6.6. (c) The circular 2D truncation window $w_c(x,y) = \text{rect}(\sqrt{x^2 + y^2}/R)$, drawn here for $R = 1$. (d) Its continuous Fourier transform is the narrow "sombrero" function $W_c(u,v) = \frac{\pi}{4}R^2 \text{somb}(R\sqrt{u^2 + v^2}) = R J_1(\pi R \sqrt{u^2 + v^2})/(2\sqrt{u^2 + v^2})$; note the circular symmetry of this function.

As we have already seen in the 1D case, the form that leakage takes depends on the form of the Fourier transform of the window being used. And indeed, by changing the shape of the window we can change the shape of the leakage [Ramirez85 p. 109]. This fact is even more striking in the multidimensional case. For example, Figs. 6.9(a),(b) show the 2D DFT of the same periodic function; notice the different shapes of the leakage effect due to the different window shapes being used.

The shapes of the leakage effect in the 2D case are indeed so remarkable, that some leakage-related terms in the literature have been coined after these shapes. For example, the term *ringing* that is often used as a synonym to ripple is inspired from 2D DFT spectra like in Fig. 6.9(b), where the ripple effect takes the form of concentric decaying rings; interestingly, the term ringing is often used instead of ripple even in 1D cases, where no rings are involved. Another term that is inspired from the 2D leakage shape in DFT spectra like Fig. 6.9(a) is *apodisation*, which is employed in some references such as [Weisstein99 pp. 55–56] to denote the windowing or tapering process used for leakage reduction. Apodisation literally means "removing the feet", referring to the protruding "feet" (or rather tails) of the 2D-sinc leakage effect that occurs due to the 2D rectangular truncation window rect($x/R,y/R$) used by default.

Remark 6.6: It is interesting to note that leakage may also affect folded-over (i.e. aliased) elements in the DFT spectrum, such as folded-over impulses that do not happen to fall exactly on a DFT output element. And conversely, aliasing may also affect leaked-out elements in the DFT spectrum if they happen to extend beyond the DFT borders (see, for example, Fig. 6.9(a)). Note also that leakage may affect different elements in the DFT spectrum in a different way; for example, in an impulsive DFT spectrum each impulse may be leaked to a different extent, depending on its precise location between elements of the DFT output array. Note in particular that the DC impulse is never affected by leakage, since it is always located precisely on the first output array element (before the reorganization), which corresponds to the spectrum origin. ∎

Let us now see what is the multidimensional counterpart of Remark 6.1, namely, what are the conditions in the 2D (or MD) case under which a given function g($\mathbf{x}$) does not have leakage artifacts in its DFT spectrum. The generalization of condition (b) in Remark 6.1 is rather straightforward, so we will concentrate here on the generalization of condition (a), which concerns periodic functions.

Figure 6.9: The shape of the 2D leakage artifact depends on the truncation window being used. This is illustrated here using the original function $g(x,y) = \cos(2\pi f[x\cos\theta + y\sin\theta])$, a rotated cosinusoidal wave of frequency f, whose CFT consists of an impulse pair. The two first rows show the effect of two different truncation windows on the DFT of the function $g(x,y)$. (a) The default truncation window $w(x,y)$, consisting of an $R\times R$ square pulse, gives a cross-like leakage artifact around each impulse that does not fall exactly on a discrete element in the DFT spectrum (compare the shape of this leakage artifact with the shape of $W(u,v)$ in Fig. 6.8(b)). Note that since each of the leakage crosses trails away *ad infinitum*, it wraps around the DFT border and re-enters from the opposite side due to aliasing (see Chapter 5). (b) The circular truncation window $w_c(x,y)$, consisting of a circular cylinder of radius $R/2$, gives a sombrero-like leakage artifact around each impulse that does not fall exactly on a discrete element in the DFT spectrum (compare the shape of this leakage artifact with the shape of $W_c(u,v)$ in Fig. 6.8(d)). (c) Leakage can be completely avoided in this case by slightly changing the angle θ and the frequency f of $g(x,y)$ so that both of its impulses fall exactly on a discrete element of the DFT output.

Signal domain

Spectral domain: Real part

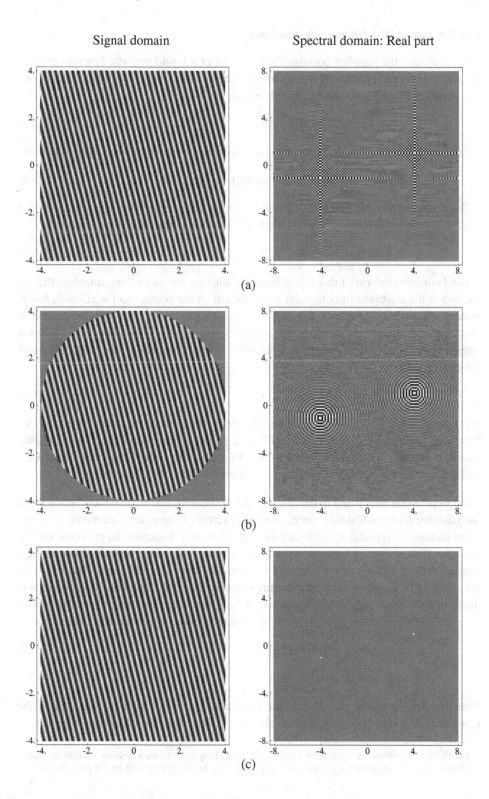

(a)

(b)

(c)

6.6.1 The case of 2D 1-fold periodic functions

We start with the simplest possible 2D case, that of a 1-fold periodic function over the x,y plane. For example, consider the cosine function with frequency of f cycles per unit (i.e. period of $P = 1/f$ units) along the x direction:

$$g(x,y) = \cos(2\pi f x) \tag{6.5}$$

After a rotation by angle θ counterclockwise this cosine function becomes (Fig. 6.9(a)):

$$g(x,y) = \cos(2\pi f[x\cos\theta + y\sin\theta]) \tag{6.6}$$

The period of this 2D function is given by the *period vector* $\mathbf{P}$ in the x,y plane which emanates from the origin and whose length is P units along the angle θ:

$$\mathbf{P} = (P\cos\theta, P\sin\theta) \tag{6.7}$$

The Fourier transform of this 1-fold periodic function consists of two impulses that are located in the u,v plane to both sides of the origin, at the points $(u,v) = \pm(f\cos\theta, f\sin\theta)$. Note that if our 1-fold periodic function $g(x,y)$ is more complex than a simple cosine, its spectrum will contain higher harmonic impulses, too, that are located in the u,v plane at integer multiples of the first harmonic impulses, i.e. at the points $(u,v) = \pm k(f\cos\theta, f\sin\theta)$ for integer values of k. The vector

$$\mathbf{f} = (f\cos\theta, f\sin\theta) \tag{6.8}$$

that connects the origin of the continuous-world spectrum to the location of the first harmonic impulse, $(f\cos\theta, f\sin\theta)$, is called the *frequency vector* of the 1-fold periodic function $g(x,y)$. Note that because in the 1-fold periodic case the vectors $\mathbf{P}$ and $\mathbf{f}$ are collinear, we also have in this case the connections $\mathbf{f} = \mathbf{P}/P^2$, $\mathbf{P} = \mathbf{f}/f^2$ and $\mathbf{P} \cdot \mathbf{f} = 1$.

Let us now consider $\mathbf{p}$ and $\mathbf{q}$, the discrete-world counterparts of the period vector $\mathbf{P}$ and the frequency vector $\mathbf{f}$, which are expressed in terms of input array elements or output array elements, respectively. We call these discrete-world vectors the *p-vector* and the *q-vector*. If the sampling interval in the signal domain is Δx along both dimensions,[15] the length of the period vector $\mathbf{P}$ is given in terms of input-array elements by $p = P/\Delta x$. Similarly, if the length of each output-array element in the DFT spectrum is Δu Hz to both directions, the length of the frequency vector $\mathbf{f}$ is given in terms of output-array elements by $q = f/\Delta u$. We therefore have, in terms of input or output array elements, respectively:

$$\mathbf{p} = (p\cos\theta, p\sin\theta) = \mathbf{P}/\Delta x \tag{6.9}$$

$$\mathbf{q} = (q\cos\theta, q\sin\theta) = \mathbf{f}/\Delta u \tag{6.10}$$

Note that since $\mathbf{p}$ and $\mathbf{q}$ are collinear the 1D relation $pq = Pf/(\Delta x \Delta u) = N$ remains valid along angle θ (using Eq. (4.5)), and we also have $\mathbf{q} = \mathbf{p}N/p^2$, $\mathbf{p} = \mathbf{q}N/q^2$ and $\mathbf{p} \cdot \mathbf{q} = N$.

---

[15] For the sake of simplicity we assume here that the array length N and the sampling interval Δx remain the same in all dimensions; but an equivalent expression can be also formulated for the general case.

$N = 16$

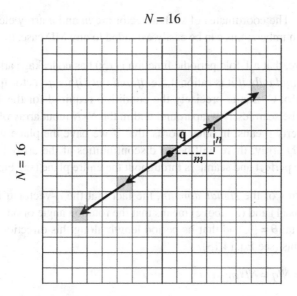

Figure 6.10: Schematic illustration of the the DFT spectrum of a 2D 1-fold periodic
function whose impulses fall exactly on elements ("pixels") of the
output array. In such cases, the q-vector **q**, which emanates from the
spectrum origin and points to the first harmonic impulse of the
spectrum, has exactly an integer number of array elements along each
of its Cartesian components: $\mathbf{q} = (m,n)$. Each array element in the DFT
spectrum represents a frequency step of Δu in the continuous-world
spectrum, so the corresponding frequency vector in the continuous-
world spectrum is $\mathbf{f} = (m\Delta u, n\Delta u)$. See also Problem 3-8.

Now, a straightforward generalization of the 1D case suggests that a 2D 1-fold periodic
function $g(x,y)$ is free from leakage artifacts *if and only if* its impulses are located precisely
on 2D output array elements, i.e. exactly on "pixels" of the 2D image that represents the
2D DFT spectrum. This happens, of course, when each of the two Cartesian components
of the frequency vector **f** is exactly an integer multiple of the frequency step Δu, or in other
words, when the two Cartesian components of the corresponding discrete-world q-vector
q are integer numbers, say, m and n (see Fig. 6.10). This means that:

(1) The rotation angle θ of $g(x,y)$ has a rational slope of $\tan\theta = \frac{n}{m}$, and

(2) The frequency of $g(x,y)$ along this angle is given, in terms of DFT spectrum pixels, by:

$$q = \sqrt{m^2 + n^2}$$

We will call 1-fold periodic functions $g(x,y)$ which satisfy these conditions *rational
1-fold periodic functions*. The q-vector $\mathbf{q} = (q\cos\theta, q\sin\theta)$ of such a function, which
connects the origin $(0,0)$ to the pixel (m,n) in the 2D DFT spectrum, will be called a

rational q-vector. The coordinates of such a vector are m and n array elements, both of which are integer numbers; this can be easily extended to any MD case, too.[16]

We see, therefore, that a 1-fold periodic function $g(x,y)$ has no leakage artifacts in the 2D DFT spectrum *if and only if* it is rational, i.e. *if and only if* its q-vector $\mathbf{q}$ is rational. As shown in Appendix C, this is precisely the condition required for the 1-fold periodic function $g(x,y)$ to be seamlessly wraparound within the $N{\times}N$ input array of the DFT. This wraparound property means, in other words, that if we pave the plane with juxtaposed replicas of this 2D array, there will be no discontinuities at the array borders and the periodicity will be perfect and seamless throughout the entire plane (see Fig. C.9).

Note that in terms of the *signal domain*, the fact that the q-vector $\mathbf{q}$ is rational (i.e. satisfies conditions (1) and (2) above) means that the rotation angle of our 1-fold periodic function $g(x,y)$ is $\tan\theta = \frac{n}{m}$, and that its period length along this direction is, in terms of input-array elements (see Eq. (4.13)):

$$p = N/q = N/\sqrt{m^2 + n^2} \tag{6.11}$$

This does not yet mean, however, that the p-vector $\mathbf{p}$ of this function is also rational (i.e. extends along an integer number of input-array pixels both horizontally and vertically). In order that the p-vector $\mathbf{p}$ be rational, it must *itself* satisfy condition (2). We will return to this point in greater detail in Apendix C, where we discuss the implications of the rationality of the p-vectors and/or of their corresponding q-vectors.

Finally, note that condition (2) can be also translated from the discrete-world language of array elements (or pixels) into the continuous-world language of true frequency units in terms of Hz (see the 1D equivalent in Sec. 4.2): The length of a single pixel along each dimension of the DFT output array is given by $\Delta u = \frac{1}{N\Delta x}$ (Eq. (4.5)). Therefore, condition (2) says that the frequency f in terms of Hz along the direction of the frequency vector $\mathbf{f}$ must have the form:

$$f = q\Delta u = \sqrt{m^2 + n^2}\,\Delta u = \sqrt{m^2 + n^2}/(N\Delta x) \tag{6.12}$$

for some integers m and n. Its components along the u and v axes are $m\Delta u$ and $n\Delta u$.

6.6.2 The case of 2D 2-fold periodic functions

We now proceed to the more general case of 2D periodic functions, that of a 2-fold periodic function over the x,y plane. This time the function $g(x,y)$ has two independent period vectors $\mathbf{P}_1$ and $\mathbf{P}_2$ rather than one in the x,y plane, and two frequency vectors $\mathbf{f}_1$ and $\mathbf{f}_2$ rather than one in the u,v plane. Therefore, its spectrum may contain impulses at all the locations $k_1\mathbf{f}_1 + k_2\mathbf{f}_2$ with integer values of k_1 and k_2. Just as in the 1-fold periodic case, here too we consider the discrete-word counterparts of the period vectors $\mathbf{P}_1$ and $\mathbf{P}_2$, the

---

[16] Note that by abuse of language we may also say that the corresponding frequency vector $\mathbf{f}$ in the continuous world is a *rational frequency vector*, although its coordinates are not integer numbers m and n but rather $m\Delta q$ and $n\Delta q$ Hz.

p-vectors $\mathbf{p}_1$ and $\mathbf{p}_2$, which are expressed in terms of input-array elements; and the discrete-world counterparts of the frequency vectors $\mathbf{f}_1$ and $\mathbf{f}_2$, the q-vectors $\mathbf{q}_1$ and $\mathbf{q}_2$, which are expressed in terms of output-array elements:

$$\mathbf{p}_i = \mathbf{P}_i / \Delta x \qquad\qquad\qquad (6.13)$$

$$\mathbf{q}_i = \mathbf{f}_i / \Delta u \qquad\qquad\qquad (6.14)$$

Now, a 2-fold periodic function $g(x,y)$ generates no leakage artifacts in the 2D DFT spectrum *if and only if* both of its q-vectors $\mathbf{q}_1$ and $\mathbf{q}_2$ are rational, namely, when $\tan\theta_1 = n_1/m_1$ and $q_1 = \sqrt{m_1^2 + n_1^2}$, and $\tan\theta_2 = n_2/m_2$ and $q_2 = \sqrt{m_2^2 + n_2^2}$. Note that in the case of 2-fold periodic functions the two main periodicity directions are not the same as the two main frequency directions, but rather orthogonal to them; see Appendix C and Fig. C.5 there. Nevertheless, the graphic interpretation of these conditions in terms of the signal domain remains the same as in the 1-fold periodic case, namely, that our 2-fold periodic function $g(x,y)$ is perfectly cyclical (seamlessly wraparound) within the $N{\times}N$ input array of the DFT. We will return to this point in greater detail in Appendix C.

6.6.3 The general MD case

The discussions above can be also extended to the MD n-fold periodic case (see Appendix C). We finally obtain the following MD generalization of Remark 6.1:

Remark 6.7: Just like in the 1D setting, in the multidimensional setting, too, there exist two particular cases in which leakage does not occur in spite of the finite-range windowing of the underlying signal: (a) when the given function $g(\mathbf{x})$ is n-fold periodic with $1 \leq n \leq M$, and it is perfectly cyclical (seamlessly wraparound) within its sampling range (truncation window) R^M;[17] or (b) when $g(\mathbf{x})$ is a finite-duration function, and it is fully included within the sampling range R^M so that it is not truncated by the rectangular window $\text{rect}(\mathbf{x}/R)$. Note that in terms of the spectral domain, case (a) means that each of the impulses of the n-fold periodic function $g(\mathbf{x})$ in the DFT spectrum is perfectly sharp, i.e. that (i) it falls precisely on an element of the output array, and (ii) its width is exactly one frequency sample (a single element of the output array). This is, indeed, the dual formulation of case (a) in terms of the spectral domain, as shown in Appendix C. In all cases other than (a) and (b) leakage is unavoidable, and the only solution is to reduce leakage as much as possible, using the same methods as in the 1D setting. ∎

Note that the conditions of Remark 6.7 do not prevent aliasing; but if aliasing does occur — for example in the form of impulses that exceed beyond the borders of the DFT spectrum and re-enter from the opposite side — it will not be accompanied by leakage, so that each impulse, aliased or not, will fall exactly on one single pixel of the DFT spectrum. The conditions on $g(\mathbf{x})$ for having neither aliasing nor leakage in the DFT spectrum are given in the following multidimensional counterpart of Remark 6.2:

---

[17] This also means that the input array N^M seamlessly paves the entire M-dimensional discrete space.

Remark 6.8: By combining Remark 5.4 and Remark 6.7 we see that the only case where both aliasing and leakage are completely absent occurs when the given signal $g(\mathbf{x})$ is both band limited and periodic, the sampling rate being used in each dimention is at least twice the largest frequency component of $g(\mathbf{x})$ in that dimension, and the function $g(\mathbf{x})$ is perfectly cyclical (seamlessly wraparound) within its sampling range (truncation window) R^M. This is, indeed, the only case where the (reorganized) DFT and the continuous Fourier transform perfectly coincide on the N^M discrete points of the DFT spectrum (within a scaling constant, as already explained above in Sec. 4.3). Note that case (b) of Remark 6.7 is excluded here, since a finite-duration function $g(\mathbf{x})$ cannot be band limited [Brigham88 p. 103], so that case (b) cannot be aliasing-free.   ■

6.7 Signal-domain leakage

Just as in the case of aliasing, it follows from the duality between DFT and the inverse DFT that all our previous discussions on the leakage phenomenon under DFT remain true under IDFT, too, except that the roles of the input and output arrays are reversed. For example, each row in Fig. 6.11 shows in the right-hand column the result obtained by applying DFT to the discrete sequence shown in the left-hand column. But the same figure can be also interpreted the other way around, where the left-hand column is the result of applying IDFT to the discrete sequence shown in the right-hand column. Now, just as in the DFT, if the truncation which occurs in the input of the IDFT (the truncation of the continuous-world spectrum $G(u)$ into the finite range length F) is inadequate, it will cause leakage in the output of the IDFT; but this time, the leakage occurs in the signal domain, and we can therefore call it *signal-domain leakage*, or in the 1D case *time-domain leakage* [Smith96]. Stated in other words, if a continuous-world frequency function $G(u)$ is truncated within a finite range length F in order to be forwarded to IDFT, and

(a) $G(u)$ is periodic (meaning that $g(x)$ is impulsive) and its truncation into the finite range length F does not give exactly an integer number of periods (see Figs. 6.11(b)–(c)); or

(b) $G(u)$ is aperiodic and it does not effectively die out by the borders of this range;

then the IDFT of the sampled version of $G(u)$ will suffer from signal-domain leakage.

However, the same signal-domain leakage in the very same figure can be also regarded from a second point of view, which does not explicitly require the use of the IDFT: Suppose we are given a continuous-world signal $g(x)$ whose continuous-world spectrum is $G(u)$, and that we try to approximate this continuous-world Fourier transform by DFT. As we already know, the DFT can only take into account a finite range of frequencies. Therefore, if the original, underlying continuous-world spectrum $G(u)$ does not die out by the borders of this range, $G(u)$ should be truncated before the transition to the discrete world (in order to avoid frequency foldover due to aliasing; see Sec. 5.2). This truncation operation can be viewed, still in the continuous world, as a multiplication of the underlying spectrum $G(u)$ by a rectangular "truncation window" $W(u)$ whose length equals the DFT frequency range length $F = N\Delta u$, namely, $W(u) = \text{rect}(u/F)$. This multiplication in the spectral domain is equivalent in the signal domain to a convolution of the input signal $g(x)$ with the inverse Fourier transform of the truncation window $W(u)$, $w(x) = F\operatorname{sinc}(Fx)$. This convolution represents, indeed, a signal-domain leakage of $g(x)$. Note, however, that the situation is easier to understand from the *spectral-domain* point of view: Signal-domain leakage occurs if the truncation of the continuous-world spectrum $G(u)$ by $W(u)$ is inadequate (non-integer multiple of the period, if $G(u)$ is periodic; or too narrow, if $G(u)$ is aperiodic). And conversely, signal-domain leakage does not occur (a) if $G(u)$ is periodic and it is truncated exactly on an integer multiple of its period; or (b) if $G(u)$ is aperiodic and its frequency range is fully included within that of the DFT spectrum (compare with conditions (a) and (b) in Remark 6.1, which speaks about *spectral-domain* leakage).[18]

Example 6.3: Signal-domain leakage when $G(u)$ is periodic:

Consider Fig. 6.3, and suppose that we are interested in the inverse situation: what input should we supply to the DFT in order to obtain in its output the truncated cosines shown in the signal domain of Fig. 6.3? In this case, shown in Fig. 6.11, the given continuous-world signal is $g(x) = \frac{1}{2}\delta(x-f) + \frac{1}{2}\delta(x+f)$, and the resulting continuous-world spectrum is $G(u) = \cos(2\pi f u)$. The 3 rows of the figure only differ in the value of the cosine frequency f. The difficulty here, when we proceed to the discrete world, is twofold: From the signal domain point of view, how should we sample the impulse pair $g(x)$ when the impulses do not fall exactly on sampling points, i.e. when they fall in-between input array elements, like in rows (b) and (c) of Fig. 6.11? From the point of view of the spectral domain the dual question is, how can we obtain in the DFT output a cosine function which is not exactly truncated into an integer number of its periods, like in rows (b) and (c)?[19] The answer to both questions is the same: Instead of sampling into the input array of the DFT the function $g(x)$ itself, we have to sample the convolution $g(x) * F\operatorname{sinc}(Fx)$, which corresponds in the spectral domain to the truncation of $G(u)$ into the frequency range length F. From the signal domain point of view, this convolution represents, indeed, signal-domain leakage, as we can see in the discrete input signal in rows (b) and (c). In row (a), however, where the impulses *do* fall on input-array elements, no leakage effect appears in the discrete input signal, since in this case the convolution simply gives back the original impulse pair $g(x)$ (up to a vertical scaling by F), as explained in Fig. 6.4(a) and Sec. 6.5. But although this convolution is practically superfluous in row (a), its effect of vertically scaling the input impulses by F should not be forgotten, in order to obtain the correct cosine amplitude in the DFT spectrum. This scaling by F (amplitude correction) is discussed in more detail in Sec. A.4 of Appendix A.

It is interesting to note, however, that even after this convolution, the resulting DFT spectrum still suffers from a small discrepancy with respect to the continuous cosine spectrum $G(u)$. This tiny difference is best visible in the leftmost point of the DFT spectrum in rows (b) and (c) of Fig. 6.11. It occurs due to the truncation of the function

---

[18] Note that from the signal-domain point of view the situation is less easy to understand, since the convolution of $g(x)$ with $F\operatorname{sinc}(Fx)$ does not always cause $g(x)$ to leak, as we have just seen in conditions (a) and (b). See also Sec. D.4 in Appendix D.

[19] The duality between these two questions is explained in detail in Sec. 6.5.

$g(x) * F\text{sinc}(Fx)$ at both ends of the input array (see the left-hand column in Fig. 6.11), i.e. due to conventional *spectral-domain* leakage, as explained in the beginning of the present chapter. ■

In conclusion, we have the following result:

Remark 6.9: (Signal-domain leakage): If the continuous-world frequency function $G(u)$ does not effectively die out by the borders of the finite range F that is taken care of by the DFT, then the input signal $g(x)$ should be convolved with $F\text{sinc}(Fx)$ before it is forwarded to the DFT. This convolution represents signal-domain leakage in $g(x)$. Ignoring this point (which is precisely what happens by default when $g(x)$ undergoes standard sampling and truncation before it is forwarded to the DFT) will result in a discrepancy between the resulting DFT and the original continuous-world spectrum $G(u)$. When $G(u)$ is aperiodic this discrepancy will occur since the DFT output will not correspond to a truncated version of $G(u)$ but rather to a folded-over (aliased) version of $G(u)$. And when $G(u)$ is periodic, this discrepancy will occur if impulses of $g(x)$ are not located exactly on input array elements, thus giving a bad sampling of $g(x)$ and hence a bad DFT output. ■

We see, therefore, that thanks to the duality between the signal and frequency domains, leakage and aliasing occur in *both* domains. Moreover, it turns out that they are very closely related — in fact, they are duals of each other, as explicitly stated by the following result (note that signal-domain aliasing has already been discussed in Sec. 5.7):

Figure 6.11: Example of signal-domain leakage that occurs when the continuous-world spectrum $G(u)$ is periodic, so that the continuous-world input function $g(x)$ is impulsive. In the present example we have $G(u) = \cos(2\pi fu)$ and therefore $g(x) = \frac{1}{2}\delta(x-f) + \frac{1}{2}\delta(x+f)$; this is exactly the inverse of the situation shown in Fig. 6.3. In row (a) the truncation at the borders of the DFT spectrum corresponds exactly to an integer number of periods of $G(u)$; and in terms of the signal domain, the two impulses of $g(x)$ fall exactly on input array elements. In this case no signal-domain leakage is present. In rows (b)–(c) the truncation at the borders of the DFT spectrum no longer corresponds to an integer number of periods; and in terms of the signal domain, the impulses of $g(x)$ fall in-between input array elements. Note that the correct cosine is obtained in the DFT spectrum because we have sampled in the input array the convolution of the impulse pair $g(x)$ with $F\text{sinc}(Fx)$ rather than the impulse pair $g(x)$ itself, to reflect the spectral-domain truncation to a frequency range of length F. This convolution precisely represents signal-domain leakage. All rows are shown here after the required adjustments as discussed in Chapters 3 and 4 and in Appendix A. In each case discrete signals are represented by black dots, while the original continuous-world functions (the impulse pair $g(x)$ in the signal domain and the cosine function $G(u)$ in the spectral domain) are drawn with

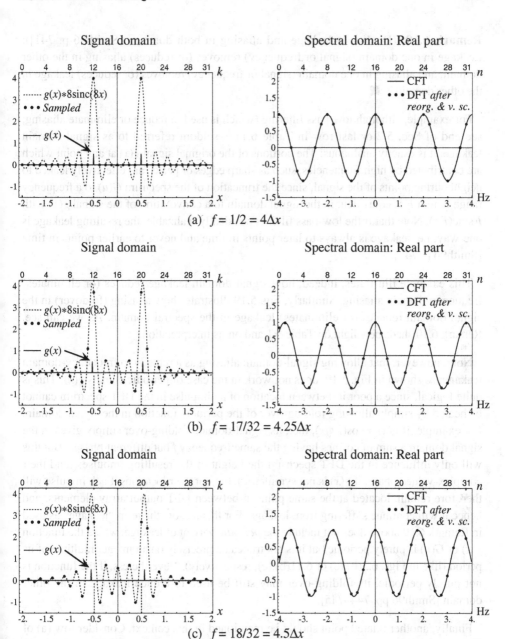

(a) $f = 1/2 = 4\Delta x$

(b) $f = 17/32 = 4.25\Delta x$

(c) $f = 18/32 = 4.5\Delta x$

continuous lines. The dashed lines in the signal domain correspond to the convolution $g(x) * F\text{sinc}(Fx)$. Note that the amplitude at the two peaks of the convolution $g(x) * F\text{sinc}(Fx)$ is $F = 8$ times higher than the amplitude $\frac{1}{2}$ of the original continuous-world impulse pair $g(x)$. A similar vertical scaling (by $R = 8$) occurs in the spectral domain of Fig. 6.3, but it is not visible there since it has been corrected by the adjustments due to Appendix A.

Remark 6.10: (Duality of leakage and aliasing in both domains [Smith96 p. 741]): Leakage in one domain (signal or frequency) removes (or reduces) aliasing in the other domain; and aliasing in one domain (signal or frequency) removes (or reduces) leakage in the other domain. ■

For example, although low-pass filtering (which is used to reduce or eliminate aliasing; see end of Sec. 5.2 or last row in Table 6.1) is seldom referred to as signal-domain leakage, it is exactly analogous: The portions of the original signal $g(x)$ at any point which are contributed by high frequencies, such as sharp edges or peaks, are effectively leaked to neighbouring points of the signal, since the truncation of the spectrum $G(u)$ to a frequency range length F is equivalent in the signal domain to a convolution of the signal $g(x)$ with $F\text{sinc}(Fx)$. Note that if the low-pass filter is physically realizable, the resulting leakage is one-way, i.e. leakage is always to later points in time and never to earlier points in time [Smith96 p. 741].

This example illustrates, indeed, how signal-domain leakage reduces (or eliminates) frequency-domain aliasing. Similarly, Fig. 5.19 illustrates how aliasing (foldover) in the signal domain reduces (or eliminates) leakage in the spectral domain. As we can see, Remark 6.10 sheds new light on Table 6.1 and on its interpretation.

Note, however, that allowing signal-domain aliasing as a remedy to frequency-domain leakage, as shown in Fig. 5.19, does not work in the case of periodic signals $g(x)$. This is quite logical, since a poor in-between location of an impulse in the DFT spectrum cannot be healed by simply allowing folding-over of the periodic function in the signal domain. For example, if $g(x) = \cos(2\pi f x)$, then its signal-domain folding-over simply gives in the signal domain a sum of cosines having the same frequency f but different phases. But this will only influence in the DFT spectrum the height of the resulting impulses (and their complex-valued phase, if $g(x)$ is not symmetric), but not their locations. Each impulse will therefore remain located at the same point in-between DFT output array elements, and hence it will continue suffering from leakage. For this reason, the term "leakage" as used in Remark 6.10 above does not include the *periodic* variant of leakage, where the function $g(x)$ or $G(u)$ is purely periodic but it is not truncated precisely on an integer multiple of its period (like in Figs. 6.3(b),(c) or 6.11(b),(c), respectively). Nevertheless, if the function is not purely periodic, its folding-over may still be used to reduce leakage in the other domain [Smith96 pp. 744–745].

Finally, another related point should be mentioned in this context. Consider row (a) of Fig. 6.5. It is obvious that in order to get, back in the signal domain, the same discrete sequence as the one shown in the left-hand column, we must provide as input to the IDFT the *leaked* signal shown in the right-hand column by dashed lines. Failing to do so will result in a distorted signal in the output array of the IDFT, that will not match the signal shown in the left-hand side of the figure. In this case, however, it would be more appropriate to use the term *input-domain leakage*, since the leakage effect in the input array of the IDFT occurs, in fact, in the frequency domain, and it cannot be considered as signal-domain leakage.

PROBLEMS

6-1. Can the DC impulse in a DFT spectrum leak out? Why? Verify your answer by checking the DFT figures in the present chapter and in other examples of your own.

6-2. Is it possible that within the same impulse comb in a 1D DFT spectrum some of the impulses be leaked out and some other impulses not? *Hint*: What happens in a 1D comb when $q = 4$? when $q = 3.5$? and when $q = 3.2$?

6-3. Part (a) of Remark 6.1 says that when the given function $g(x)$ is periodic, and the range length R of the truncation window is an exact integer multiple of the period of $g(x)$, no leakage occurs. A student claims that Fig. 6.2 contradicts this result, since in this figure the cosine function has been truncated at exactly an integer multiple of the period, and yet, the impulses in the right-hand part of row (c) are clearly smeared and leaked-out (compare with row (a) of the figure). What is your opinion?

6-4. *Rational vectors*. Fig. 6.12 shows a table with all the 2D rational vectors (rational combinations of integers m and n; see Fig. 6.10) for $m = 1,...,16$ and $n = 1,...,11$. Each cell in the table gives the ratio n/m, the corresponding rational angle $\theta = \arctan(n/m)$, and the value of $q = \sqrt{m^2+n^2}$. What is the significance of this table? *Hint*: This table may come in handy when you wish to design a 2D discrete periodic example with rational q-vectors, like Fig. C.9(b). Note that in Fig. C.9(b) we have used $m = 7$, $n = 2$ so that the rotation angle of our 2D cosine function was $\theta = \arctan(2/7) = 15.9454°$ and its frequency was, by Eq. (6.12), $f = \sqrt{m^2+n^2} / R = 7.2801 / 16 = 0.455$ Hz. This table only shows the possible rational vectors for small values of m and n; but it can be extended to any desired values of m and n.

6-5. Is it possible to find a rational vector whose angle is exactly 15°? 30°? 45°? Explain.

6-6. Suppose you need to plot for a new publication a 2D cosine function with frequency $f = 1$ and rotation angle of $\theta = 15°$, along with its spectrum. In order to obtain using your 2D DFT a clear spectral representation of the two impulses without awkward leakage effects, you agree to use a close rational approximation of the specified parameters $f = 1$ and $\theta = 15°$. Suppose that you wish to plot the function within the range $-8...8$ along both axes, so that $R = 16$. This means that in terms of array elements (pixels) you have $q = f / \Delta u = fR = 16$. Which combination of m and n would you choose in the table? Explain.

6-7. What would our student from Problem 5-1 obtain in his DFT plot if he modified the parameters of the given line grating ($P = 1$, $\theta = 15°$) according to the results of the previous problem? *Hint*: See Fig. 6.13. Once the impulses in the DFT spectrum are sharp and no longer suffer from leakage, would you expect their heights to be the same as in the CFT? Explain.

6-8. If our student chooses in the table another close rational approximation, he will get again sharp impulses in the DFT spectrum, but their locations will be somewhat different than before. Will the heights of the impulses remain unchanged?

6-9. Several figures in this book have been designed using the table of Fig. 6.12, in order to avoid leakage artifacts in the DFT spectrum. Consider, for example, Fig. 8.7, in which all the impulses in the DFT spectrum are sharp and clean. Given that the array size used in this figure is $N = 64$ to both directions, can you find out the values of m and n in this case?

6-10. How can the table in Fig. 6.12 be extended to the 3D case?

6-11. *Identification of leakage in DFT spectra*. How can you tell if a certain frequency in the DFT spectrum is indeed a true spectral frequency of the given spectrum, or just a leakage artifact? Remark 5.3 in Chapter 5 provides a simple method that often allows to distinguish between true and aliased frequencies in the multidimensional case, based on the rotation theorem (or on the dilation theorem). Can you think of a similar method for the leakage artifact? Consider, for example, Fig. 6.9(a); how will it be modified if the original continuous-world function $g(x,y)$ is slightly rotated before sampling? Will the cross-like structure around each impulse be rotated, too, like any true spectral element of $G(u,v)$?

6-12. Having obtained the DFT spectrum of a single rotated b/w line grating, our student now wishes to study the spectrum of a superposition of two such gratings having slightly different rotation angles.

(a) What is the CFT of such a grating superposition? *Hint*: See, for example, [Amidror09 Secs. 2.6, 6.5].

(b) What would you expect to see in the DFT spectrum of such a grating superposition? In particular, is it possible to avoid leaked-out impulses?

6-13. *Limitations of the DFT*. Based on his experience so far, our student comes to the conclusion that after all, for his specific needs, the use of DFT is not the best option: Due to the various artifacts inherent to the DFT (such as aliasing and leakage), impulses in the DFT spectrum do not always represent true spectral entities in the continuous-world spectrum; and even when they do, their values (heights) do not necessarily give a reliable estimation of their true values in the CFT. Even after applying all the required adjustments as explained in Chapters 3, 4 and Appendix A, a perfect match with the CFT is only possible in simple cases where the DFT artifacts can be avoided or easily mastered. In more complex situations, the DFT can offer a good *qualitative* approximation to the continuous-world spectrum, but not a fully reliable *quantitative* evaluation. What is your opinion on this conclusion?

6-14. Can you think of an alternative method, not based on DFT, that could help our student in finding the impulse locations and heights in the spectrum of a superposition of two (or even more) periodic gratings? *Hint*: Due to the 1-fold periodicity of each grating, the superposition of n gratings corresponds in the spectral domain to a convolution of the n respective impulse combs. The impulse locations and heights in such cases can be easily determined using the properties of impulse comb convolutions. This can be done, for example, by a short computer program that simulates the behaviour of the continuous-world comb convolution. See, for example, Problem 5-22 in [Amidror09 p. 148].

6-15. Can you think of other Fourier or spectral analysis situations where the DFT is not the optimal tool? *Hint*: In general, in most situations where the computation of the CFT is feasible — be it by directly using the definition formulas, by using Fourier-pair tables in the literature, by applying transformation and combination rules to already known Fourier pairs, or by any other means (such as the comb-convolution approach in the last problem) — the results you obtain will be more accurate than by using DFT. But there also exist various other circumstances where the DFT approach is not optimal. This is the case, for example, in the spectral analysis of sound, music, radar or video signals, etc., where the spectral content varies with time or with spatial location, either continuously or discretely. In such cases an approach based on joint-domain analysis (such as joint time/frequency analysis in the 1D case or joint spatial/spatial-frequency analysis in the 2D case) may be preferred [Gröchenig01; Poularikas96, Chapter 12]. In applications where compression or multiresolution are sought, the wavelet approach may be more judicious [Mallat91; Strang93; Rao98; Poularikas96, Chapter 10; Bovik05, Chapters 4.2 and 5.4]. Thus, for each application one should use the method

$\frac{n}{m}$	1	2	3	4	5	6	7	8	9	10	11
1	1.0 45.0° 1.4142	2.0 63.4349° 2.2361	3.0 71.5651° 3.1623	4.0 75.9638° 4.1231	5.0 78.6901° 5.099	6.0 80.5377° 6.0828	7.0 81.8699° 7.0711	8.0 82.875° 8.0623	9.0 83.6598° 9.0554	10.0 84.2894° 10.0499	11.0 84.8056° 11.0454
2	0.5 26.5651° 2.2361	1.0 45.0° 2.8284	1.5 56.3099° 3.6056	2.0 63.4349° 4.4721	2.5 68.1986° 5.3852	3.0 71.5651° 6.3246	3.5 74.0546° 7.2801	4.0 75.9638° 8.2462	4.5 77.4712° 9.2195	5.0 78.6901° 10.198	5.5 79.6952° 11.1803
3	0.3333 18.4349° 3.1623	0.6667 33.6901° 3.6056	1.0 45.0° 4.2426	1.3333 53.1301° 5.0	1.6667 59.0362° 5.831	2.0 63.4349° 6.7082	2.3333 66.8014° 7.6158	2.6667 69.444° 8.544	3.0 71.5651° 9.4868	3.3333 73.3008° 10.4403	3.6667 74.7449° 11.4018
4	0.25 14.0362° 4.1231	0.5 26.5651° 4.4721	0.75 36.8699° 5.0	1.0 45.0° 5.6569	1.25 51.3402° 6.4031	1.5 56.3099° 7.2111	1.75 60.2551° 8.0623	2.0 63.4349° 8.9443	2.25 66.0375° 9.8489	2.5 68.1986° 10.7703	2.75 70.0169° 11.7047
5	0.2 11.3099° 5.099	0.4 21.8014° 5.3852	0.6 30.9638° 5.831	0.8 38.6598° 6.4031	1.0 45.0° 7.0711	1.2 50.1944° 7.8102	1.4 54.4623° 8.6023	1.6 57.9946° 9.434	1.8 60.9454° 10.2956	2.0 63.4349° 11.1803	2.2 65.556° 12.083
6	0.1667 9.4623° 6.0828	0.3333 18.4349° 6.3246	0.5 26.5651° 6.7082	0.6667 33.6901° 7.2111	0.8333 39.8056° 7.8102	1.0 45.0° 8.4853	1.1667 49.3987° 9.2195	1.3333 53.1301° 10.0	1.5 56.3099° 10.8167	1.6667 59.0362° 11.6619	1.8333 61.3895° 12.53
7	0.1429 8.1301° 7.0711	0.2857 15.9454° 7.2801	0.4286 23.1986° 7.6158	0.5714 29.7449° 8.0623	0.7143 35.5377° 8.6023	0.8571 40.6013° 9.2195	1.0 45.0° 9.8995	1.1429 48.8141° 10.6301	1.2857 52.125° 11.4018	1.4286 55.008° 12.2066	1.5714 57.5288° 13.0384
8	0.125 7.125° 8.0623	0.25 14.0362° 8.2462	0.375 20.556° 8.544	0.5 26.5651° 8.9443	0.625 32.0054° 9.434	0.75 36.8699° 10.0	0.875 41.1859° 10.6301	1.0 45.0° 11.3137	1.125 48.3665° 12.0416	1.25 51.3402° 12.8062	1.375 53.9726° 13.6015
9	0.1111 6.3402° 9.0554	0.2222 12.5288° 9.2195	0.3333 18.4349° 9.4868	0.4444 23.9625° 9.8489	0.5556 29.0546° 10.2956	0.6667 33.6901° 10.8167	0.7778 37.875° 11.4018	0.8889 41.6335° 12.0416	1.0 45.0° 12.7279	1.1111 48.0128° 13.4536	1.2222 50.7106° 14.2127
10	0.1 5.7106° 10.0499	0.2 11.3099° 10.198	0.3 16.6992° 10.4403	0.4 21.8014° 10.7703	0.5 26.5651° 11.1803	0.6 30.9638° 11.6619	0.7 34.992° 12.2066	0.8 38.6598° 12.8062	0.9 41.9872° 13.4536	1.0 45.0° 14.1421	1.1 47.7263° 14.8661
11	0.0909 5.1944° 11.0454	0.1818 10.3048° 11.1803	0.2727 15.2551° 11.4018	0.3636 19.9831° 11.7047	0.4545 24.444° 12.083	0.5455 28.6105° 12.53	0.6364 32.4712° 13.0384	0.7273 36.0274° 13.6015	0.8182 39.2894° 14.2127	0.9091 42.2737° 14.8661	1.0 45.0° 15.5563
12	0.0833 4.7636° 12.0416	0.1667 9.4623° 12.1655	0.25 14.0362° 12.3693	0.3333 18.4349° 12.6491	0.4167 22.6199° 13.0	0.5 26.5651° 13.4164	0.5833 30.2564° 13.8924	0.6667 33.6901° 14.4222	0.75 36.8699° 15.0	0.8333 39.8056° 15.6205	0.9167 42.5104° 16.2788
13	0.0769 4.3987° 13.0384	0.1538 8.7462° 13.1529	0.2308 12.9946° 13.3417	0.3077 17.1027° 13.6015	0.3846 21.0375° 13.9284	0.4615 24.7751° 14.3178	0.5385 28.3008° 14.7648	0.6154 31.6075° 15.2643	0.6923 34.6952° 15.8114	0.7692 37.5686° 16.4012	0.8462 40.2364° 17.0294
14	0.0714 4.0856° 14.0357	0.1429 8.1301° 14.1421	0.2143 12.0948° 14.3178	0.2857 15.9454° 14.5602	0.3571 19.6538° 14.8661	0.4286 23.1986° 15.2315	0.5 26.5651° 15.6525	0.5714 29.7449° 16.1245	0.6429 32.7352° 16.6433	0.7143 35.5377° 17.2047	0.7857 38.1572° 17.8045
15	0.0667 3.8141° 15.0333	0.1333 7.5946° 15.1327	0.2 11.3099° 15.2971	0.2667 14.9314° 15.5242	0.3333 18.4349° 15.8114	0.4 21.8014° 16.1555	0.4667 25.0169° 16.5529	0.5333 28.0725° 17.0	0.6 30.9638° 17.4929	0.6667 33.6901° 18.0278	0.7333 36.2538° 18.6011
16	0.0625 3.5763° 16.0312	0.125 7.125° 16.1245	0.1875 10.6197° 16.2788	0.25 14.0362° 16.4924	0.3125 17.354° 16.7631	0.375 20.556° 17.088	0.4375 23.6294° 17.4642	0.5 26.5651° 17.8885	0.5625 29.3578° 18.3576	0.625 32.0054° 18.868	0.6875 34.5085° 19.4165

Figure 6.12: A table showing all the 2D rational combinations for $m = 1,...,16$ and $n = 1,...,11$. Each cell gives the ratio n/m, the corresponding rational angle $\theta = \arctan(n/m)$, and the value of $q = \sqrt{m^2 + n^2}$. The diagonal guiding lines show the angles of 15°, 30°, 45°, 60° and 75°. In order to compare with Fig. 6.10, it may be helpful to rotate this table by 90° counterclockwise, so that m increases to the right and n increases going upward.

which is best adapted to the problem. You may have a glimpse at the various alternative analysis methods in [Poularikas96] or in other handbooks on signal or image processing.

6-16. Is it possible for an impulse in a 2D or MD DFT spectrum to leak in one dimension only? *Hint*: Consider Fig. 6.9(c). How would the DFT spectrum look like if each of the two cosine impulses fell between two neighbouring pixels in the horizontal direction? and if it fell between two neighbouring pixels in the vertical direction? Part (a) of the figure shows what happens when the impulses fall between two pixels both horizontally and vertically.

6-17. As we have seen in Sec. 6.6.1, a rotated line grating such as in Fig. 5.14(a) can be made fully leakage-free by slightly modifying its angle and frequency to be rational. Suppose that we have obtained such a leakage-free line grating, whose impulses in the 2D DFT spectrum are therefore perfectly sharp (like in Fig. 8.7(a)). Let us take now only the central line of this grating, the one that passes through the origin (like in Fig. 8.8(a)). This line is a subset of the leakage-free line grating, and it clearly has a rational angle. Will it be leakage free? *Hint*: As we can clearly see, Fig. 8.8(a) is not leakage free, although the line has a rational angle; does this contradict our results in Sec. 6.6? Will the 2D DFT input array containing this line segment pave seamlessly the entire x,y plane? In which cases it will, and in which cases it will not? Consider in particular the rotation angle of the line.

6-18. A student looks at Figs. 8.6(b),(c) just after studying Sec. 3.6 in Chapter 3, and comes to the conclusion that the checkerboard-like nature of the spectra in these figures is due to a reorganization problem in the input array of the DFT (either missing reorganization or superfluous reorganization).

(a) What is your opinion?

(b) How will the cross-shaped leakage artifact of an impulse in a 2D DFT spectrum look like in case of missing or superfluous input array reorganization? Generate and plot a few such cases using your favorite DFT software package and verify your answer. *Hint*: Compare Fig. 4.8 with its non-reorganized counterpart, Fig. 3.11 (note that in Fig. 3.11 no output array reorganization was performed, either).

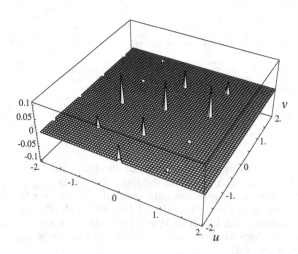

Figure 6.13: The DFT plot obtained by the student in Problem 6-7.

6-19. Student A applies DFT to a cosine function and obtains the result shown in Fig. 6.14. How do you explain this result? And how do you explain the scale along the vertical axis?

6-20. Student B applies DFT to the same cosine function as Student A, but obtains the result shown in Fig. 6.15. How do you explain this result? *Hint*: See Chapter 3.

6-21. Student C applies DFT to the same cosine function as student A, but obtains the result shown in Fig. 6.16. How do you explain this result? *Hint*: See Sec. 6.4.

6-22. What are the main adverse effects of leakage? *Hint*: Limitation of the frequency resolution; significant alteration of the spectrum amplitudes; a strong spectral peak may hide behind its leaked tails other weaker peaks, thus limiting the possibility of detecting spectral peaks; wrong graphical representation of the underlying continuous-world function (if the aim of using DFT is a visualization of the CFT).

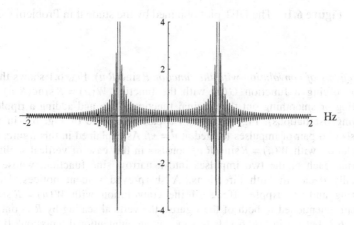

Figure 6.14: The DFT plot obtained by the student in Problem 6-19.

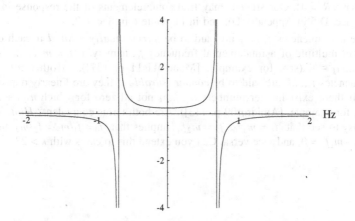

Figure 6.15: The DFT plot obtained by the student in Problem 6-20.

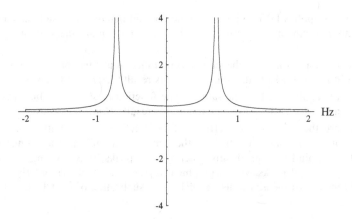

Figure 6.16: The DFT plot obtained by the student in Problem 6-21.

6-23. *The effects of convolution with the function* $R\,\mathrm{sinc}(R\,u)$. Fig. 6.1 shows that the effect of convolving a function $G(u)$ with the function $W(u) = R\,\mathrm{sinc}(R\,u)$ consists of smearing or smoothing out the original function $G(u)$, and adding a ripple effect (i.e. decaying oscillations). Fig. 6.2 shows a particular case of Fig. 6.1 in which $G(u)$ consists of a pair of impulses located at $u = \pm f$. As explained in this figure, the effect of convolution with $W(u) = R\,\mathrm{sinc}(R\,u)$ consists in this case of vertical scaling by R and smearing each of the two impulses into a narrow sinc function, whose oscillations gradually decay to both directions. A sharp-eyed student notices that while the smearing and the ripple effects of the convolution with $W(u) = R\,\mathrm{sinc}(R\,u)$ are explicitly mentioned in both of the figures, the vertical scaling by R is only mentioned in Fig. 6.2, but not in Fig. 6.1. Is this merely an unintentional omission? If not, how do you explain this difference? *Hint*: Looking at Fig. D.6(a) in Appendix D we see, indeed, that the convolution of a non-impulsive function $G(u)$ with $W(u) = R\,\mathrm{sinc}(R\,u)$ does not cause a vertical scaling of $G(u)$ by R even when $R \neq 1$ (note that in Fig. D.6(a) we have $R = 4$). Our student may find some elements of the response in Remark D.1 (see Sec. D.5 in Appendix D), and in Footnote 6 in Sec. 6.2.

6-24. The k frequencies $f_1,...,f_k$ are said to be *harmonically related* if each of them is an integer multiple of a fundamental frequency f_0, namely, if $f_1 = m_1 f_0, \ ... \ f_k = m_k f_0$ with $m_1,...,m_k \in \mathbb{Z}$ (see, for example, [Manolakis11 p. 137]). In other references, the k frequencies $f_1,...,f_k$ are said to be *commensurable* if they are linearly dependent over $\mathbb{Z}$, i.e. if there exist integer numbers $n_1,...,n_k$ not all zero for which $n_1 f_1 + ... + n_k f_k = 0$ (see, for example, [Amidror09 p. 113]). Are both terms equivalent? *Hint*: When $k = 2$ it is easy to see that $f_1 = m_1 f_0$, $f_2 = m_2 f_0$ implies that $f_0 = f_1/m_1 = f_2/m_2$ and hence that $m_2 f_1 - m_1 f_2 = 0$, and vice versa. Can you extend this to cases with $k > 2$?

Chapter 7

Issues related to resolution and range

7.1 Introduction

In this chapter we explain how to choose optimal values for the various parameters that are related to the resolution and the range of the DFT data, both in the signal and in the spectral domains. We discuss the significance of these parameters, and show the interconnections between them. Although the discussion is provided here for the 1D case, it can be readily extended to any multidimensional case, too.

7.2 The choice of the array size

It is clear that a larger array size N allows us to increase the length R of the range within which our given signal $g(x)$ is sampled, and/or to increase the sampling rate of the signal (i.e. to reduce the sampling interval Δx being used). This is clearly expressed by Eq. (4.2): $R = N\Delta x$. Both of these options are, of course, welcome; and moreover, as we have seen in Chapters 5 and 6, they can also contribute to the reduction of DFT artifacts such as aliasing and leakage.

Since the array size N is common to both input and output arrays of the DFT, an increase in the size of N also affects the DFT spectrum. And indeed, as we can see from Eqs. (7.1)–(7.4) below, a larger array size N allows for an increase in the frequency range length F of the DFT spectrum, and/or for an increase in the resolution of the DFT spectrum (i.e. a smaller frequency step Δu). These are, indeed, the frequency-domain counterparts of a reduction in the sampling interval Δx and of an increase in the sampling range length R, respectively (see Eqs. (7.3) and (7.4) below).

In general, if the array size N is not limited by extraneous computational, storage or other constraints, it should be taken sufficiently large to allow for the optimal values of R and Δx (or their spectral-domain counterparts, Δu and F). For example, if the input function $g(x)$ has a finite duration, it would be desirable that the entire span of the function be covered by R, in order to avoid truncation; and if $g(x)$ is rather jumpy, it would be advantageous to use a small sampling interval Δx. In any case, using a large size of N has no major shortcomings (except, possibly, in terms of processing time or storage), so it is almost always advisable to choose a sufficiently large value of N.

It should be remembered, however, that the choice of N may also have some consequences in terms of the graphic representation of the input and output data (on a computer display, a printed plot, etc.). For instance, in the 2D case, in order to reduce

185

jaggies along slanted lines in the signal domain (for example, when the input signal consists of a slightly rotated 0/1-valued line grating like in Fig. 5.14) it may be desirable to use an array of at least 512×512 elements. But on the other hand, if the displayed elements ("pixels") become too small, single-element details such as isolated impulses in the DFT spectrum (like in the spectral domain of Fig. 5.14) may become too small and therefore hardly visible in the 2D plot. If we want to make visible fine-detailed features such as isolated impulses in the spectrum or isolated pixels in a 2D image, we need to choose the array size accordingly, taking into account the physical dimensions of the final displayed plot.

It is useful to note at this point that in many cases we do not really need to plot the entire range used by the DFT, and we can, instead, only plot the most interesting part of the data. This is indeed a good practice, since it allows us to show in greater detail the most important part of the data and to omit all the rest; this can be done both in the input and in the output arrays of the DFT. (Note however that for didactic reasons we prefer not use this technique in the figures of this book, and unless mentioned otherwise we always show the entire ranges of the input and output data.)

A final point to remind here is that in many FFT algorithms N must be a power of 2, so if the actual data length N does not satisfy this requirement it may be necessary to increase N to the next closest power of 2. Various possible ways of filling the input array when increasing N, as well as their effects on the DFT results, are discussed in Appendix B.

7.3 The choice of the sampling interval

The smaller the sampling interval Δx, the larger the frequency range covered by the DFT (see Eq. (4.9) or Eq. (7.4) below). On the other hand, for any given value of N, the larger the sampling interval Δx, the higher the frequency resolution in the spectrum (meaning that Δu is smaller) and hence, the better the ability to distinguish between closely spaced peaks or impulses in the spectrum. This is a direct consequence of Eq. (4.4).

Note also that the choice of the sampling interval Δx may influence the aliasing in the resulting DFT spectrum (see Chapter 5). If Δx is chosen small enough so that the sampling frequency $f_s = 1/\Delta x$ (Eq. (4.6)) is at least twice the highest frequency present in the input signal, no aliasing will occur. But even when such a choice is not possible (e.g. if the input signal is not band limited), still the smaller Δx, the less significant is the resulting aliasing.

As we can see, we are confronted here with conflicting considerations: On the one hand, there are good reasons for choosing Δx as small as possible. The tradeoff is, however, that for a fixed N, smaller Δx values imply a higher value of Δu (i.e. a coarser spectral resolution; see Eq. (4.4)), as well as a smaller sampling range length R (see Eq. (4.2)), which may, in turn, increase the leakage effect. To solve this conflicting situation, N can be increased as Δx is decreased in order to keep R and Δu unchanged.

Finally, it is sometimes useful to make a first estimation of the optimal sampling interval empirically: If possible, one can perform a series of DFT computations with successively reduced sampling interval Δx, and choose the case which provides the best compromise [Brigham88 p. 172].

7.4 The choice of the sampling range

Another parameter that can be considered is the sampling range of the original function, i.e. the range $x_{min}...x_{max}$ within which we sample the original function with a sampling interval Δx, in order to fill the N-elements long input array of the DFT. If we denote the length of the sampling range by $R = x_{max} - x_{min}$, we get (see Eq. (4.2)):

$$R = N\Delta x$$

(Note that since we have N samples, one could have argued that there exist only $N-1$ sampling intervals between them, so that we actually have $R = (N-1)\Delta x$. But if we remember that the input array is considered by the DFT as cyclical, and admit that each of the N samples occupies along the continuous x axis an identical segment of width Δx, we see indeed that $R = N\Delta x$.)

This means that any two of the three parameters N, Δx and R can be freely chosen, fixing automatically the third one. And yet, even if N and Δx have already been chosen (thus fixing R as well), we still have the freedom of choosing which R-units long part of the original function, along the continuous x axis, will be sampled into our N-element long input array. For example, as we have seen in Sec. 4.2, if the original function is symmetric about the origin, it is natural to sample it within the almost symmetric range $[-R/2...R/2)$; but other choices can be also justified depending on the function at hand. In all cases, as we have seen in Chapter 2, the sampled part of the function will be considered by the DFT as cyclical, i.e. as an N-element period from a periodic sequence which repeatedly extends to both directions *ad infinitum*, even if the original underlying function was not periodic.[1]

It should be noted that if the chosen sampling range is not sufficiently wide, so that it truncates significant parts of the input function, a typical leakage artifact may occur in the DFT spectrum (such as the ripple effect shown in Fig. 6.5). Therefore, if the input signal $g(x)$ is known to die out at a certain distance from the origin, it is highly recommended to choose R such that most of the non-zero span of $g(x)$ be included, as far as possible without truncation. Similarly, if the input signal $g(x)$ is known to be periodic and its period is known, it is highly recommended to choose R to be equal to this period. As we have seen in Chapter 6, this guarantees that no leakage artifacts occur in the resulting DFT

---

[1] Note that choosing a different range having the same length R can be sometimes assimilated with a cyclic shift in the input data. According to the shift theorem (see point 6 in Sec. 2.4.4 or [Brigham88 pp. 107, 114]), this results in a change in the *phase* of the spectrum, while its *magnitude* (absolute value) remains unchanged (see also Secs. 3.3 and 8.3). However, this is no longer true if the new range implies a different truncation of the input function so that the input data is not purely shifted.

spectrum, so that all the impulses will be sharp. Finally, if the input signal $g(x)$ is not periodic and it does not die out outside a certain range, the best we can do is to choose a sampling range that is as far as possible representative of what $g(x)$ looks like along the x axis. Intuitively, our aim when choosing the sampling range should be to capture the most of the given signal, or at least the most significant parts of it.

Remark 7.1: In fact, if $g(x)$ is periodic, Remark 6.1 guarantees that leakage does not occur whenever the sampling range R is *any exact integer multiple* of the period. But in reality, there is usually no advantage in choosing a sampling range R that consists of several full periods of the input signal, since all of the periods are identical, and the additional periods will provide no new information about the signal. Instead, it would usually be better to decrease the sampling step Δx and get within the same N-element input array a more detailed sampling of a smaller range R consisting of a *single* period of $g(x)$. Sampling m full periods of $g(x)$ rather than one (within the same input array of length N) will cause a loss of input information, because the new repeatedly sampled periods will come to the detriment of a much denser sampling which could be obtained within a single period. In terms of the DFT spectral domain, sampling m full periods of $g(x)$ rather than one will result in packing $m-1$ zeroes after each element in the output array (see the repeat theorem in point 11a of Sec. 2.4.4). But since the output array length remains N we are only left in the output array with N/m rather than N significant (i.e. non zero-packed) spectral values. In other words, when sampling m full periods rather than one (within the same input array of length N) we lose spectral information, since the higher Fourier series coefficients will be pushed outside of the output array and lost.[2]

On the other hand, having several periods of $g(x)$ within the sampling range R may have some *graphical* advantages when plotting the input and output arrays of the DFT of $g(x)$. Consider, for example, Figure 7.1(a) which shows the input and the output arrays of the DFT in the case of a periodic triangular function $g(x)$. As we can see, the input array of the DFT contains exactly 4 sampled periods of $g(x)$, and indeed, in the output array each impulse (Fourier series coefficient) is followed by 3 zero elements. In the signal domain, having more than one period allows to clearly convey the periodic nature of $g(x)$; and in the spectral domain, the resulting zero packing allows to clearly see the impulsive nature of the resulting spectrum. Had we chosen to include only one sampled period in the same N-element input array, as shown in Fig. 7.1(c), the detail we would supply to the DFT about $g(x)$ would be 4 times finer, and hence we would get 4 times more Fourier series coefficients in the DFT output. But on the other hand, the DFT output would then seem to be compact and "continuous", and it would not as clearly convey graphically the impulsive nature of the spectrum (compare the DFT spectra in Figs. 7.1(a)–(c)). See also Sec. D.9 in Appendix D on the interpretation of the DFT results as an approximation to a CFT or as an approximation to a Fourier series decomposition, and the relations between

---

[2] Note, however, that if sampling m full periods rather than one is also accompanied by an m-fold extension of the input and output arrays from N to mN elements, so that we have no loss in terms of the sampling density of $g(x)$, then the resulting zero packing in the output array will not cause any loss of information in the DFT spectrum, either; see Sec. B.5 in Appendix B, and row (e) in Fig. B.1.

them. Appendix C provides further information on the number of periods in the sampling range and its interconnections with the DFT spectrum. ■

Let us now consider the influence of widening the range R on the output of the DFT. Doubling R (and Δx) while N remains fixed reduces by half the frequency range obtained in the output array of the DFT (see Eq. (4.9)) but also the frequency step Δu (see Eq. (4.4)). In other words, the N discrete frequencies included in the output array will be twice denser but they will only cover half of the original frequency range. But if required (and possible), N can be increased as R is increased in order to keep Δx unchanged.

Note, however, that doubling R has no influence on the amplitude of the DFT spectrum (if the standard amplitude scaling correction explained in Sec. 4.4 is adapted to the new Δx value). The only exceptions are cases involving an impulsive spectrum (where the impulse heights do depend on R, as explained in Appendix A); and cases where doubling the sampling interval Δx may introduce aliasing artifacts, as explained in Chapter 5.

As we can see, the choice of the sampling range is not always trivial, and it may involve several conflicting considerations. Furthermore, it may also have some unexpected consequences on the DFT output, as we have seen, for example, in Secs. 3.3 and 3.5.

Finally, it is worthwhile to note here, just as we did above in the case of Δx, that sometimes an estimation of the optimal sampling range can be also obtained empirically: If possible, one can perform a series of DFT computations with successively increased sampling range length R, and choose the case which provides the best compromise.

7.5 The choice of the frequency step and of the frequency range

The frequency-domain counterparts of the sampling interval Δx and the sampling range length R are, respectively, the frequency step (or frequency interval) Δu and the frequency range length F.

As we have seen in Eq. (4.4) above, the frequency step Δu is related to the sampling step (sampling interval) Δx by the relationship $\Delta u = \frac{1}{N\Delta x}$.

As for the frequency range length $F = f_{max} - f_{min}$ of the DFT spectrum, it is given by the frequency-domain counterpart of Eq. (4.2) (see [Briggs95 p. 21] or [Gonzalez87 p. 96]):

$$F = N\Delta u \qquad (7.1)$$

(Note that from Eq. (4.9) we have $F = f_{max} - f_{min} = 2\frac{1}{2\Delta x} - f_1 = N\Delta u - \Delta u = (N-1)\Delta u$, but using the same considerations as in the beginning of Sec. 7.4 we can indeed say that $F = N\Delta u$.)

The frequency range length F is also related to the sampling range length R of the input signal by (see, for example, [Briggs95 p. 22]):

$$RF = N \tag{7.2}$$

This can be obtained from Eq. (4.5) by substituting therein Δx and Δu as obtained from Eqs. (4.2) and (7.1); or alternatively, by multiplying Eqs. (4.2) and (7.1) and then using Eq. (4.5). Inserting this result into Eqs. (7.1) and (4.2), respectively, we also obtain:

$$\Delta u = 1/R \tag{7.3}$$

and: $$\Delta x = 1/F \tag{7.4}$$

This means that the *resolution* in each of the two domains, the signal domain and the spectral domain, is inversely determined by (i.e. reciprocal to) the *range length* of the other domain. Note also that Eq. (4.5) means that the resolutions of the two domains are inversely proportional. Similarly, Eq. (7.2) means that the range lengths of the two domains (R and F) are inversely proportional.

Thanks to this duality between the signal-domain and frequency-domain parameters it is clear that the frequency step Δu and the frequency range length F are fully determined by their signal-domain counterparts, the sampling interval Δx and the sampling range length R. However, in some cases it may be advantageous to consider things in terms of the frequency domain parameters, or even in terms of both domains. Table 7.1 summarizes the connections between the main signal-domain and frequency-domain parameters and shows how they influence each other. As an illustration, one may refer to Figs. A.4–A.7 in Appendix A.

For example, suppose we wish to refine (decrease) the frequency step Δu. As we can see in Table 7.1, this can be done in two different ways: either by keeping N unchanged and increasing R and Δx, ot by keeping Δx unchanged and increasing N and R (see also [Briggs95 pp. 55–56]). In fact, as we can see in the table, such a double choice exists for *any* paramerer change we may wish to perform — since any parameter change (increase or decrease) appears in the column of the respective parameter twice, in two different rows.

Figure 7.1: The effect of choosing a sampling range length R that includes several full periods of the given periodic function $g(x)$. This is illustrated here using the periodic triangular signal $g(x) = \text{triangwave}(2x)$, which has the period $P = 2$ (i.e. frequency $f = \frac{1}{2}$). Each row shows the sampled version of $g(x)$ as well as its resulting DFT (after all the required adjustments as discussed in Chapters 3, 4 and Appendix A). For the sake of comparison, the original continuous function $g(x)$ and its CFT $G(u)$ are also shown in the same plots, using continuous lines. As explained in Example 5.5, the CFT of $g(x)$, $G(u)$, is an impulse train with impulse interval $\Delta u = f = \frac{1}{2}$ whose amplitude is modulated by *half* the CFT of a single triangular period tri($2x$), namely by $\frac{1}{4}\text{sinc}^2(\frac{1}{2}u)$. (a) When the sampling range length is $R = 8$ (i.e. 4 periods of $g(x)$). (b) When the sampling range length is $R = 4$ (i.e. two periods of $g(x)$).

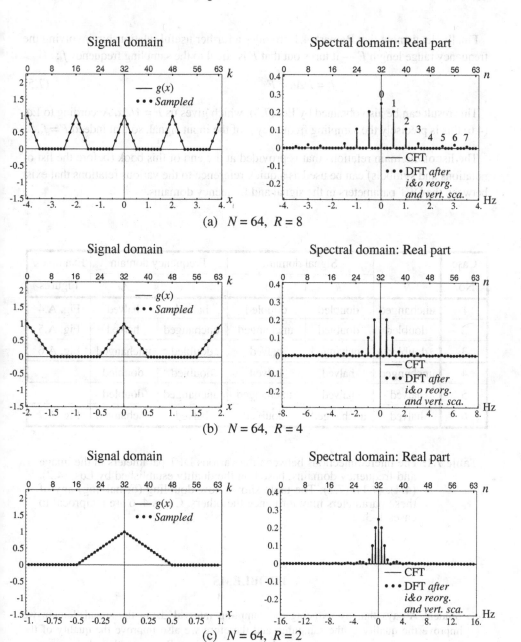

(a) $N = 64, \ R = 8$

(b) $N = 64, \ R = 4$

(c) $N = 64, \ R = 2$

(c) When the sampling range length is $R = 2$ (i.e. a single period of $g(x)$). Note that if row (c) is understood as the DFT of the non-periodic function $h(x) = \mathrm{tri}(2x)$, whose CFT is $H(u) = \frac{1}{2}\mathrm{sinc}^2(\frac{1}{2}u)$, then the impulse height correction of Appendix A (division by $R = 2$) should not be applied to the DFT output. This agrees, indeed, with Remark D.3 in Sec. D.9 of Appendix D.

Finally, let us note that Remark 4.1 provides a further useful relationship involving the frequency range length F — it turns out that F is equal to the sampling frequency f_s:

$$f_s = N\Delta u = F \tag{7.5}$$

This result can be also obtained by Eq. (7.4), which gives us $F = 1/\Delta x$. According to Eq. (4.6) this is precisely the sampling frequency f_s of the input signal, so that indeed $F = f_s$.

The list of the main relations that is provided at the end of this book (before the list of notations and symbols) can be used as a quick reference to the various relations that exist between the DFT parameters in the signal and frequency domains.

Case		Signal domain			Frequency domain		Examples/
No.	N	R	Δx	F	Δu		Figures
1	unchanged	doubled	doubled	halved	halved		Fig. A.4
2	doubled	doubled	unchanged	unchanged	halved		Fig. A.5
3	doubled	unchanged	halved	doubled	unchanged		Fig. A.6
4	unchanged	halved	halved	doubled	doubled		
5	halved	halved	unchanged	unchanged	doubled		
6	halved	unchanged	doubled	halved	unchanged		Fig. A.7

Table 7.1: The interconnections between the various DFT parameters in the image and frequency domains, based on the duality established by Eqs. (4.2), (4.4), (7.1), (7.2). The table shows how doubling (or halving) any of these parameters may influence the others. Cases 4–6 are reciprocal to cases 1–3.

PROBLEMS

7-1. Student A says that by reducing the sampling interval Δx in the signal domain we improve the quality of the sampling, and therefore we also improve the quality of the frequency resolution in the spectral domain. Student B argues that this reasoning contradicts Eq. (4.4), $\Delta x \Delta u = 1/N$, according to which improving Δx gives a worse Δu, and vice versa. Can you help them find a way out of this contradiction?

7-2. *The interconnections between the DFT parameters* Δx, Δu, R, F *and* N. Make a list of the various relations you can think of between the DFT parameters Δx, Δu, R, F and N. Note that some of these relations involve three parameters, like $R = N\Delta x$, but others involve only two, like $\Delta u = 1/R$. Explain the significance of each of the relations in your list. (See also the list of the main relations at the end of this book, before the list of notations and symbols.)

7-3. Consider Eq. (7.3), $\Delta u = 1/R$. How can it be that the DFT spectrum resolution Δu does not depend on N, but only on R? Is it possible that for all values of N we get the same spectrum resolution Δu if R remains unchanged? *Hint*: See row 3 in Table 7.1.

7-4. What are the 2D and MD counterparts of each of the relations between the DFT parameters Δx, Δu, R, F and N? Are there any new relations in the 2D or MD cases that have no equivalents in the 1D case?

7-5. Explain Remark 7.1 and Footnote 2 therein in light of the stretch and the repeat theorems (see points 11 and 11a in Sec. 2.4.4). What is the equivalent in the 2D case?

7-6. We have seen in Remark 7.1 that when sampling m full periods of an original periodic signal $g(x)$ rather than one (within the same input array of length N) we lose spectral information, since the higher Fourier series coefficients will be pushed outside of the output array and lost (see also Fig. 7.1). A student argues that because the output array of the DFT is cyclical, the Fourier series coefficients that are pushed outside the end of the output array are not really lost: Rather, they re-enter from the opposite end of the output array, just as we have seen in Chapter 5 (see, for example, Fig. 5.11). What is your opinion?

7-7. Looking at Fig. 7.1, Student A comes to the conclusion that given a periodic signal $g(x)$, by increasing the number of its periods in the input array of the DFT one obtains a better spectral resolution in the DFT output. For example, in row (b) of Fig. 7.1 the 64-element DFT spectrum shows the frequency range between –8...8 Hz, while in row (a) the same 64-element DFT spectrum shows in a higher resolution the smaller frequency range between –4...4 Hz. This also agrees with row 1 in Table 7.1. However, Student B claims that no gain of spectral information can be obtained by sampling a larger number of periods which are all strictly identical. Furthermore, since the sampling resolution in row (a) of Fig. 7.1 is lower than in row (b), no *gain* of spectral information occurs when passing from (b) to (a), but rather a *loss* of spectral information. What is your opinion?

7-8. Student A claims that Fig. 7.1 does not respect the repeat theorem as given in point 11a of Sec. 2.4.4, since in this figure the number of repetitions L does not influence the impulse heights in the spectral domain. Student B says that this could be explained if Fig. 7.1 were based on the alternative DFT definition, as in the column "Alt DFT" in Sec. 2.4.4, but student A reminds him that in the present book we are not using the alternative DFT definition. Is Fig. 7.1 flawed by an error?

7-9. Most rows in Table 7.1 provide in the last column an example (a figure in this book) which illustrates the case in question. Complete the table by providing figures of your own to illustrate the remaining rows of the table.

7-10. From row 4 of Table 7.1 we learn that by improving the sampling resolution in the signal domain (using a smaller sampling interval Δx) we *lose* in spectral resolution, since the resulting frequency interval Δu is larger. How can it be? See also Problem 7-1.

7-11. Suppose you wish to find the DFT spectrum of an unknown signal that you obtained from some astronomical measurements. How would you choose the various DFT parameters?

7-12. Suppose you suspect that there exists some hidden periodicity in the astronomical signal of the previous problem.

(a) How can you check if your assumption is correct?

(b) How can you determine the periodicity in question (or, equivalently, the corresponding frequency)?

(c) How can you optimize your DFT plot if you know the periodicity (or the frequency) in question?

(d) How can you optimize the DFT plot if there exist in the given signal several different periodicities?

7-13. *Twin peaks.* Suppose that you observe in the DFT output a peak having two very close summits. While this "M-shaped" peak may truly be part of the desired spectrum, one may also suspect that such a spectral peak reflects in fact the existence of two closely located individual impulses that are simply merged together due to insufficient spectral resolution (see, for example, Fig. 9-3 in [Smith03 p. 173]). How would you increase the spectral resolution in order to resolve this uncertainty?

7-14. *The interconnections between the sampling frequency f_s and the other DFT parameters.* Make a list of the various relations you can think of between the sampling frequency f_s and the DFT parameters Δx, Δu, R, F and N. Explain the significance of each of the relations in your list.

7-15. What are the 2D and MD counterparts of each of the relations between the sampling frequency f_s and the DFT parameters Δx, Δu, R, F and N? Are there any new relations in the 2D or MD cases that have no equivalents in the 1D case?

7-16. Is it possible to sample a 1D signal using a non-uniform sampling interval (for example, logarithmic sampling) and apply DFT to the resulting sampled signal? What continuous-world spectrum would you expect this DFT to match? *Hint*: On the extension of the 1D DFT to irregularly spaced data see, for example, [Press02, Sec. 13.8], [Bagchi99], [Marvasti01, Chapter 7], [Henson92] and [Dutt93].

7-17. Is it possible to sample a 2D signal using a circular or radial sampling pattern, and apply DFT to the resulting sampled signal? *Hint*: On the computation of 2D DFT on irregular grids, and more particularly in the case of polar coordinates, see [Briggs95 pp. 292–299]. On the extension of the sampling theorem to polar coordinates see [Stark79], [Stark82]. See also [Marvasti01], which is an excellent source of information on non-uniform sampling and non-uniform DFT in 1D and 2D settings, and includes many further references.

7-18. The CFT is said to *preserve a given coordinate transformation* if the application of that coordinate transformation to any CFT input affects the CFT output (the spectrum) by exactly the same coordinate transformation. For example, the 2D CFT *preserves rotations* since rotating the CFT input $g(x,y)$ by angle θ results in an equal rotation of the spectrum $G(u,v)$. The 2D CFT is therefore a rotation-preserving transform (see in the Glossary the difference between *rotation-preserving* and *rotation-invariant*).

(a) Does the CFT (1D, 2D or MD) generally preserve linear (or non-linear) coordinate transformations? *Hint*: Consider, for example, the dilation rule in Secs. 2.4.1–2.4.3. See also Sec. 10.3 in [Amidror09].

(b) What about the DFT? *Hint*: Unfortunately, even in the rare cases where the DFT *does* preserve a given coordinate transformation, the DFT usually loses this property. For example, although 2D CFT preserves *all* rotations, DFT only preserves rotations through integer multiples of 90° (see, for example, [Park09]).

(c) What happens under 2D CFT or 2D DFT to the polar coordinate transformation?

7-19. Suppose you are given a continuous 2D signal having a circular symmetry (such as the functions shown in Fig. 6.8(c) or Fig. 6.8(d)). A student suggests to apply to this signal a polar coordinate transformation, then sample the signal in the r,θ space using a standard orthogonal sampling grid along the r and θ axes, apply DFT to this 2D discrete signal, and finally apply an inverse coordinate transformation to obtain the spectral samples of the original 2D signal. What is your opinion?

Chapter 8

Miscellaneous issues

8.1 Introduction

In this chapter we group together some further questions and potential pitfalls that may await the unaware DFT users, but which do not fit into any of the previous chapters. In Sec. 8.2 we discuss the representation of discontinuities in the discrete world, in the input and output arrays of the DFT. In Sec. 8.3 we briefly review issues and artifacts that are related to the phase when using the polar (i.e. magnitude and phase) representation of the complex-valued spectrum. Then, in Sec. 8.4 we discuss symmetry related issues and the DFT artifacts that may result thereof. In Sec. 8.5 and 8.6 we discuss two artifacts that may occur due to sampling and reconstruction of continuous-world signals, namely, jaggies and sub-Nyquist artifacts, respectively. These two discrete-world artifacts are not specific to DFT, but they do occur in DFT applications, too, and we seize the opportunity offered by their explanation here to illustrate some questions related to DFT spectra and their continuous-world counterparts. Then, in Sec. 8.7 we review the main displaying considerations that should be taken into account in order to obtain a correct graphic representation of the DFT results, and finally in Sec. 8.8 we review very briefly some numeric precision considerations.

8.2 Representation of discontinuities

Consider the function $g(x) = \mathrm{sinc}(x)$, whose spectrum is the unit square pulse $G(u) = \mathrm{rect}(u)$. If we look carefully at the DFT of $g(x)$ (see Fig. 8.1), we note that at both of the pulse discontinuities (i.e. the transitions from 0 to 1 and then from 1 back to 0) the DFT output contains a point having the midvalue height $\frac{1}{2}$. It turns out that this is, indeed, the general rule in discontinuities: the value of a function at a discontinuity is simply taken to be the midvalue [Brigham88 p. 167], both in the signal and in the spectral domains.

The "midvalue rule" which defines the behaviour at discontinuities is fundamental and omnipresent in the Fourier theory. This rule already exists in the two main branches of the continuous-world Fourier theory, namely, in the CFT and in the Fourier series development of periodic functions. In the case of CFT this rule is required, when $g(x)$ contains discontinuities, in order for the inverse Fourier transform to hold [Brigham88 pp. 22, 167; Bracewell86 p. 7]. For this reason, if a continuous-world function $g(x)$ has a discontinuity at the point x, it is always recommended to explicitly define its value there to be equal to the midvalue $\frac{1}{2}[g(x^+) + g(x^-)]$, even if this subtlety seems to be superfluous as it does not really affect the evaluation of the Fourier integral in Eq. (2.1).[1] This midvalue rule is even more significant in the case of Fourier series, since it turns out that the Fourier series associated with a periodic function $p(x)$ converges at any discontinuity point x of $p(x)$ to the midvalue $\frac{1}{2}[p(x^+) + p(x^-)]$ (see [Bracewell86 p. 205; Briggs95 pp. 37–38]).

It is therefore not really surprising that a similar rule exists in the discrete world, too, in the case of DFT. And indeed, as already mentioned in references such as [Briggs95 pp. 38, 93–95] and [Brigham88 p. 167], the midvalue rule should always be observed when a function is sampled for input to the DFT. For example, when sampling the function $g(x) = \text{rect}(x)$ (see Fig. 8.2(a)), the two sampling points that fall on its discontinuities $x = -\frac{1}{2}$ and $x = \frac{1}{2}$ should be assigned the midvalue $\frac{1}{2}$, and only their neighbouring samples will actually take the values of 0 and 1, respectively.[2] The same is also true in the 2D and MD cases. For example, as shown in Fig. 3.6(a), the value of the 2D unit rectangle function $g(x,y) = \text{rect}(x)\text{rect}(y)$ (i.e. the 2D counterpart of $g(x) = \text{rect}(x)$) at the discontinuities along its four edges is neither 0 nor 1 but rather the midvalue $\frac{1}{2}$, except at the four corners where the value is $\frac{1}{2} \cdot \frac{1}{2} = \frac{1}{4}$ (since $g(x,y)$ is the product of the two perpendicular bars $g_1(x,y) = \text{rect}(x)$ and $g_2(x,y) = \text{rect}(y)$, each of which has the value $\frac{1}{2}$ along its two edges).

Furthermore, because the DFT input is considered to be cyclical (i.e. one N-element period from a periodic discrete signal that repeats itself to both directions *ad infinitum*), the first element and the last element of the input array are in fact neighbours. For this reason, the midvalue rule should also be applied to a discontinuity which occurs between the two ends of the input array. For example, when sampling the function $g(x) = e^{-x}\text{step}(x)$ (see Figs. 3.1, 3.3), the sampled value at the discontinuity point $x = 0$ should be defined as the midvalue $\frac{1}{2}[g(x^+) + g(x^-)] = \frac{1}{2}$, even if the discontinuity is located at the left end of the input array [Brigham88 p. 168]. This is also true in the 2D and MD cases. And due to the reversibility of the DFT, it is clear that the same midvalue rule applies both to the signal and to the spectral domains, whenever a discontinuity occurs.

Note, however, that the midvalue rule is not systematically followed in the DFT literature: Some references such as [smith03 pp. 212–217] do not apply it, while others use it selectively (for example, [Brigham88] applies this rule in the 1D case [Brigham88 pp. 122, 168], but not in the 2D case [Brigham88 p. 242]). And indeed, because this rule only concerns the value of few points of the sampled signal, it may be legitimately asked whether this subtlety is really so important, and what happens if we forget to apply it. Let us answer this question with the help of a concrete example.

Consider Fig. 8.2. Row (a) of this figure shows a sampled version of the unit square pulse function $g(x) = \text{rect}(x)$, which correctly takes into account the midvalue rule at both

---

[1] Note that in the context of CFT some references such as [Bracewell86 pp. 52–53] do not really insist on this midvalue rule, while others do, like [Brigham88 pp. 13–18] or [Gaskill78 pp. 42–43] for the 1D case and [Gaskill78 pp. 67, 71] for the 2D case.

[2] Note that this gradual transition at a discontinuity has nothing to do with the ramp-like graphic presentation of a sharp transition in the signal when drawing it as a connected line plot (Sec. 1.5.1). While the former is a true property of the discrete signal, the latter is just a graphical plotting artifact.

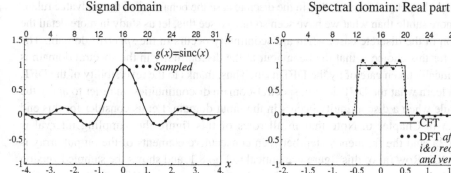

Figure 8.1: The continuous-world function $g(x) = \mathrm{sinc}(x)$ and its spectrum $G(u) = \mathrm{rect}(u)$, and their discrete counterparts in the input and output arrays of the DFT. Two interesting phenomena can be observed here in the DFT spectrum: (1) The pulse discontinuities in the DFT spectrum (both the transitions from 0 to 1 and from 1 to 0) contain an array element having the midvalue height $\frac{1}{2}$. (2) Because the sampling range length $R = 8$ in the signal domain is not sufficiently large, it truncates significant parts of the sinc function and leaves them outside the scope of the input array. Consequently, a typical ripple effect (similar to the Gibbs phenomenon) occurs about both sharp transitions of the square pulse in the spectral domain, due to leakage. This ripple effect is discussed in Sec. 5.7 (see Fig. 5.19) and in Example 6.2.

of the discontinuity points $x = \pm\frac{1}{2}$. As we can see in the spectral domain of row (a), the resulting DFT output (after the corrections of Chapters 3 and 4) is a discrete function that closely approximates the CFT $G(u) = \mathrm{sinc}(u)$, up to some aliasing error (which is unavoidable since our function $g(x)$ is not band limited).[3] Row (b) of the figure shows what happens to the DFT output when we do not apply the midvalue rule in the input array, and the discontinuity jumps directly from 0 to 1 and vice versa. As we can see, the resulting DFT spectrum in this case differs from the CFT $G(u)$ much more than in row (a): Note the significant error in the height of the resulting DFT and in the location of its zero crossing points. Row (c) of the figure shows another discrete variant of $g(x)$ having direct jumps from 0 to 1 and back, but this time, unlike in row (b), the discrete samples at the discontinuities have the value 0 rather than 1. Note that in this case, too, the resulting DFT spectrum significantly differs from the CFT $G(u) = \mathrm{sinc}(u)$. Another example illustrating the errors that are introduced to the DFT output when we do not apply the midvalue rule can be found in [Briggs95 pp. 93–95].

---

[3] In order to obtain in the DFT spectrum the exact, unaliased function $G(u) = \mathrm{sinc}(u)$, we have to take in the DFT input the *rippled* version of the unit pulse, which is precisely its band-limited (i.e. low-pass filtered) version. The DFT of this function is simply a truncated (and therefore unaliased) version of the true function $G(u) = \mathrm{sinc}(u)$. This can be illustrated by Fig. 8.1 if we reverse the roles of its signal and spectral domains.

It should be noted, however, that in the discrete case the behaviour of the midvalue rule is often more subtle than what we have seen so far. To see this, let us study in more detail the situation in the discrete case, when a discontinuity occurs in the *spectral* domain. The reason for this choice is that the behaviour at the discontinuity in the spectral domain is automatically taken care of by the DFT itself. Thus, thanks to the reversibility of the DFT, we can learn what the DFT does in spectral-domain discontinuities, in order to apply the same rule when a discontinuity occurs in the input domain. Let us consider for this end Fig. 6.5 in Chapter 6. Note that in all rows of this figure the sampling interval is $\Delta x = 0.25$, and the frequency step between consecutive elements of the output array is $\Delta u = 0.125$. Row (a) of this figure is identical to Fig. 8.1, and shows the sampled version of the function $g(x) = \text{sinc}(x)$, as well as the corresponding DFT output array (after applying the corrections of Chapters 3 and 4). This output array is the DFT approximation to the continuous-world spectrum of $g(x)$, the unit square pulse $G(u) = \text{rect}(u)$. And indeed, we see that both of the pulse discontinuities (i.e. the transitions from 0 to 1 and then from 1 back to 0) fall on an array point whose height is the midvalue $\frac{1}{2}$. More precisely, if the cutoff frequency of the square pulse, $\frac{1}{2}$Hz, falls exactly on the n-th element of the output array, then the n-th element itself will contain the midvalue $\frac{1}{2}$, and only its two neighbours, the elements $n-1$ and $n+1$, will contain the values 0 and 1.[4] Consider now row (c) of Fig. 6.5, which shows the function $g(x) = \text{sinc}(2ax)$ and its CFT $G(u) = \frac{1}{2a}\text{rect}(\frac{u}{2a})$ (a square pulse of height $\frac{1}{2a}$ extending between $-a$ and a), as well as their discrete counterparts, when $a = 18/32$. In this case the cutoff frequency is not exactly $a = \frac{1}{2} = 4\Delta u$ as in row (a), but rather $a = 18/32 = 4.5\Delta u$. This means that in this case the cutoff frequency does not fall exactly on an element of the output array but rather halfway between two consecutive output array elements. Similarly, in rows (b) and (d), where a equals $17/32 = 4.25\Delta u$ and $19/32 = 4.75\Delta u$, respectively, the cutoff frequency falls again

Figure 8.2: The continuous-world function $g(x) = \text{rect}(x)$ and its spectrum $G(u) = \text{sinc}(u)$, and their discrete counterparts in the input and output arrays of the DFT. (a) Using the correct discrete representation of the input function, where each of the pulse discontinuities (the transitions from 0 to 1 and from 1 to 0) contains an array element having the midvalue height $\frac{1}{2}$. Note that the DFT spectrum is slightly corrupted by aliasing, as we can see by comparing it with the true CFT, $G(u) = \text{sinc}(u)$, which is drawn by a continuous line. (b) Using a wrong discrete representation of the input function $g(x) = \text{rect}(x)$ where both of the pulse discontinuities take the value 1, so that the transitions are done directly from 0 to 1 and from 1 to 0. (c) Using another wrong discrete representation of the input function where both of the pulse discontinuities take the value 0. Note that the DFT spectra obtained in rows (b) and (c) are significantly more corrupted than

---

[4] Possibly with some discrepancies due to a typical ripple effect, similar to the Gibbs phenomenon, that occurs in both sharp transitions of the square pulse in the spectral domain, due to leakage. This ripple effect is discussed in Sec. 5.7 (see Fig. 5.19) and in Example 6.2.

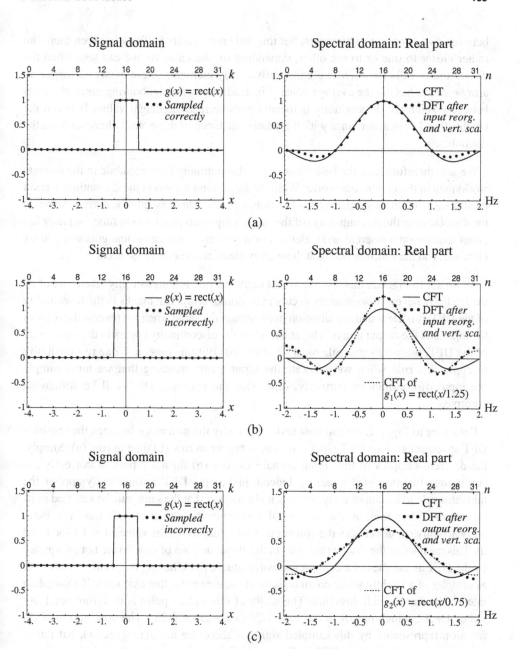

(a)

(b)

(c)

in row (a): their heights as well as their zero crossing points are incorrect. The array length (number of samples) being used in all of the rows is $N = 32$ and the sampling range length is $R = 8$, so that the sampling interval is $\Delta x = 8/32 = 0.25$. The frequency step in the DFT output is, according to Eq. (7.3), $\Delta u = 1/R = 0.125$.

between two consecutive elements, but this time not exactly halfway between them but rather closer to one or to the other, depending on the case. As we can see, when the discontinuity falls *between* two consecutive array elements, there exists no longer an array element having the average value $\frac{1}{2}$; instead, the two neighbouring array elements between which the discontinuity is located get intermediate height values between the heights 0 and 1, in accordance with the relative distance of these array elements from the discontinuity.

We see, therefore, that the behaviour near a discontinuity is more subtle in the discrete world than in the continuous world: While in the continuous world the discontinuity point itself is included in the domain of the function, and takes there exactly the average value, in the discrete case the discontinuity (of the underlying continuous-world function) may fall either exactly on a discrete array element or anywhere between two neighbouring array elements, and the function height at these array elements is determined accordingly.

As we see in Fig. 6.5, this discrete-world midvalue rule is automatically taken care of by the DFT whenever a discontinuity occurs in its output array. But thanks to the reversibility of the DFT, it is clear that the situation must remain the same when we reverse the roles of the input and the output arrays. Therefore, when the discontinuity occurs in the *input* array of the DFT, the very same rule must be observed. But this time, it is *our* responsibility to apply this rule when we generate the input array, meaning that we must sample the input discontinuities correctly; otherwise the resulting DFT will be somewhat corrupted.

Returning to Fig. 8.2, we can now understand why the difference between the resulting DFT spectrum and the CFT $G(u) = \text{sinc}(u)$ is bigger in row (b) than in row (a): Simply, the discrete samples in the signal domain of row (b) do not represent correctly the continuous function $g(x) = \text{rect}(x)$. Indeed, just as the DFT automatically applies the midvalue rule to its output array, so does it also expect that the same rule be satisfied in its input array. Therefore, in row (b), the DFT assumes that the discontinuities in its input array actually occur *between* the consecutive samples having the values 0 and 1 (or 1 and 0). This means that the discrete samples in the signal domain of row (b) are not interpreted by the DFT as samples of a continuous-world square pulse having the width 1, but rather as samples of a slightly wider continuous-world square pulse that extends half a sampling interval further to each direction. The width of this square pulse is therefore not 1 but rather $1 + \Delta x$, which means in our case 1.25 (since $\Delta x = 0.25$). The continuous-world function represented by this sampled signal is therefore not $g(x) = \text{rect}(x)$, but rather $g_1(x) = \text{rect}(x/1.25)$, whose CFT is $G_1(u) = 1.25\text{sinc}(1.25u)$ rather than $G(u) = \text{sinc}(u)$. And indeed, the DFT spectrum in row (b) corresponds well to the function $G_1(u)$, which is plotted in our figure using a dashed curve. Note that the small discrepancy between our DFT and $G_1(u)$ is basically due to aliasing, which remains unavoidable in the DFT spectrum because $g_1(x)$, too, is not band limited. Similarly, in row (c) the discrete samples in the signal domain are not interpreted as a continuous-world square pulse having the width 1, but rather as a slightly narrower continuous-world square pulse that extends half a sampling interval less to each direction. The width of this square pulse is therefore $1 - \Delta x = 0.75$. Consequently, the continuous-world function represented by this sampled signal is not $g(x) = \text{rect}(x)$, but rather $g_2(x) = \text{rect}(x/0.75)$, whose CFT is $G_2(u) = 0.75\text{sinc}(0.75u)$ rather than $G(u) = \text{sinc}(u)$. And indeed, the DFT spectrum in row (c) matches well (up to some minor error mainly due to aliasing) the function $G_2(u)$, which is plotted in our figure using a dashed curve.

Note, however, that this explanation of the situation in rows (b) and (c) of Fig. 8.2 is only a close approximation. In fact, the correct sampling of a square pulse of width $1 + \Delta x$ or $1 - \Delta x$, whose discontinuities fall between input array elements, should not assign to the two array elements adjacent to a discontinuity the values 0 and 1. Rather, these elements should get some intermediate values, depending on their relative distance from the true discontinuity point (see Fig. 6.5). This suggests that in rows (b) and (c) of Fig. 8.2 the small discrepancy between the DFT results and the continuous spectra $G_1(u)$ or $G_2(u)$ is not only due to aliasing, but also due to this rather negligible sampling inaccuracy.

In conclusion, the *discrete counterpart of the midvalue rule* can be formulated as follows: If a discontinuity of a continuous-world spectrum $G(u)$ falls exactly on the n-th discrete element of the DFT output array, then this element takes the midvalue height, and only the elements $n-1$ and $n+1$ really reflect the spectrum heights $G(u^-)$ and $G(u^+)$ to both sides of the jump. But if the discontinuity of the continuous-world spectrum falls *between* the discrete elements n and $n+1$, then these discrete elements take intermediate values between $G(u^-)$ and $G(u^+)$ depending on the relative distance of these two elements from the true discontinuity point. This rule is automatically applied by the DFT in the output array whenever a discontinuity occurs in the spectrum (as we have seen in Fig. 6.5). But when it is *our* role to sample a continuous-world function having one or more discontinuities as input to the DFT, the application of this midvalue rule during the sampling process is clearly less obvious. The best we can do in this case is to check where the discontinuity falls: If it falls very close to an input array element, then we should assign to this element the midvalue height. But if the discontinuity falls between the input array elements n and $n+1$, we should reasonably estimate (interpolate) the intermediate heights at these two input array elements in accordance with their respective distance from the discontinuity point between them. Brute-force sampling without taking the midvalue rule into account will assuredly infect the DFT results with some errors, which may be more or less significant depending on the case.

Remark 8.1: It is important to note, however, that the midvalue rule does not concern the case of impulses. If we carefully observe the DFT spectrum of a function such as $g(x) = \cos(2\pi f x)$ (see Fig. 4.2 or Fig. A.2 in Appendix A) we will see that its impulses jump directly up and down, with no midvalues to the left and to the right of the peak. This could be indeed expected, since the width of an impulse (that falls precisely on an array element) is a single array element; see Appendix A for more details on impulses in the continuous and discrete worlds. ∎

8.3 Phase related issues

Like any other complex-valued entity, the spectrum $G(u)$ of a function $g(x)$ can be represented either by its real-valued and its imaginary-valued parts:

$$G(u) = \text{Re}[G(u)] + i\,\text{Im}[G(u)] \tag{8.1}$$

or, using the polar notation, by its *magnitude* (also called *absolute value* or *modulus*) and its *phase* (also called *argument*):

$$G(u) = \text{Abs}[G(u)] \cdot e^{i\,\text{Arg}[G(u)]} \tag{8.2}$$

where:
$$\text{Abs}[G(u)] = \sqrt{\text{Re}[G(u)]^2 + \text{Im}[G(u)]^2} \tag{8.3}$$

$$\text{Arg}[G(u)] = \arctan\frac{\text{Im}[G(u)]}{\text{Re}[G(u)]} \tag{8.4}$$

(Note that $\text{Re}[G(u)]$, $\text{Im}[G(u)]$, $\text{Abs}[G(u)]$ and $\text{Arg}[G(u)]$ are all real-valued functions of the real variable u.) Both of these representations can be used, of course, in the continuous case as well as in its discrete counterpart.

Although the representation of the spectrum in terms of its real-valued part and its imaginary-valued part is rather straightforward, often the polar representation in terms of magnitude and phase is more convenient and easier to understand [Smith03 pp. 163–164]. For example, a shift of a in the signal-domain function $g(x)$ usually changes both $\text{Re}[G(u)]$ and $\text{Im}[G(u)]$, in a rather obscure way; but in the polar representation, thanks to the shift theorem, the magnitude-spectrum $\text{Abs}[G(u)]$ remains unchanged and only the phase-spectrum $\text{Arg}[G(u)]$ is affected by a linear increment of $\varphi(u) = 2\pi ua$ [Smith03 p. 188–190], [Smith07 pp. 149–150].[5]

However, in spite of its usefulness, the graphic presentation of the phase spectrum suffers from several shortcomings. For example, we can mention its 2π ambiguity, and its high numeric instability in zones where the magnitude is very small, which causes violent but meaningless oscillations. These and other pitfalls related to the phase have already been explained in detail in the literature, and the interested readers are encouraged to consult references such as [Smith03 pp. 164–168] or [Smith07 pp. 180–181].

Finally, let us mention here one more point that is also related to the phase of the DFT spectrum. As we have seen in Chapter 2, there exist in the literature many variants of the CFT and DFT definitions, which may differ, among other things, in the sign of the exponential part. Using a DFT definition with reversed signs in the exponential part will result in a reversed list of values in the output [Wolfram96 p. 868]; more precisely, the positive and negative frequencies will be interchanged (since a sign reversal of the

---

[5] Note that the phase of the spectrum $G(u)$ is a straight line through the origin $u = 0$ *if and only if* the signal-domain function $g(x)$ is symmetric around some point x_0. If the point of symmetry is $x_0 = 0$ (i.e. the origin) the slope of the phase line is zero, and the phase of the spectrum is identically zero; if $x_0 > 0$ the slope of the phase line is negative, and if $x_0 < 0$ the slope is positive.

exponent can be assimilated with a sign reversal in the index n within the exponent, where due to the cyclic nature of the DFT we have $-n = N - n$ [Smith07 pp. 118–119]).

Note that if the original input signal is real-valued (which is most often the case for a real-world signal), then its spectrum is Hermitian, meaning that the real-valued part of the spectrum is even while its imaginary-valued part is odd. In this case, using a DFT definition with the opposite exponential sign convention will simply change the sign of the imaginary-valued part of the resulting spectrum, without affecting its real-valued part (this is a straightforward consequence of the output frequency reversal mentioned above, when the spectrum is Hermitian). And if we prefer to use the polar representation of the spectrum, the sign inversion in the exponent will only cause a sign inversion in the phase of the resulting spectrum, without affecting its magnitude (this simply results from Eqs. (8.3)–(8.4)). Thus, when the DFT implementation being used is based on a different exponential sign convention the user must remember to change the sign of the phase (or the imaginary part) of the resulting DFT, in addition to all the other required corrections according to Chapters 3 and 4. This is the case, for example, when the DFT algorithm we use is based on a DFT definition with a positive exponent sign in the direct transform (like in the Mathematica® software package [Wolfram96 p. 868]), while the corresponding continuous-world results being used have been obtained using a CFT definition with a negative exponent sign in the direct transform (as in [Bracewell86 p. 6]).

8.4 Symmetry related issues

The continuous Fourier transform has some very useful symmetry properties (see, for example, [Bracewell86 pp. 14–16]). In particular, if the given function $g(x)$ is even (i.e. symmetric about the origin: $g(-x) = g(x)$) then its spectrum $G(u)$ is also even, and if $g(x)$ is odd (antisymmetric about the origin: $g(-x) = -g(x)$) then $G(u)$ is odd, too. Furthermore, if $g(x)$ is real-valued (which is often the case for a real-world signal), then its spectrum is Hermitian (meaning that its real-valued part is even and its imaginary-valued part is odd, which is succinctly expressed by $G(-u) = G^*(u)$); and in particular:

(a) If $g(x)$ is real-valued and even, so is its spectrum $G(u)$;

(b) If $g(x)$ is real-valued and odd, its spectrum $G(u)$ is imaginary-valued and odd.

These useful symmetry rules are also preserved in the discrete world, when we consider the DFT (after the usual reorganizations) rather than the continuous Fourier transform; see, for example, [Bracewell86 p. 366] or [Smith07 pp. 142–148]. Interestingly, however, the situation in the discrete case is somewhat more subtle. We illustrate this by the following two examples:

As a first example, note that a symmetric rect function in the discrete world must have an odd number of 1-valued elements across its width, namely, the center of the rectangle (which corresponds to $x = 0$), and to each of its sides an equal number of neighbouring

1-valued elements (usually followed by an element with the midvalue $\frac{1}{2}$, as we have seen in Sec. 8.2). Any other possibility will result in a non purely-real DFT.[6]

As a second example, note that in the continuous world when $g(x)$ is real valued, $G(0)$ is real valued, too (since the Hermitian nature of $G(u)$ implies that $\mathrm{Im}[G(0)] = -\mathrm{Im}[G(0)]$, which means that $\mathrm{Im}[G(0)] = 0$). However, in the discrete world it turns out that when the input signal of the DFT is real valued, not only the output element corresponding to the zero frequency is real valued, but also the output element representing the highest (or lowest) frequency in the spectrum, which is located $N/2$ elements away from the zero-frequency element [Briggs95 pp. 77, 120]; see, for example, the leftmost spectral element in Figs. 3.1(d) and 3.3(f). This is explained in Problem 8-18. It follows therefore that if the DFT spectrum does not have a discontinuity on its border, its imaginary part must gradually go down to zero when approaching the spectrum borders, even if its continuous counterpart behaves differently (see, for example, Fig. 5.8(b), where this deviation of the DFT with respect to the CFT is explained by aliasing).

Let us now proceed to the multidimensional case. Once again, as we have done above in the 1D case, we will first consider the continuous world, and only then return to the discrete world and the DFT.

From the point of view of symmetry, the 2D and MD cases are much more interesting than the 1D case, since they have many more possible types of symmetry. To start with, a multidimensional function may have a different line-symmetry along each of its axes. For example, in the 2D case, the function $g(x,y) = y$ is even in the x direction (symmetric with respect to the y axis) and odd in the y direction (antisymmetric with respect to the x axis): $g(-x,y) = g(x,y)$, $g(x,-y) = -g(x,y)$; while the function $g(x,y) = e^y$ has only one symmetry (it is even in the x direction). However, in the multidimensional case there may also exist line-symmetries with respect to lines other than the main axes; and in addition, there may also

Figure 8.3: Symmetry properties of the multidimensional Fourier transform of a real-valued signal $g(\mathbf{x})$. This figure was generated by 2D DFT with $N = 64$. (a),(b) If $g(\mathbf{x})$ is real-valued and centrosymmetric, so is its spectrum $G(\mathbf{u})$. (c) If $g(\mathbf{x})$ is real-valued and anticentrosymmetric, its spectrum $G(\mathbf{u})$ is imaginary-valued and anticentrosymmetric. (d) If $g(\mathbf{x})$ is real-valued but has any other symmetry (or no symmetry at all), its spectrum $G(\mathbf{u})$ has a centrosymmetric real-valued part as well as an anticentrosymmetric imaginary-valued part (this is the multidimensional generalization of a Hermitian function). The functions used here are as follows (with $f = 1$): (a) $g(x,y) = \cos(2\pi f[x\cos\theta + y\sin\theta])$; (b) $g(x,y) = \cos(2\pi f\sqrt{x^2+y^2})$; (c) $g(x,y) = \sin(2\pi f[x^3 - y^3]/16)$; (d) $g(x,y) = \cos(2\pi f[y - 0.15x^2])$. The spectra of these 2D functions are explained in [Amidror09, Chapter 10].

[6] Interestingly, this is true independently of the array length N, be it an even or an odd number (see Problem 8-5).

| Signal domain | Spectral domain: Real part | Spectral domain: Imaginary part |

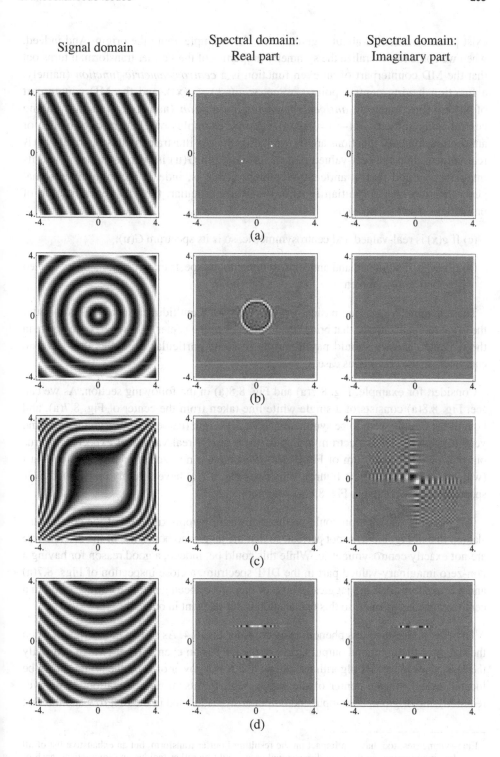

(a)

(b)

(c)

(d)

exist point-symmetries about a given point, for example about the origin. And indeed, when we come to generalize the symmetry properties of the Fourier transform, it turns out that the MD counterpart of an even function is a *centrosymmetric function* (namely, a function having central point-symmetry: $g(-\mathbf{x}) = g(\mathbf{x})$), and the MD counterpart of an odd function is an *anticentrosymmetric function* (namely, a function having central antisymmetry: $g(-\mathbf{x}) = -g(\mathbf{x})$).[7] A few examples of 2D centrosymmetric or anticentrosymmetric functions are shown in Fig. 8.3. As illustrated in this figure, if $g(\mathbf{x})$ is real valued, then the real-valued part of its spectrum $G(\mathbf{u})$ is centrosymmetric and its imaginary-valued part is anticentrosymmetric (this is, indeed, the multidimensional generalization of a Hermitian function); and in particular, the MD generalization of properties (a) and (b) above is:

(c) If $g(\mathbf{x})$ is real-valued and centrosymmetric, so is its spectrum $G(\mathbf{u})$;

(d) If $g(\mathbf{x})$ is real-valued and anticentrosymmetric, its spectrum $G(\mathbf{u})$ is imaginary-valued and anticentrosymmetric.

These symmetry rules are basically preserved by the multidimensional DFT, too (after the usual reorganizations, that bring the origin back to the center). However, once again, in the discrete case we should pay attention to some particular subtleties that have no equivalent in the continuous case.

Consider, for example, Fig. 8.7(a) and Fig. 8.8(a) in the following section. As we can see, Fig. 8.8(a) consists of a single white line taken from the center of Fig. 8.7(a), and both of the figures seem to be symmetric about the origin (namely, centrosymmetric). And yet, although the DFT spectrum of Fig. 8.7(a) is purely real-valued, as expected, it turns out that the DFT spectrum of Fig. 8.8(a) *does* have a non-zero imaginary-valued part (which is not shown in this figure). Why does Fig. 8.7(a) have a purely real-valued DFT spectrum, as expected, but Fig. 8.8(a) does not?

One possible explanation could be that our discrete approximation of the continuous, slanted line in Fig. 8.8(a) is not perfect, so that the jaggies to both sides of the discrete line are not exactly centrosymmetric. While this could be, indeed, a good reason for having a non-zero imaginary-valued part in the DFT spectrum, a close inspection of Figs. 8.7(a) and 8.8(a) shows that the jaggies in both of them have been generated correctly and in a centrosymmetric manner, so this explanation is not relevant in our case.

To better understand this phenomenon, consider Fig. 8.4. As shown in this figure, when the size N of the input and output arrays of the DFT is an even number (which is usually the case, since most FFT algorithms assume that N is a power of two), the origin cannot be located exactly in the center of the array. After the usual reorganizations (wherever required), the origin in both input and output arrays is located in the first row and the first

---

[7] Line-symmetries, too, have an impact on the resulting Fourier transform, but an exhaustive list of all possible combinations in the multidimensional case would be rather tedious and impractical; each of these cases can be elaborated individually whenever needed [Gaskill78 p. 313]. See also [Komrska79].

column of the first quadrant, i.e. in the position $N/2 + 1$ along each of the array's axes. This means that along each of the axes the origin has $N/2$ negative neighbours, but only $N/2 - 1$ positive neighbours. Therefore, a perfect symmetry (or antisymmetry) about the origin can be obtained if we keep the most negative row and the most negative column unused, i.e. filled with zeroes (see Fig. 8.4(e),(g)). But because of the periodic nature of the DFT, the most negative row or column can be also identified with the missing row or column in the positive end. Therefore, as shown in Fig. 8.4(c),(d), if the input array containing the centrosymmetric stucture is perfectly wraparound, so that the most negative row is indeed identical to the expected (but missing) $N+1$-th row and the most negative column is identical to the expected (but missing) $N+1$-th column, then the signal is still centrosymmetric and has a purely real-valued DFT spectrum even if the most negative row and column are not zero. (Note, however, that this does not work in the anti-centosymmetric case, in which the only solution for having a purely imaginary-valued DFT spectrum is to zero the most negative row and column of the input array).

Figure 8.4: Subtleties in the discrete counterpart of the multidimensional Fourier symmetry rules. This figure was generated by 2D DFT with $N = 32$. (a) A centrosymmetric discrete signal consisting of two 1-valued samples (white pixels) to the left of the origin and two 1-valued samples to its right, and the corresponding 2D DFT. Note that the sample at the origin of the signal domain has been assigned the value of -1 in order that its location be clearly indicated by a black pixel. (b) A similar centrosymmetric discrete signal, but this time with 15 1-valued samples to each side of the origin. Because $N = 32$, the 15-th sample to the right of the origin is already on the border, while to the left of the origin there still remains room for one additional sample. (c) If we set this additional sample, too, to 1, our signal contains 16 1-valued samples to the left of the origin but only 15 1-valued samples to its right. And yet, the resulting discrete signal is still considered as centrosymmetric, and indeed its DFT spectrum has a zero imaginary-valued part. The reason is that thanks to the periodic nature of the DFT, this leftmost sample corresponds simultaneously to the -16-th sample to the left of the origin and to the (missing) 16-th sample to the right of the origin. (d) A more complex 2D discrete signal that is still considered as centrosymmetric thanks to the periodic nature of the DFT. Note that its DFT spectrum has a zero imaginary-valued part. (e),(f) Signals similar to (b),(c) except that they are not purely horizontal. In this case, as we can see in (f), if we add the leftmost sample, too, the resulting signal is no longer centrosymmetric, and it clearly contains one more sample to the left. Unlike in (c), this additional sample cannot be identified here with a sample that exceeds to the right, since these two samples would not match in the vertical direction. Because the discrete signal in (f) is not centrosymmetric, its DFT spectrum does have a non-zero imaginary-valued part. (g),(h) Similar to (e),(f), but this time the problem occurs both in the horizontal and in the vertical directions. More information on the pattern in row (d) is provided in Problem 8-7.

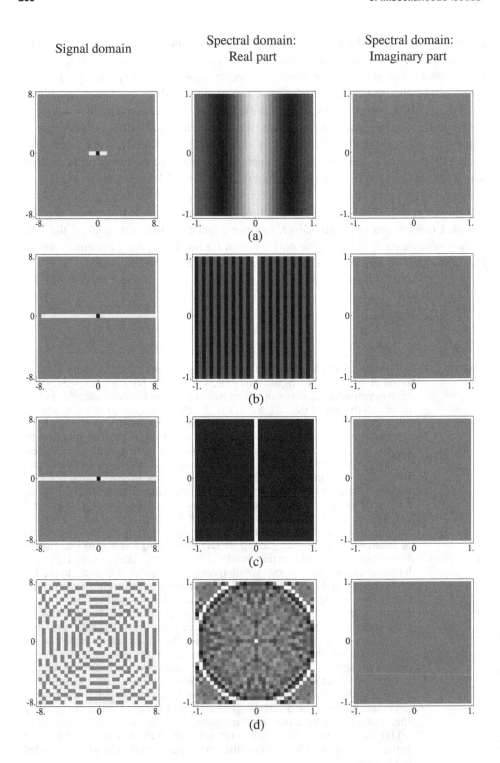

Signal domain
Spectral domain: Real part
Spectral domain: Imaginary part

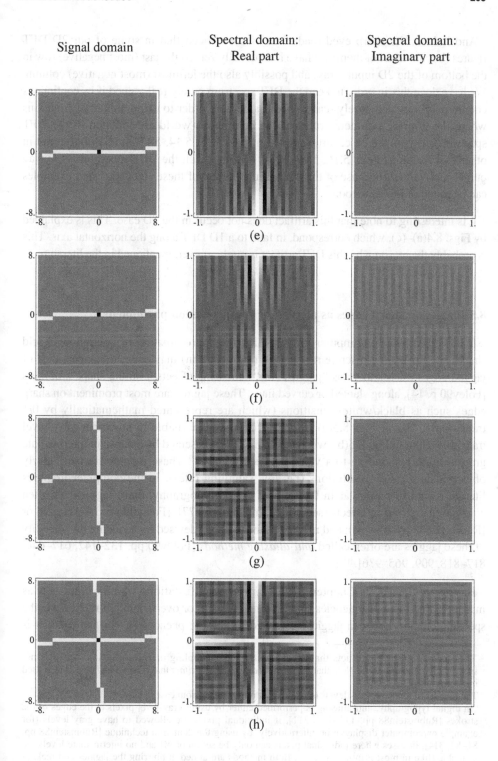

And indeed, the sharp-eyed reader may have noticed that in some of our 2D DFT figures (but not in all of them) we have intentionally zeroed the last (most negative) row in the bottom of the 2D input array, and possibly also the leftmost (most negative) column. We have done this in order to keep the DFT spectrum purely real-valued if its continuous counterpart is indeed purely real-valued (namely, in order to avoid abnormal situations where the Fourier symmetry rules of the continuous-world are violated in the DFT spectrum). This is the case, for example, in Figs. 5.13, 5.14, 5.15 and 8.8(a), but not in other cases such as Figs. 5.16, 5.17 and 8.7(a), in which the DFT spectrum is anyway purely real-valued (because of the wraparound nature of these signals). Other examples can be found in Fig. C.8, too.

It is interesting to note that this artifact does not occur in the 1D case. This is explained by Figs. 8.4(a)–(c), which correspond, in fact, to a 1D DFT along the horizontal axis. This is probably the reason why this DFT artifact is rarely, if ever, mentioned in the literature.

8.5 Jaggies on sharp edges as aliasing or reconstruction phenomena[8]

It is well known in computer graphics that the approximation of continuous-world shapes by pixels on a discrete raster grid (discretization) may cause the appearance of jagged edges (also known as "jaggies" or "staircasing effects") on the shape's borders [Foley90 p. 14], along slanted or curved lines. These jaggies are most prominent on sharp edges such as black/white transitions (which are represented mathematically by 0/1 transitions), like in Fig. 8.5(a). But they can be also visible in smoother gray-level transitions, like in Fig. 8.5(b), where each pixel is represented by a constant intermediate gray level corresponding to a value between 0 and 1.[9] These jaggies are particularly objectionable at low resolutions (i.e. low sampling rates). In the computer graphics literature, and in particular in the field of digital typography, these jaggies are often considered as aliasing effects (see, for example, [Crow77], [Foley90 pp. 14–15, 628] or [Rubinstein88 pp. 45–48]); and moreover, the methods devised for reducing the visibility of these jaggies are often called *anti-aliasing methods* [Foley90 pp. 132–142, 617–646, 817–818, 909, 965–976].[10]

Having reviewed in Chapter 5 the various manifestations of aliasing, such as masquerading lower frequencies (in the signal domain) or overlapping of replicas (in the spectral domain), how can jaggies be related to aliasing phenomena? In particular, it is

---

[8] This section concerns artifacts that occur due to signal sampling or reconstruction, and it is not specifically related to DFT. Note that the term *reconstruction artifact* that is explained here is also used in the following section.

[9] The number of available gray levels is limited, but it may differ from case to case.

[10] In digital typography, anti-aliasing is performed either by using gray-level pixels at the edges of the strokes [Rubinstein88 pp. 111–115, 311], if individual pixels are allowed to have gray levels (for example on computer displays); or, alternatively, by using the dentation technique [Rubinstein88 pp. 81–82, 314], in cases where individual pixels can only be set on or off and no intermediate levels are available (like in most printing devices). Both methods are aimed at blurring the jaggies and making them less visible when viewed from a reasonable distance.

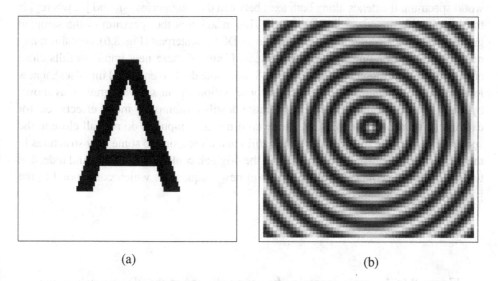

(a)   (b)

Figure 8.5: Jaggies in discrete-world images: (a) Along sharp edges (here:
slanted black lines forming the letter A). (b) Along smooth
gray-level shapes (here: a circular cosinusoidal wave).

clear that if jaggies are indeed aliasing phenomena, they must be represented in some way
or another in the spectral domain, too; but how exactly? In the present section we will try
to eludidate these and other similar questions. Note, however, that throughout this section
we assume that the jaggies in question are sufficiently big to be perceived by the naked
eye, and we do not discuss issues related to the human visual system, such as visual angle
resolution, viewing distance, modulation transfer functions (MTFs), etc. More details
about human vision and its modelization can be found in references such as [Wandell95],
[Daly92], [Pratt91, Sec. 2.4], [Boff86], [Cornsweet70], etc.

Let us consider, as a simple illustrative example, a periodic line grating that is slightly
rotated by angle θ (see Example 5.7). The spectrum of the original continuous-world
grating, before sampling, is an infinite comb whose impulses are located along a straight
line passing through the origin at the angle θ, where the impulse interval along this line
corresponds to the frequency of the grating. This spectrum is obviously not band limited.
After sampling the original continuous-world grating, as we have seen in Chapter 5, the
continuous-world spectrum of the resulting sampled grating consists of the original comb
plus infinitely many replicas thereof, that are centered about each impulse of a 2D impulse
nailbed (this nailbed being itself the Fourier transform of the sampling nailbed).[11] The
DFT spectrum, on its part (see Fig. 8.6), contains only the central part of this continuous-

---

[11] More precisely, the replicas are centered about the points (kf_S, lf_S) where f_S is the sampling frequency
in both directions and $k,l \in \mathbb{Z}$. See Fig. 5.12.

world spectrum: it extends along both axes between the frequencies $-\frac{1}{2}f_s$ and $\frac{1}{2}f_s$, where f_s is the sampling frequency (see Sec. 4.3).[12] This means that the spectrum of the sampled grating, both in the continuous-world and in its DFT counterpart (Fig. 8.6), contains many new impulses belonging to the new replicas. If any of these new impulses falls close enough to the spectrum origin, like in Fig. 8.6(c) (note the two encircled impulses), a new low-frequency structure (moiré effect) becomes visible in our sampled grating, as shown by the arrows in the image domain (for more details on sampling moiré effects see, for example, [Amidror09, Sec. 2.13]). But even if the new impulses do not fall close to the spectrum origin, like in Fig. 8.6(b), they still must correspond to some new structures in the sampled grating, which did not exist in the original, continuous grating. And indeed, it turns out that these new impulses represent new frequencies which correspond to the jaggedness of the sampled grating.

Figure 8.6: A simple example illustrating aliasing in the 2D case, where the spectrum consists of a 1D impulse-comb (see also Fig. 5.14). The original continuous-world function in the signal domain is a rotated line grating, whose spectrum is a rotated infinite impulse comb through the spectrum origin whose impulse interval corresponds to the frequency of the grating (see, for example, [Amidror09 pp. 23–25]). After sampling, the spectrum of the resulting sampled grating consists of the original comb plus infinitely many replicas of this comb, that are centered about the points (kf_s, lf_s), where f_s is the sampling frequency in both directions and $k,l \in \mathbb{Z}$. This means that the continuous-world spectrum of the sampled grating contains infinitely many new impulses belonging to the new replicas. But the DFT spectrum only includes the center of this continuous-world spectrum, between the frequencies $-\frac{1}{2}f_s$ and $\frac{1}{2}f_s$ to both directions (see also Problems 8-10 and 8-11). In each of the rows of this figure the left-hand column shows the sampled, discrete-world signal and the right-hand column shows its spectrum as obtained by DFT. The only difference between the three rows is in the rotation angle θ: (a) $\theta = 0°$; (b) $\theta = 14.036°$ (so that $\tan\theta = 1/4$); and (c) $\theta = 3°$. In case (c), due to the small rotation angle, impulses of the replicas fall very close to the spectrum origin (see the two impulses surrounded by circles, which are very slightly rotated with respect to the vertical axis). And indeed, these impulses correspond to a new low-frequency parasite structure (sampling moiré) that is clearly visible in the sampled line grating (see the arrows in the signal domain). The cross-like oscillations which surround impulses in the spectra of (b) and (c) are due to the leakage artifact (see Chapter 6). Note that both in the image and in the spectral domains mid-gray represents zero, black is negative and white is positive (see Footnote 5 in Sec. 1.5.2).

---

[12] As explained in Chapter 5, the impulses of the other replicas which fall within the range of the DFT spectrum can be also viewed as a cyclical continuation of the main comb, that folds-over each time it exceeds the border of the DFT spectrum and re-enters from the opposite side.

Signal domain   Spectral domain

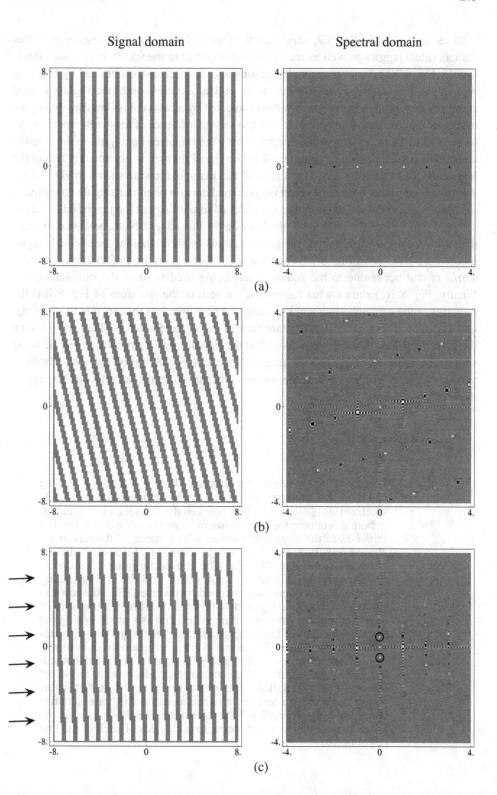

(a)

(b)

(c)

To see this, consider Fig. 8.7, which is intentionally drawn at a lower resolution so that the individual jaggies as well as the individual impulses in the spectrum become clearly visible. Fig. 8.7(a) shows the sampled grating and its spectrum as obtained by DFT. As we can see, the spectrum contains the main, original impulse comb (the slightly rotated comb passing through the origin, which is indicated by arrows) plus impulses belonging to its new replicas due to sampling. In order to see the influence of these new impulses, let us zero all of them leaving in the spectrum only the impulses belonging to the original comb. The corresponding structure back in the signal domain is obtained by taking the inverse DFT of this spectrum. The results of this manipulation are shown in Fig. 8.7(b). As we can see in this figure, the effect on the signal domain of eliminating the impulses of the new replicas (the aliased impulses) consists of smoothing out the jagged edges of the sampled grating (see the gray level pixels along the line edges). Note, however, that this does not yet completely eliminate the jaggies; as will be shown below, the residual jaggies that we can still see in Fig. 8.7(b) under a magnifying glass are in fact *reconstruction artifacts* that occur due to the square pixels being used to draw the sampled signal. Finally, Fig. 8.7(c) shows what happens if we zero in the spectrum of Fig. 8.7(a) the impulses of the original comb and keep only the new impulses that are due to sampling. Once again, the corresponding structure back in the signal domain is obtained by taking the inverse DFT of this spectrum. Note that the signal-domain structure in Fig. 8.7(c) is simply the difference between the jagged grating of Fig. 8.7(a) and its smoothed-out version of Fig. 8.7(b); this difference corresponds, indeed, to the net effect of the jaggies themselves on the line edges.

Figure 8.7: Jaggies along the edges of a sampled, slightly rotated line grating and their spectral representation. (a) The sampled grating and its spectrum, as obtained by DFT, between the frequencies $-\frac{1}{2}f_s$ and $\frac{1}{2}f_s$ in both directions; the two arrows indicate the main comb. (b) The impulses of the main comb alone, after zeroing all the remaining aliased impulses, and the corresponding signal-domain structure obtained by inverse DFT. (c) The aliased impulses alone, after zeroing the impulses of the main comb, and the corresponding signal-domain counterpart obtained by inverse DFT. As shown in (b), the effect on the signal domain of eliminating the impulses of the other replicas (the aliased impulses) consists of smoothing out the jagged edges of the sampled grating (see the gray level pixels along the line edges). The signal-domain structure in (c) is simply the difference between the jagged grating of (a) and its smoothed-out version of (b); this difference corresponds, indeed, to the net effect of the jaggies themselves on the line edges. Note that in this figure there are no leakage artifacts because this case satisfies condition (a) of Remark 6.7. Note also the overshoots (like in the Gibbs phenomenon) which occur to both sides of each of the line edges in (b).

Signal domain   Spectral domain

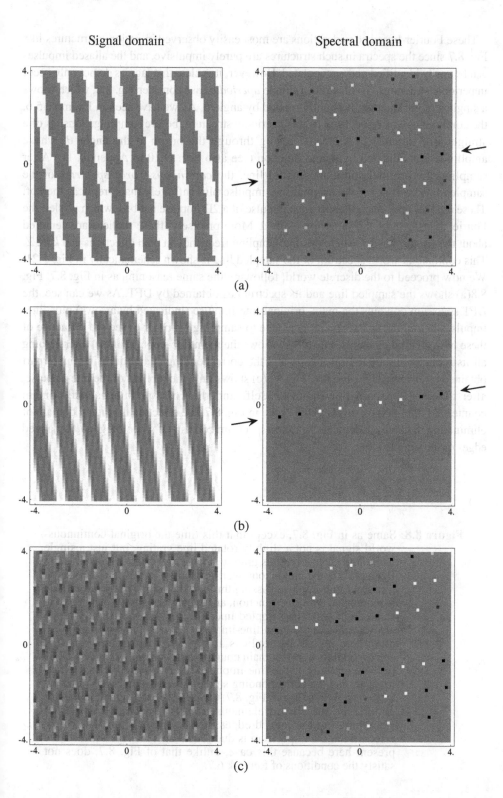

These Fourier-based considerations are most easily observed in periodic structures like Fig. 8.7 since the spectra in such structures are purely impulsive, and the aliased impulses can be easily identified and manipulated. However, it is clear that jaggies do not only occur in periodic structures. To illustrate a simple *aperiodic* case consider Fig. 8.8, which shows a single straight line that is slightly rotated by angle θ. As we have seen in Example 5.6, the continuous-world spectrum of this aperiodic structure, before sampling, consists of a straight line-impulse (a "blade") passing through the origin at the angle θ, whose amplitude is modulated by a sinc function (see also Sec. D.10 in Appendix D). After sampling the original continuous-world line, the *continuous-world spectrum* of the sampled line consists of the original line-impulse plus infinitely many replicas thereof. These replicas are centered about each impulse of a 2D impulse nailbed, which is islelf the Fourier transform of the sampling nailbed. More precisely, these replicas are centered about the points (kf_s, lf_s), where f_s is the sampling frequency in both directions and $k,l \in \mathbb{Z}$. This continuous-world spectrum of the sampled line is shown schematically in Fig. 5.21. We now proceed to the discrete world, following the same reasoning as in Fig. 8.7. Fig. 8.8(a) shows the sampled line and its spectrum as obtained by DFT. As we can see, the DFT spectrum contains the main, original line-impulse (indicated by arrows) plus line-impulses belonging to its new replicas due to sampling. In order to see the influence of these new spectral elements, Fig. 8.8(b) shows the main line-impulse alone, after zeroing all its replicas in the DFT spectrum, and the corresponding signal-domain counterpart obtained by inverse DFT. Finally, Fig. 8.8(c) shows only the replicas of the line-impulse, after zeroing the main line-impulse itself, and the corresponding signal-domain counterpart obtained by inverse DFT. Like in Fig. 8.7, the effect on the signal domain of eliminating the aliased elements from the spectrum consists of smoothing out the jagged edges of the sampled line.

Figure 8.8: Same as in Fig. 8.7, except that this time the original continuous-world signal is not a slightly rotated line grating, but just a single white line taken from this grating (the central one). As explained in Example 5.6, the continuous-world spectrum of this line consists of a straight line-impulse passing through the origin whose amplitude is modulated by a sinc function, and which is slightly rotated by the same angle. (a) The sampled line and its DFT spectrum; the two arrows indicate the main line-impulse. (b) The main line-impulse, after zeroing all its replicas in the DFT spectrum, and the corresponding signal-domain counterpart obtained by inverse DFT. (c) The replicas of the line-impulse, after zeroing the main line-impulse, and the corresponding signal-domain counterpart obtained by inverse DFT. Like in Fig. 8.7, the effect on the signal domain of eliminating the aliased elements from the spectrum consists of smoothing out the jagged edges of the sampled line. The ripple effect in the DFT spectrum is due to leakage; leakage artifacts are present here because this case, unlike that of Fig. 8.7, does not satisfy the conditions of Remark 6.7.

Signal domain   Spectral domain

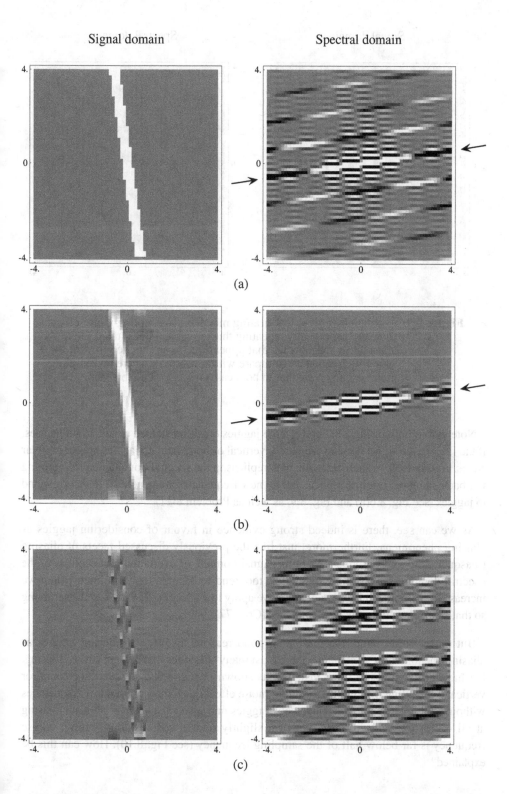

(a)

(b)

(c)

Signal domain   Spectral domain

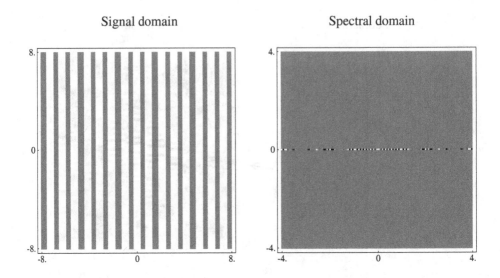

Figure 8.9: A sampled vertical line grating may have a sampling moiré effect
(note the periodically repeating thicker vertical lines) thanks to new
aliased low frequencies that appear due to an interaction with the
sampling frequency. Compare with Fig. 8.6(a), where no sampling
moiré occurs. Note that in both cases no jaggies are present.

Note that in both of Figs. 8.7 and 8.8 the jaggies occur on slanted edges. In both cases,
if the lines in the signal domain are purely vertical or horizontal no jaggies appear on their
edges. And indeed, in such cases the new replicas in the spectral domain due to sampling
fall back along the original replica, and no new frequencies are generated that correspond
to jaggies (see Fig. 8.6(a) and Fig. 8.9, as well as Problem 8-11).

As we can see, there is indeed strong evidence in favour of considering jaggies as
aliasing phenomena. Furthermore, just like the previously discussed facets of aliasing
(masquerading lower-frequencies in the signal domain, or overlapping of replicas in the
spectral domain; see Chapter 5), jaggies, too, tend to become less prominent when we
increase the sampling resolution or when we apply low-pass filtering prior to the sampling
so that sharp transitions become smoother [Crow77].

But on the other hand, there are also good reasons against considering jaggies as
aliasing phenomena. For example, there exist many structures that present strong aliasing,
but have no jaggies at all (for instance, as shown in Fig. 8.9, a sampled horizontal or
vertical line grating may have a sampling moiré effect due to new aliased low frequencies
without having jaggies). And conversely, jaggies may also exist when there is no aliasing
at all — for example in a 2D plot of a slightly rotated cosinusoidal grating, whose
frequency is far below half of the sampling frequency (see Fig. 8.10). How can this be
explained?

Signal domain   Spectral domain

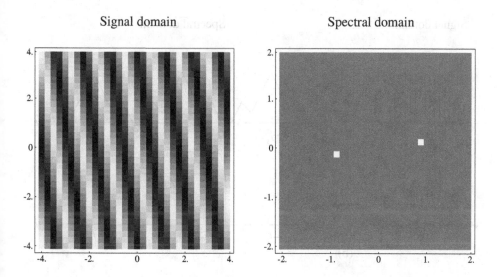

Figure 8.10: Low-resolution figure showing a slightly rotated cosine wave of
frequency $f = 0.88$ and its spectrum. Note that jaggies are visible
although the sampling frequency $f_s = 4$ is higher than $2f$, and no
aliasing occurs in the sampling process.

The key for understanding this question resides in the ambiguity of the term aliasing.
This term is used in the literature for two different effects (see, for example, pp. 343–344
in [Glassner95]): (1) the distortion that is produced by *poor sampling*, and which causes
the sampled signal to become indistinguishable from a lower frequency signal (an *alias*);
and (2) the distortion that is produced by *poor reconstruction*, which causes the signal
that is reconstructed from the samples to be different than the original continuous signal
(again, an *alias*).[13] To resolve this ambiguity, we will henceforth call the first, classical
effect *aliasing due to sampling* or *sampling aliasing*, while the second effect will be called
aliasing due to reconstruction or simply *reconstruction error* (other terms being
sometimes used in the literature are *prealiasing* and *postaliasing*, respectively [Mitchell88
p. 222]). Let us explain this in more detail.

While *Sampling* is the process that converts a continuous signal to a discrete one,
reconstruction is the process that recreates a continuous signal from its samples [Foley90
Sec. 14.10.5; Mitchell88].[14] According to the sampling theorem, all the information in the
original continuous signal is preserved in its sampled version if the sampling frequency is

---

[13] As shown below, both of these problems originate from failures in the correct application of the
sampling theorem: the failure to fulfil the required condition on the frequencies leads to sampling
aliasing; while the failure to approximate ideal reconstruction leads to reconstruction error.
[14] Note that theoretically a sampled signal consists of zero-width impulses (of varying heights), and not
of real-world "pixels" having square or circular shapes. It is precisely the reconstruction process that
brings back the "flesh" arround each of the sampled "bones".

Signal domain   Spectral domain

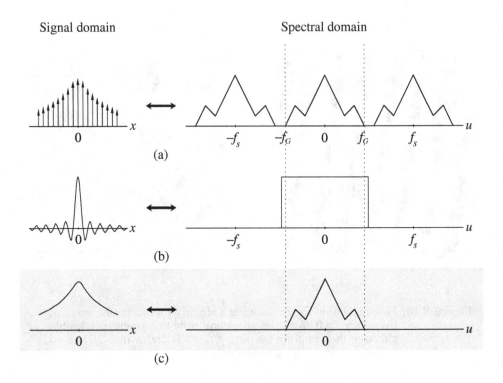

Figure 8.11: Continuation of Fig. 5.3, showing schematically the signal- and spectral-
domain representations of an *ideal* reconstruction process, that gives back
exactly the original continuous signal we had before sampling. (a) The
sampled signal, as in Fig. 5.3(c). (b) The ideal reconstruction function
(a narrow sinc function) according to the sampling theorem and its
spectrum, which is a rect function (a 1-valued pulse) that extends from
$-\frac{1}{2}f_s$ to $\frac{1}{2}f_s$. (c) The perfectly reconstructed signal (convolution of the
signals (a) and (b)) and its spectrum (product of the spectra (a) and (b)).
Note that the reconstructed signal in row (c) is indeed identical to the
original continuous signal shown in row (a) of Fig. 5.3.

at least twice that of the heighest frequency contained in the signal. Under this condition,
the theorem guarantees that the original continuous signal can be perfectly reconstructed
by multiplying the spectrum of the sampled signal with a rect function that cuts off all
frequencies beyond half of the sampling frequency, or equivalently, by convolving the
sampled signal with the Fourier transform of this rect function, i.e. with the corresponding
sinc function [Brigham88 p. 83]. This is shown in Figs. 5.3 and 8.11; note that the
explanation is easier to understand in the spectral domain (the right-hand side of the
figures). However, in practice, the reconstruction of a sampled signal is never done by
convolving its impulses with an infinite-range sinc function, as stipulated by the sampling

Signal domain   Spectral domain

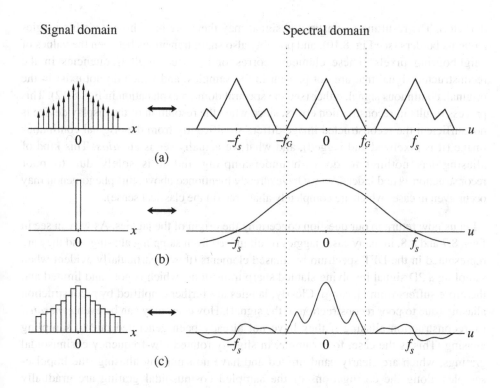

Figure 8.12: Continuation of Fig. 5.3, showing schematically the signal- and spectral-
domain representations of a *non-ideal* reconstruction process, that causes
reconstruction artifacts. (a) The sampled signal, as in Fig. 5.3(c). (b) A
non-ideal reconstruction function (a rect function representing a square
pixel) and its spectrum, which is a sinc function that extends ad infinitum.
(c) The reconstructed signal (convolution of the signals (a) and (b)) and
its spectrum (product of the spectra (a) and (b)). Row (c) is not identical
to row (a) in Fig. 5.3, and it contains reconstruction artifacts (note the
debris from the neighbouring replicas that give new high-frequency noise
in the resulting reconstructed signal).

theorem.[15] Instead, each impulse of the sampled signal is typically represented (i.e.
convolved) by a pulse function whose width equals the distance between two consecutive
samples [Foley90 p. 641]. For example, in the 2D case (like in computer displays or in
digital printing devices) each impulse of the sampled signal to be displayed is convolved
with a single pixel shape (which may be, depending on the device, a square dot, a circular

---

[15] Note that the condition required by the sampling theorem, that the signal being sampled is band
limited, implies that in the signal domain the signal has infinite duration [Zwillinger96 p. 535].
Similarly, the reconstruction of the continuous signal according to the sampling theorem is done by
convolving the sampled signal with an unrealizable, infinite sinc function. These assumptions yield a
mathematical model that is only an idealized approximation to the real world. The sampling theorem's
claim that perfect reconstruction is possible is mathematically correct for the model, but only an
approximation for real-world signals (see, for example, [Chu08 p. 182] or [Wikipedia12]).

dot, etc.). The resulting reconstructed signal may therefore have highly visible jaggies along its borders (see Fig. 8.10), and possibly also sharp transitions between the values of neighbouring pixels. These elements correspond to new high frequencies in the reconstructed signal that are not present in the samples, and which do not exist in the original, continuous signal, either (see the spectral domain explanation in Fig. 8.12). This process results in reconstruction errors, since when the resolution of the display device is not sufficient the reconstructed image differs significantly from the original continuous image (it is pixelized and jagged), and what we actually see is an *alias*. This kind of aliasing has nothing to do with undersampling and it is solely due to poor reconstruction;[16] and indeed, as we have already mentioned above, this phenomenon may occur even in cases which are completely alias-free (in the classical sense).

Let us now return to our question concerning the origin of the jaggies. As we can see in Figs. 8.7 and 8.8, in many cases jaggies result indeed from sampling aliasing, and they are represented in the DFT spectrum by aliased elements (this is particularly evident when sampling a 2D signal involving slanted sharp transitions, which is not band limited and therefore suffers from aliasing). Clearly, jaggies are further amplified by reconstruction aliasing (due to poor reconstruction of the signal). However, they can be even *generated* by reconstruction aliasing if they have not already been generated due to sampling aliasing. This is the case, for example, in slightly rotated low-frequency cosinusoidal gratings, which are clearly band limited and have no sampling aliasing (the impulse-samples along the corrugations of the sampled cosinusoidal grating are gradually attenuated in such a way that no jaggies are apparent). In cases like this, if reconstruction is correctly done, as stipulated by the sampling theorem, no jaggies should appear in the reconstructed signal; and if jaggies do appear — like in Fig. 8.10 — this is only due to reconstruction aliasing. And indeed, this observation led some sources to the conclusion that jaggies are not aliasing artifacts as often claimed in the literature, but rather reconstruction artifacts (see Problem 8-12).

As we can see, it follows from our discussion that jaggies have mixed origins: in some cases like in Figs. 8.7 and 8.8 they are due to sampling aliasing (although even then they may be further amplified by poor reconstruction); while in other cases, like in Fig. 8.10, they are a pure product of poor reconstruction. In the first case jaggies clearly manifest themselves in the DFT spectrum, but in the latter case they do not. This is, indeed, the origin of the confusion that prevails in the literature with regards to jaggies and aliasing.

Remark 8.2: It is important to stress that the new high frequencies due to reconstruction aliasing are not represented in the DFT spectrum of the sampled signal, since they only occur in a later stage, during the reconstruction of the continuous signal (whence the name *postaliasing*).[17] And yet, the issue of reconstruction *does* have an impact on our

---

[16] Note, in particular, that the new high frequencies that are generated by poor reconstruction are not corrupting the low frequency components, so no sampling aliasing is actually taking place.

[17] Of course, they *would* have been represented in the continuous-world spectrum of the reconstructed signal, had we cared to produce such a spectrum.

discussion on DFT and its artifacts: Each time we plot a discrete function or its DFT, be it in the 1D or MD case, we implicitly make a choice regarding the precise way the discrete values should be graphically plotted (i.e. reconstructed). For example, in the 1D case we may represent individual samples by isolated dots, by vertical spikes or bars emanating from the horizontal axis, or even as a curve that connects neighbouring points and somewhat approaches the familiar look of the original continuous function. In the 2D case, too, sampled functions may be represented in various ways. The most usual way (that we tacitly utilize in all our 2D figures here) uses adjacent square pixels, each having a distinct constant gray level, to represent the sampled values. These square pixels are clearly visible in cases where the array size N is low, but when N is sufficiently high (with respect to the figure's dimensions and the viewing distance) we may have the illusion of dealing with a continuous function (see, for example, Fig. 8.3). However, the various graphical techniques that may be used to plot the sampled functions and their DFT spectra only influence the visual presentation of the plots, and they do not add or remove frequencies from the DFT spectra themselves. A more detailed discussion on the graphic presentation of sampled signals can be found in Sec. 1.5. ■

Finally, let us return to the question with which we opened this section: What is the connection between the sampling-induced jaggies and the classical manifestations of aliasing (masquerading lower-frequencies in the signal domain, or spectral replications in the spectral domain)? Indeed, the most notorious signal-domain manifestation of aliasing consists of cases where new low-frequency structures become visible due to the sampling process, as shown in Fig. 5.1(b) for the 1D case and in Fig. 8.6(c) for the 2D case. These new low-frequency structures are simply sampling moirés. But the spectral-domain replicas due to the sampling process may also generate new frequencies that are much higher than this. These new high frequencies contribute to the microstructure details of the sampled signal and give it its typical jagged look, as shown for example in Figs. 8.6(b) or 8.7.

8.6 Sub-Nyquist artifacts[18]

We have seen in Chapter 5 that when an aliased frequency due to sampling falls close enough to the center of the signal's Fourier spectrum, a low-frequency sampling moiré effect may become visible in the signal domain, i.e. in the sampled signal.

Interestingly, however, a conspicuous low-frequency structure may also appear in the sampled signal when no aliasing occurs, so that no aliased frequencies fall close to the center of the signal's spectrum. It is even more surprising to realize that such artifacts occur even within the frequency range in which sampling is supposed to be safe by virtue of the sampling theorem. And to crown it all, these conspicuous low frequencies are not

---

[18] This section concerns artifacts that occur due to signal sampling or reconstruction, and it is not specifically related to DFT. Note that the term *reconstruction artifact* that is being used here was already explained in the previous section.

even represented in the spectral domain. This happens, for example, when the signal being sampled is a cosine function $g(x) = \cos(2\pi f x)$ whose frequency f is slightly below half of the sampling frequency ($\frac{1}{2}f_s$), i.e. close to the maximum frequency that can be represented in the DFT spectrum (see Eq. 4.10). But because this phenomenon occurs here when the cosine frequency f is *below* $\frac{1}{2}f_s$ (see Fig. 8.13(f)) it is clear by virtue of the sampling theorem that this artifact is *not* caused by aliasing. For this reason it is sometimes called a *sub-Nyquist distortion* [Williams00], where the Nyquist frequency means half of the sampling frequency.[19] This phenomenon has already been reported in several references [Mitchell88 pp. 222, 225; Foley90 p. 642; Gomes97 pp. 206–207; Jain98 pp. 126–127; Williams00; Fielding06], but it still remains poorly understood, sometimes just being dismissed as a moiré effect or as a *reconstruction artifact*, without further explanations.

To better understand this phenomenon, let us consider Fig. 8.13. The rows of this figure show a series of cosine functions $g(x) = \cos(2\pi f x)$ that only differ in their frequency f. All of these cosines have been sampled with the same sampling frequency f_s. When the frequency f of the continuous cosine precisely equals $\frac{1}{2}f_s$ (or in other words, when the cosine period equals two sampling intervals; see row (g) of the figure), the sampling is done exactly twice per period of the cosine; hence, the cosine is sampled on its successive maxima and minima, giving the sampled values 1, -1, 1, -1... But when the original continuous function $g(x) = \cos(2\pi f x)$ has a frequency f just slightly below $\frac{1}{2}f_s$, say $f = \frac{1}{2}f_s - \varepsilon$ (so that no aliasing occurs; see row (f) of the figure), sampling is done slightly more often than twice per period of the cosine. Hence, the successive sampling points fall at first close to the extrema of the cosine and have values close to 1, -1, 1, -1... (for example, around the origin), but farther away, due to the slight frequency difference, the successive sampling points fall near the zero-crossing points of the cosine and have values close to 0, 0, 0, 0... These different regions of the sampled signal repeat themselves periodically[20] at a much lower frequency than f or f_s, and indeed they look in the signal domain like a moiré effect, as shown in row (f) of the figure. Let us study this phenomenon in more detail, using a concrete example.

Example 8.1: A 1D sub-Nyquist artifact:

Consider the case shown in Fig. 8.13(f). Here, the sampling frequency is $f_s = 8$ Hz, so that the maximum frequency allowed by the sampling theorem without causing aliasing is $\frac{1}{2}f_s = 4$ Hz (note that this is also the maximum frequency representable in the DFT spectrum, as we have seen in Eq. (4.10)). The frequency of the continuous cosine function being sampled in this case is $f = 3.75$ Hz, so that $\varepsilon = \frac{1}{2}f_s - f = 0.25$ Hz. As we can see in the figure, the resulting beating effect has a period of 2, i.e. a frequency of $\frac{1}{2}$ Hz, which is precisely 2ε. Why is the low frequency of the beating effect *twice* the frequency difference $\varepsilon = \frac{1}{2}f_s - f$ rather than simply ε? To better understand this consider Fig. 8.14, which shows

---

[19] As already mentioned in Footnote 4 in Sec. 4.3, we usually prefer to avoid using the term "Nyquist frequency" because of its ambiguity. The name of the present artifact is an exception, but it should not lead to any confusions. See also Problem 8-14 about the name of this artifact.

[20] Note that the distance between successive zero-crossing points of the cosine is the same as the distance between successive extrema.

in the *continuous-world spectrum* what happens when we sample a continuous cosine function $g(x) = \cos(2\pi f x)$ having the frequency $f = \frac{1}{2}f_s - \varepsilon$ using the sampling frequency f_s. The reason we prefer to use here continuous-world spectra is that they are easier to understand than DFT spectra (such as in Fig. 8.13), since their frequency range is not limited, and they do not suffer from wraparound and foldover of higher frequencies as DFT spectra do.

Fig. 8.14(a) shows the continuous-world spectrum of our original, unsampled cosine function, which consists of an impulse pair at the frequencies $\frac{1}{2}f_s - \varepsilon$ and $-\frac{1}{2}f_s + \varepsilon$, while Fig. 8.14(b) shows the continuous-world spectrum of its sampled counterpart. As we already know from Chapter 5, if the CFT of $g(x)$ is $G(u)$, then the continuous-world spectrum of the sampled version of $g(x)$ consists of infinitely many replicas of the original spectrum $G(u)$, which are centered about all the integer multiples of the sampling frequency f_s (see Eq. (5.2) and Fig. 5.9). Fig. 8.14(b) shows only three of these replicas: the original one (which is identical, up to a certain amplitude scaling factor, to row (a) of the figure) plus its two nearest neighbours, that are centered about f_s and $-f_s$. As we can see in Fig. 8.14(b), these new replicas, that are introduced due to the sampling, add to the continuous-world spectrum a new impulse pair that is located very close to the impulse

Figure 8.13: A series of cosine functions $g(x) = \cos(2\pi f x)$ having gradually increasing frequency f, and their spectra (as obtained by DFT, after all the required adjustments as explained in Chapters 3, 4 and Appendix A). For the sake of clarity, consecutive samples (dots) are connected by straight line segments; the original, continuous cosines are drawn by thinner curves. The array length (number of samples) being used is $N = 64$, the sampling range length is $R = 8$, and therefore the sampling interval is $\Delta x = 8/64 = 0.125$ and the sampling frequency is $f_s = 8$ Hz. The only difference between the rows is in the cosine frequency f: (a) $f = 0.25$ Hz; (b) $f = 0.5$ Hz; (c) $f = 2$ Hz; (d) $f = 3$ Hz; (e) $f = 3.5$ Hz; (f) $f = 3.75$ Hz; (g) $f = 4$ Hz, namely $f = \frac{1}{2}f_s$; (h) $f = 4.25$ Hz; (i) $f = 7.5$ Hz; (j) $f = 7.75$ Hz. In row (g) the cosine frequency equals half of the sampling frequency, i.e. the maximum frequency that can be represented in the DFT spectrum. Note that in this case the positive frequency impulse at 4 Hz is already aliased and falls on top of the negative frequency impulse at –4 Hz. This happens because, as we have seen in Sec. 3.2, when N is even the DFT spectrum contains in its positive end one element less than in its negative end, so that the minimum frequency $-\frac{1}{2}f_s$ is indeed included in the DFT spectrum, but the maximum frequency $\frac{1}{2}f_s$ is already folded over. In rows (h)–(j) the cosine frequency f exceeds $\frac{1}{2}f_s$, meaning that aliasing occurs (note that the corresponding impulses exceed the boundaries of the DFT spectrum and re-enter from the opposite end). In particular, when the folded-over impulses are sufficiently close to the spectrum origin (see rows (i),(j)) a strong low-frequency sampling moiré becomes more prominent and visible than the original cosine function itself. Note, however, that the low-frequency beating effects which appear in rows (e),(f),(h) are not true moirés (there are no corresponding low-frequency impulses in the respective spectra).

Signal domain   Spectral domain

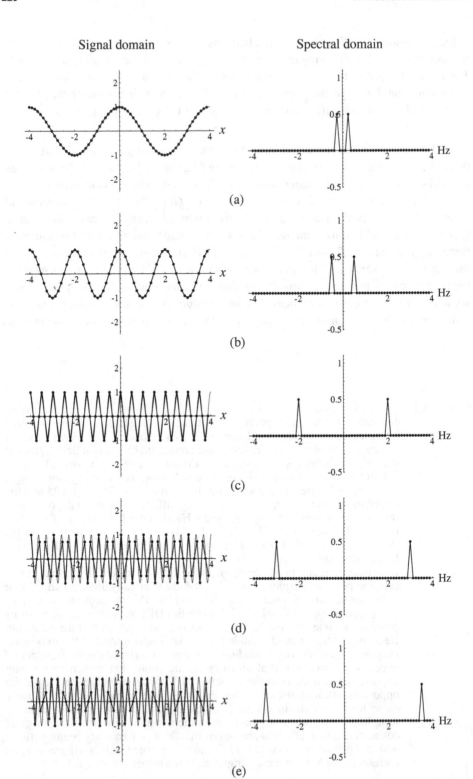

Signal domain   Spectral domain

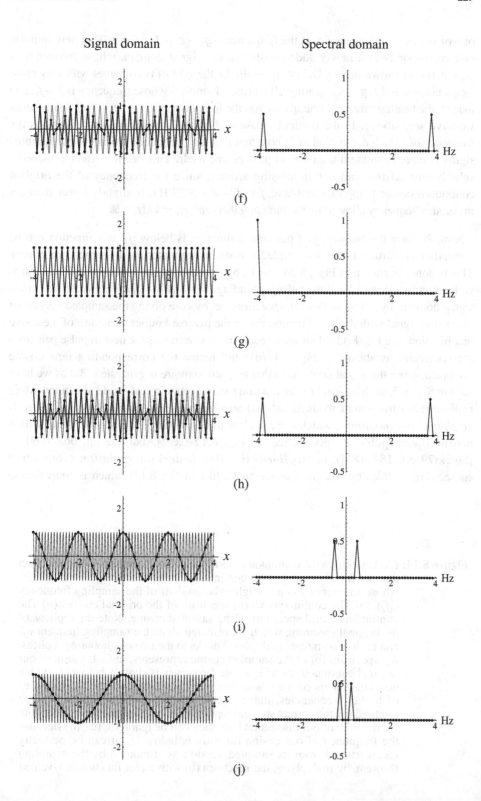

(f)

(g)

(h)

(i)

(j)

pair of our original cosine, i.e. at the frequencies $\frac{1}{2}f_s + \varepsilon$ and $-\frac{1}{2}f_s - \varepsilon$. This new impulse pair corresponds to a newly added cosine in the signal domain, whose frequency is $\frac{1}{2}f_s + \varepsilon$. But as shown in Sec. D.3 of Appendix D, the sum of two cosines with very close frequencies f_1 and f_2 gives a beating effect (pseudo moiré) whose frequency is $f_2 - f_1$; and indeed, the beating effect we obtain here has the frequency of $(\frac{1}{2}f_s + \varepsilon) - (\frac{1}{2}f_s - \varepsilon) = 2\varepsilon$, as we have seen above. In the particular case of Fig. 8.13(f) this beating effect has the frequency of $\frac{1}{2}$ Hz, i.e. a period of 2; note that this low-frequency artifact in our sampled signal is more prominent than the original cosine itself. This beating effect is indeed a sub-Nyquist artifact and not an aliasing artifact, since the frequency of the original continuous cosine being sampled here, $f = \frac{1}{2}f_s - \varepsilon = 3.75$ Hz, is slightly lower than the maximum frequency allowed by the sampling theorem, $\frac{1}{2}f_s = 4$ Hz. ∎

Now, because the frequency of our cosine function is below $\frac{1}{2}f_s$, this function can be perfectly reconstructed from its sampled version, as stipulated by the sampling theorem. This is done, as shown in Figs. 8.14(c)–(d), by multiplying the spectrum of Fig. 8.14(b) with a rect function (a 1-valued pulse) extending from $-\frac{1}{2}f_s$ to $\frac{1}{2}f_s$ (or equivalently, in the signal domain, by *sinc-function interpolation*, i.e. by convolving the sampled version of the cosine signal with the sinc function that is the inverse Fourier transform of the above rect function; see Fig. 8.11). This ideal reconstruction removes the new impulse pair from the spectrum, as shown in Fig. 8.14(d), and hence the corresponding new cosine disappears from the signal domain, so that no pseudo moiré is generated. But as we have seen in Sec. 8.5, such an ideal reconstruction rarely occurs in real-world situations. More realistic reconstruction methods include: (i) *zero-order interpolation* (also called *nearest-neighbour interpolation*), which is equivalent to convolving the sampled signal with a narrow square pulse or "pixel", like in Fig. 8.12 (see [Abdou82 pp. 668–669] or [Stucki79 pp. 184–185]); or (ii) *linear* (i.e. *first-order*) *interpolation* (connecting consecutive samples by straight line segments, like in Fig. 8.13), which is equivalent to

Figure 8.14: Explanation in the continuous-world spectrum of the sub-Nyquist artifact that occurs when sampling a continuous cosine function $g(x) = \cos(2\pi f x)$ whose frequency f is just slightly below half of the sampling frequency ($\frac{1}{2}f_s$). (a) The continuous-world spectrum of the original cosine. (b) The continuous-world spectrum of the sampled cosine. Note the replicas of the original spectrum, which are centered about the sampling frequency f_s and each of its integer multiples. Thanks to the two neighbouring replicas, the spectrum (b) of the sampled cosine represents, in fact, a sum of our original continuous cosine, whose frequency is slightly below $\frac{1}{2}f_s$, with a new continuous cosine whose frequency is slightly above $\frac{1}{2}f_s$. (Impulses of higher frequencies, that are also present in the spectrum (b), are required for the correct spectral representation of the sampled cosine, but for the sake of our discussion here they can be ignored). (c), (d) Because the frequency of our cosine function is below $\frac{1}{2}f_s$, it can be perfectly reconstructed from its sampled version as stipulated by the sampling theorem, by multiplying the spectrum (b) with a rect function (a 1-valued

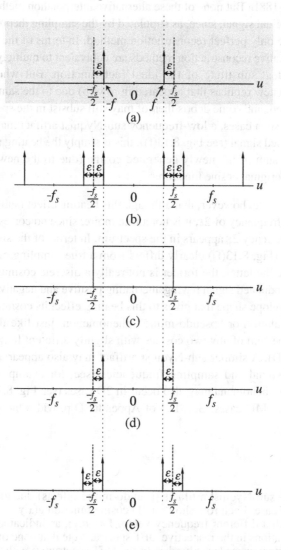

Spectral domain

pulse) extending from $-\frac{1}{2}f_s$ to $\frac{1}{2}f_s$. (e) When reconstructed by multiplying the spectrum (b) with a non-ideal substitute of the ideal rect function (c), debris of the new replicas that appeared in (b) due to the sampling may still subsist in the spectrum. The sub-Nyquist artifact that appears in the sampled signal (see Fig. 8.13(f)) is simply the beating modulation effect that occurs in the sum of the original cosine having the frequency $f = \frac{1}{2}f_s - \varepsilon$ and the new cosine having the very close frequency of $\frac{1}{2}f_s + \varepsilon$. As shown in (b), this beating effect is generated by the sampling operation; but it becomes actually visible due to the non-ideal reconstruction. Note that the low beating frequency itself is not present in the spectrum, meaning that it is not a true moiré effect but rather a pseudo-moiré effect.

convolving the sampled signal with a narrow triangular pulse (see [Abdou82 p. 669] or [Stucki79 pp. 186–188]). But none of these alternative interpolation methods can perfectly reconstruct the original signal, since, as stipulated by the sampling theorem, sinc-function interpolation is the only perfect reconstruction method. In terms of the spectral domain, both of these alternative reconstruction methods are equivalent to multiplying the spectrum (b) with a non-ideal substitute of the ideal rect function. But when doing such a multiplication, the new replicas that appeared in row (b) due to the sampling will not be completely removed, and some debris thereof may still subsist in the spectrum, as shown in Fig. 8,14(e). In such cases, a low-frequency sub-Nyquist artifact may indeed become visible in the sampled signal (see Fig. 8.13(f)); this is simply the beating modulation effect that occurs in the sum of the newly generated cosine (due to the new impulses in the spectrum) and the original cosine function.

It is important to note, however, that although this beating effect (sub-Nyquist artifact) has a visible, low frequency of 2ε, it is not a true moiré, since no corresponding impulse having the low frequency 2ε appears in the spectrum. In terms of the signal domain, too, our beating effect (Fig. 8.13(f)) clearly differs from a true sampling moiré effect (Figs. 8.13(i),(j)): Unlike the latter, the former is not really a discrete cosinusoidal signal, but rather a discrete modulated signal having alternating positive and negative values; and it is only its global envelope shape that gives to this beating effect its cosinusoidal look. This is, in fact, a modulation or "pseudo-moiré" phenomenon, just like the beating effect which occurs in the sum of any two cosines with slightly different frequencies (see Sec. D.3 in Appendix D). A similar sub-Nyquist artifact may also appear at other alias-free combinations of signal and sampling frequencies (see, for example, [Williams00]). Furthermore, such phenomena may also occur in 2D cases (see Fig. 8.15 or [Fielding06 pp. 132–134]) or in MD cases. Sec. D.3 of Appendix D provides the generalization of

Figure 8.15: 2D sub-Nyquist artifacts (pseudo-moiré effects) due to sampling in the 2D case. Each row shows a 2D cosine function $g(x,y) = \cos(2\pi[ux + vy])$ with a different frequency vector $\mathbf{f} = (u,v)$, as indicated by the impulse locations in the respective DFT spectra. Note that none of these frequency vectors exceeds the borders of the DFT spectrum (i.e. the frequency range $-\frac{1}{2}f_s \dots \frac{1}{2}f_s$ in each dimension), meaning that no aliasing occurs in any of the cases. And yet, as shown in rows (b)–(d), when the frequency vectors are just slightly below half of the sampling frequency (which is also the maximum frequency that can be represented in the DFT spectrum), a strong low-frequency beating effect may appear, which is even more prominent than the original sampled cosine itself. This beating effect has a much lower frequency (i.e. a much larger period) than the original cosine, and sometimes, like in row (c), even a different orientation. These artifacts are not true moirés (note that there are no corresponding low-frequency impulses in the respective spectra). The sampling frequency used in all of the rows is identical (4 Hz in each of the two directions); the array size is 64×64 and the sampling range is –8...8 to each direction.

Signal domain   Spectral domain

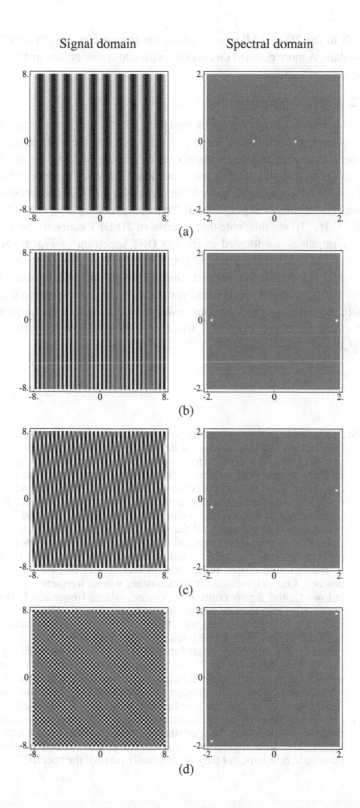

(a)

(b)

(c)

(d)

pseudo moirés to the 2D and MD cases, where the frequencies are 2D or MD vectors rather than scalars. A more detailed discussion on pseudo-moiré effects in 1D, 2D or MD can be found in [Amidror09a].

Example 8.2: A 2D sub-Nyquist artifact:

To take a concrete 2D example, let us study Fig. 8.15(c) in more detail. Here, the sampling frequency to both directions is $f_s = 4$ Hz, so that the maximum frequency allowed by the 2D sampling theorem without causing aliasing is $\frac{1}{2}f_s = 2$ Hz in each of the two directions (note that this is also the maximum frequency representable in each direction of the 2D DFT spectrum, as we have seen in Eq. (4.10)). The frequency of the 2D continuous cosine function $g(x,y) = \cos(2\pi[ux + vy])$ being sampled in this case is $\mathbf{f}_1 = (1.875, 0.25)$ Hz. To see this, note that in terms of 2D DFT elements (pixels) the two corresponding impulses are located in the 2D DFT spectrum of Fig. 8.15(c) in the discrete locations $\pm(30,4)$ pixels. Since 32 pixels (i.e. $N/2$) correspond to the maximum frequency $\frac{1}{2}f_s = 2$ Hz, it follows, indeed, that 30 pixels in the horizontal direction correspond to 1.875 Hz, and 4 pixels in the vertical direction correspond to 0.25 Hz. Let us now denote the difference between our cosine frequency $\mathbf{f}_1$ and the frequency $(\frac{1}{2}f_s, 0)$ by $(\varepsilon_u, \varepsilon_v)$. We have, therefore, $(\varepsilon_u, \varepsilon_v) = (\frac{1}{2}f_s, 0) - \mathbf{f}_1 = (2,0) - (1.875, 0.25) = (0.125, -0.25)$, where $\mathbf{f}_1 = (\frac{1}{2}f_s, 0) - (\varepsilon_u, \varepsilon_v)$.

Figure 8.16: Explanation in the continuous-world spectrum of the sub-Nyquist artifact that occurs in Fig. 8.15(c) when sampling a continuous 2D cosine function $g(x,y) = \cos(2\pi[ux + vy])$ whose frequency $\mathbf{f}_1 = (u,v)$ is just slightly below half of the sampling frequency $\frac{1}{2}f_s$ in the horizontal direction. (a) The continuous-world spectrum of the original cosine. (b) The continuous-world spectrum of the sampled cosine. Note the impulse pair of the original cosine, at the frequencies $\pm\mathbf{f}_1$, and its replicas which are centered about all the points (kf_s, lf_s) where f_s is the sampling frequency in both directions and $k,l \in \mathbb{Z}$. Thanks to the two neighbouring replicas to the right and to the left of the original replica, the spectrum (b) of the sampled cosine represents, in fact, a sum of two very close 2D cosines: Our original continuous cosine, whose frequency $\mathbf{f}_1$ is slightly below $\frac{1}{2}f_s$, and a new continuous cosine, whose frequency $\mathbf{f}_2$ is slightly above $\frac{1}{2}f_s$. (Impulses of higher frequencies, that are also present in the spectrum (b), are required for the correct spectral representation of the sampled signal, but for the sake of our discussion here they can be ignored). The sub-Nyquist artifact that appears in the sampled signal (see Fig. 8.15(c)) is simply the 2D beating modulation effect that occurs in the sum of the two very close 2D cosines having the frequencies $\mathbf{f}_1$ and $\mathbf{f}_2$. (c) Schematic view of the frequency vectors $\pm\mathbf{f}_1$, $\pm\mathbf{f}_2$, their sums and their differences. Note that the low beating frequency $\mathbf{f}_2 - \mathbf{f}_1$ itself, which is clearly visible in the signal domain of Fig. 8.15(c), has no corresponding impulse (black dot) in the spectrum (b), meaning that it is not a true moiré effect. The line segments in (a)–(c) have been added to clarify the geometric relations, but they are not really parts of the spectra.

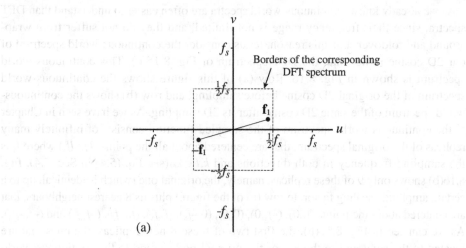

(a)

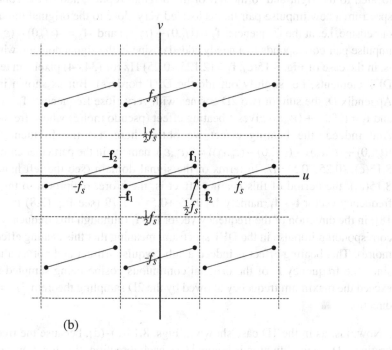

(b)

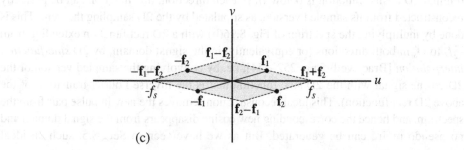

(c)

As we already know, continuous-world spectra are often easier to understand than DFT spectra, since their frequency range is not limited, and they do not suffer from wrap-around and foldover. For this reason, let us consider the continuous-world spectrum of our 2D cosine rather than the DFT spectrum of Fig. 8.15(c). This continuous-world spectrum is shown in Fig. 8.16. Row (a) of this figure shows the continuous-world spectrum of the original 2D cosine before sampling, and row (b) shows the continuous-world spectrum of the same 2D cosine after its 2D sampling. As we have seen in Chapter 5, the continuous-world spectrum of the sampled function consists of infinitely many replicas of the original spectrum, that are centered about all the points (kf_s, lf_s) where f_s is the sampling frequency in both directions and $k,l \in \mathbb{Z}$ (see Eq. (5.4) in Sec. 5.4). Fig. 8.16(b) shows only 9 of these replicas, namely, the original one (which is identical, up to a certain amplitude scaling factor, to row (a) of the figure) plus its 8 nearest neighbours, that are centered about the points $(f_s,0)$, $(-f_s,0)$, $(0,f_s)$, $(0,-f_s)$, (f_s,f_s), $(f_s,-f_s)$, $(-f_s,f_s)$ and $(-f_s,-f_s)$. As we can see in Fig. 8.16(b), the first two of these 8 new replicas, the ones that are located to the right and to the left of the original replica, add to the continuous-world spectrum a new impulse pair that is located very close to the original impulse pair $\pm\mathbf{f}_1$ of our cosine, i.e. at the frequencies $\mathbf{f}_2 = (\frac{1}{2}f_s,0) + (\varepsilon_u,\varepsilon_v)$ and $-\mathbf{f}_2 = -(\frac{1}{2}f_s,0) - (\varepsilon_u,\varepsilon_v)$. This new impulse pair corresponds to a newly added cosine in the signal domain, whose frequency is, in the case of Fig. 8.15(c), $\mathbf{f}_2 = (2.125,-0.25)$ Hz (or $(34,-4)$ pixels, in terms of the 2D DFT elements, i.e. slightly outside the DFT borders). But as shown in Sec. D.3 of Appendix D, the sum of two 2D cosines with very close frequencies $\mathbf{f}_1 = (\frac{1}{2}f_s,0) - (\varepsilon_u,\varepsilon_v)$ and $\mathbf{f}_2 = (\frac{1}{2}f_s,0) + (\varepsilon_u,\varepsilon_v)$ gives a beating effect (pseudo moiré) whose frequency is $\mathbf{f}_2 - \mathbf{f}_1$. And indeed, the beating effect we obtain here has the frequency of $\mathbf{f}_2 - \mathbf{f}_1 = ((\frac{1}{2}f_s,0) + (\varepsilon_u,\varepsilon_v)) - ((\frac{1}{2}f_s,0) - (\varepsilon_u,\varepsilon_v)) = 2(\varepsilon_u,\varepsilon_v)$, namely, in the particular case shown in Fig. 8.15(c), $(0.25,-0.5)$ Hz. In terms of the signal domain (see the left hand side of Fig. 8.15(c)), the period of this beating effect is, therefore, reciprocal to the length of the frequency vector $\mathbf{f}_2 - \mathbf{f}_1$, namely $1/\sqrt{0.25^2+0.5^2} \approx 1.79$ (see Eq. (2.8) in [Amidror09 p. 18]) in the direction of the frequency vector $\mathbf{f}_2 - \mathbf{f}_1$ (although this frequency vector has no corresponding impulse in the DFT spectrum, meaning that this beating effect is not a true moiré). This beating effect is indeed a sub-Nyquist artifact and not an aliasing artifact, since the frequency $\mathbf{f}_1$ of the original continuous cosine being sampled here does not exceed the maximum frequency allowed by the 2D sampling theorem, $\frac{1}{2}f_s = 2$ Hz to each direction. ■

Now, just as in the 1D case shown in Figs. 8.14(c)–(d), because the frequency of our original 2D cosine function is below $\frac{1}{2}f_s$ in each direction, this function can be perfectly reconstructed from its sampled version, as stipulated by the 2D sampling theorem. This is done by multiplying the spectrum of Fig. 8.16(b) with a 2D rect function extending from $-\frac{1}{2}f_s$ to $\frac{1}{2}f_s$ in both directions (or equivalently, in the signal domain, by 2D *sinc-function interpolation* [Bracewell95 pp. 252–254], i.e. by convolving the sampled version of the 2D cosine signal with the 2D sinc function that is the inverse Fourier transform of the above 2D rect function). This ideal reconstruction removes the new impulse pair from the spectrum, and hence the corresponding new cosine disappears from the signal domain and no pseudo moiré can be generated. But as we have seen in Sec. 8.5, such an ideal reconstruction rarely occurs in real-world situations. A more realistic reconstruction method may consist of 2D *zero-order* (i.e. *nearest-neighbour*) *interpolation*, which is equivalent to convolving the sampled signal with a narrow 2D pulse or "pixel" (see [Bracewell95 pp. 247–248] or [Stucki79 pp. 184–185]). This alternative reconstruction method means, in terms of the spectral domain, multiplication of the spectrum (b) with a non-ideal substitute of the ideal rect function. But when doing such a multiplication, the new replicas that appeared in row (b) due to sampling will not be completely removed, and some debris thereof may still subsist in the spectrum. In such cases, a low-frequency sub-Nyquist artifact may indeed become visible in the sampled signal (see Fig. 8.15(c)); this is simply the beating modulation effect that occurs in the sum of the newly generated cosine (due to the new impulses in the spectrum) and the original cosine function.

As we have seen above, the beating effects (sub-Nyquist artifacts) which are visible in Figs. 8.15(b)–(d) are not true moiré effects, since no corresponding low-frequency impulses appear in their spectra. These beating effects are, in fact, modulation or pseudo-moiré phenomena, just like the beating effect which occurs in the sum of any two cosines with slightly different frequencies (Sec. D.3 in Appendix D). However, just as shown in Fig. 8.13 for the 1D case, here, too, by judiciously modifying the frequency of the 2D cosine being sampled we can obtain a true sampling moiré effect, having exactly the same period and orientation as our beating effect. Figs. 8.17(b)–(d) show, indeed, for each of the 2D sub-Nyquist artifacts of Figs. 8.15(b)–(d), its corresponding true sampling moiré. This is obtained by sampling 2D cosines with higher frequencies than in Fig. 8.15, as we will see below, while keeping the sampling frequency unchanged.

Example 8.3: A true 2D sampling moiré:

Let us study in more detail the true sampling moiré effect shown in Fig. 8.17(c), and compare it with its pseudo-moiré counterpart of Fig. 8.15(c) that we have already analyzed in Example 8.2.

Here, too, we prefer to use the continuous-world spectrum rather than the DFT spectrum of Fig. 8.17(c). This continuous-world spectrum is shown in Fig. 8.18. Row (a) of this figure shows the continuous-world spectrum of our original 2D cosine before sampling. Note that the frequency vector $\mathbf{f}_1$ of this original 2D cosine is exactly twice the frequency vector $\mathbf{f}_1$ of the original 2D cosine we used in Fig. 8.16(a); as we will see below, this choice is not accidental. Row (b) of our figure shows the continuous-world spectrum of the same 2D cosine as in row (a), after its sampling (note that we are using here the same sampling frequency f_s as in Fig. 8.16 to both directions). As we already know, this spectrum consists of infinitely many replicas of the original spectrum, that are centered about all the points (kf_s, lf_s) where f_s is the sampling frequency in both directions and $k, l \in \mathbb{Z}$ (see Eq. (5.4) in Sec. 5.4). Just like in Fig. 8.16(b), row (b) of our figure shows only 9 of these replicas, namely, the original one (which is identical, up to a certain amplitude scaling factor, to row (a)) plus its 8 nearest neighbours, that are centered about the points $(f_s,0)$, $(-f_s,0)$, $(0,f_s)$, $(0,-f_s)$, (f_s,f_s), $(f_s,-f_s)$, $(-f_s,f_s)$ and $(-f_s,-f_s)$. But unlike in Fig. 8.16(b), this time the impulse pair belonging to each of the replicas is not confined inside the replica itself. This happens because in our present case the frequency $\mathbf{f}_1$ of the original 2D cosine is larger than half of the sampling frequency ($\frac{1}{2}f_s$), meaning that we are confronted here with aliasing. In terms of the signal domain this means that our sampling rate is not sufficient for capturing all details of our given 2D cosine, so that the sampled result will be mimicked by a lower frequency alias, like in Fig. 5.15(d).

If we look now at the main replica in Fig. 8.18(b), which also represents the territory of the corresponding DFT spectrum, we see that it contains, in fact, only two impulses, both of which exceed from the horizontal neighbouring replicas. These two impulses penetrate into the very center of the main replica, and fall there close to the origin. These new low-frequency impulses (that are due to the aliasing which occurs while sampling our original 2D cosine), correspond, indeed, to a new low-frequency moiré effect, which is clearly seen in Fig. 8.17(c). It is important to note, however, that unlike the low frequency which is visible in Fig. 8.15(c), the sampling artifact that we see in Fig. 8.17(c) is indeed a true sampling moiré, and not a sub-Nyquist artifact: Unlike its counterpart of Fig. 8.15(c), it *does* have a corresponding low-frequency impulse pair in the spectrum (the impulses indicated in Fig. 8.18 by $\pm\mathbf{f}_M$); and furthermore, in terms of the signal domain, it consists of a true 2D low-frequency cosinusoidal signal rather than of a modulated signal made of rapid alternating positive and negative values, as in Fig. 8.15(c). Compare also to the 1D equivalent shown in Fig. 8.13, in which row (f) shows a sub-Nyquist artifact, and row (i) shows its true sampling-moiré counterpart having the same frequency (and periodicity).

It is interesting to note, however, that in spite of the different nature of the low-frequency sampling artifacts in Figs. 8.15 and 8.17 (respectively, sub-Nyquist artifacts or true sampling moiré effects), their frequency (or period) as well as their orientation are exactly the same in both figures. The reason is that when designing Figs. 8.17(b)–(d), we have used in each of them an original 2D continuous cosine whose frequency vector $\mathbf{f}_1$ is exactly twice the frequency vector of the corresponding cosine in Fig. 8.15 (namely, twice

Figure 8.17: True 2D sampling moirés due to sampling in the 2D case (compare with the similar 2D sub-Nyquist artifacts shown in Fig. 8.15). Just like in Fig. 8.15, each row shows a 2D cosine function $g(x,y) = \cos(2\pi[ux + vy])$ with a different frequency vector $\mathbf{f} = (u,v)$. But this time, in each of the rows (b)–(d) the impulses of the cosine function *do* exceed the borders of the DFT spectrum (i.e. the frequency range $-\frac{1}{2}f_s \ldots \frac{1}{2}f_s$ in each dimension), and thus they fold over and re-enter into the opposite end of the DFT spectrum as aliased impulses. Since these aliased impulses fall at low frequencies close to the origin of the DFT spectrum, they are indistinguishable in the signal domain from the true signals having these low-frequency impulses, and give there highly visible low-frequency sampling moiré effects. Note that just like in Fig. 8.15 here, too, the sampling frequency is identical in all of the rows (4 Hz in each of the two directions); the array size is 64×64 and the sampling range is –8...8 to each direction.

Signal domain   Spectral domain

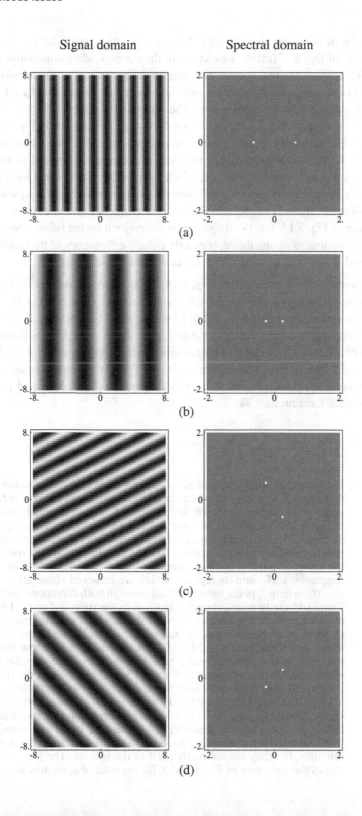

(a)

(b)

(c)

(d)

the frequency in the same orientation). To see how this works, let us consider the particular case of Fig. 8.17(c). As we can see in the corresponding continuous spectrum shown in Fig. 8.18(c), the two horizontally neighbouring replicas of the impulse pair $\pm\mathbf{f}_1$ give close to the spectrum origin two new low-frequency impulses, $\mathbf{f}_M = \mathbf{f}_2 - \mathbf{f}_1$ and $-\mathbf{f}_M = \mathbf{f}_1 - \mathbf{f}_2$, where $\mathbf{f}_2 = (f_s,0)$ represents the sampling frequency in the horizontal direction. If we compare Fig. 8.18(c) with its sub-Nyquist counterpart in Fig. 8.16(c), we see that the basic geometry in both cases remains the same, with one major difference: While in Fig. 8.18(c) the low-frequency vector $\mathbf{f}_2 - \mathbf{f}_1$ corresponds indeed to an impulse, in Fig. 8.16(c) this low-frequency vector, which falls exactly in the same location, is not endowed with an impulse of its own, and it therefore only represents there a pseudo-moiré effect (sub-Nyquist artifact) having exactly the same frequency and orientation. Note that in the 1D case of Fig. 8.13, too, the frequency of the original cosine being sampled in row (i), which gives a true sampling moiré, is exactly twice the frequency of the original cosine being sampled in row (f), and which gives the equivalent sub-Nyquist artifact.

Note that in the present moiré effect (Fig. 8.18) the frequency $\mathbf{f}_2$ does not correspond to a second cosine signal, as it does in the sub-Nyquist artifact (Fig. 8.16) or in a classical superposition moiré effect (see Fig. 2.2 in [Amidror09 p. 14]). Rather, $\mathbf{f}_2$ corresponds here to the sampling frequency vector in the horizontal direction, $(f_s,0)$. For this reason our present moiré effect is said to be a *sampling* moiré effect [Amidror09 p. 432]. Note that the frequency $\mathbf{f}_2$ itself does not appear in the DFT spectrum; as we can see in the continuous-world spectrum (Fig. 8.18(b)) this frequency is indeed located beyond the borders of the DFT spectrum. ■

Figure 8.18: Explanation in the continuous-world spectrum of the true moiré effects that occur in Fig. 8.17(c) when sampling a continuous 2D cosine function $g(x,y) = \cos(2\pi[ux + vy])$ whose frequency $\mathbf{f}_1 = (u,v)$ is exactly twice than in Fig. 8.15(c). Compare with the continuous-world spectrum explanation of Fig. 8.15(c), which is provided in Fig. 8.16. (a) The continuous-world spectrum of the original cosine. (b) The continuous-world spectrum of the sampled cosine. Note the impulse pair of the original cosine, at the frequencies $\pm\mathbf{f}_1$, and its replicas which are centered about all the points (kf_s, lf_s) where f_s is the sampling frequency in both directions and $k,l \in \mathbb{Z}$. Thanks to the two neighbouring replicas to the right and to the left of the original replica, the spectrum (b) of the sampled cosine includes very close to its origin two new impulses that are only due to the sampling. These new impulses, noted by $\pm\mathbf{f}_M$, correspond to a new low-frequency structure in the signal domain, which is indeed a true sampling moiré effect, and not a sub-Nyquist artifact. (c) Schematic view of the frequency vectors $\pm\mathbf{f}_1$, $\pm\mathbf{f}_2$, their sums and their differences. Note that unlike in Fig. 8.16(c), the low frequency $\mathbf{f}_2 - \mathbf{f}_1$ itself, which is clearly visible in the signal domain of Fig. 8.17(c), *does* have a corresponding impulse (black dot) in the spectrum (b), meaning that in this case it *is* a true moiré effect. The line segments in (a)–(c) have been added to clarify the geometric relations, but they are not really parts of the spectra. The gray area in (c) shows the geometry of Fig. 8.16(c), for the sake of comparison.

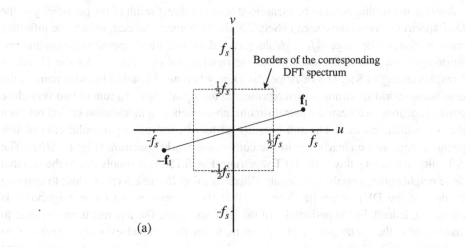

(a)

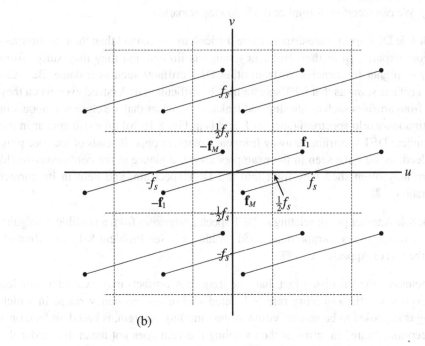

(b)

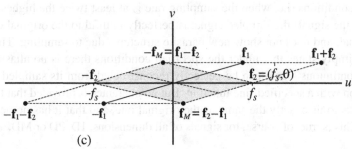

(c)

Another interesting point to be mentioned here is a direct result of the periodicity of the DFT spectrum. As we have seen in Sec. 2.5, this periodicity reflects, in fact, the infinitely many replicas of the range $-\frac{1}{2}f_s \ldots \frac{1}{2}f_s$ along each direction, which appear in the continuous-world spectrum due to the sampling of the input signal (see Fig. 2.1 for the 1D case). Now, according to Sec. D.3 of Appendix D, any two neighbouring impulse pairs in the continuous-world spectrum (which represent in the signal domain a sum of two very close cosine functions) may cause a pseudo-moiré phenomenon (a modulation effect) between the two similar cosines. The sub-Nyquist artifact is, indeed, a particular case of this phenomenon, as we clearly see in the continuous-world spectrum (Fig. 8.16(b)). The difficulty in realizing this in the DFT spectrum (Fig. 8.15(c)) is simply due to the fact that these neighbouring impulses are located there far away from each other, close to opposite borders of the DFT spectrum. The fact that these impulses are close neighbours is explained, indeed, by the periodicity of the DFT spectrum. But this fact is much easier to understand in the continuous-world spectrum, where these impulses really appear next to each other (although they fall there to both sides of a boundary between two neighbouring replicas). We can therefore formulate the following remark:

Remark 8.3: DFT spectra are usually more difficult to understand than their continuous-world counterparts, since their frequency range is limited and they may suffer from foldover of higher frequencies, or from other DFT artifacts such as leakage. But sub-Nyquist artifacts show us that DFT spectra may be difficult to understand even when they are free from artifacts such as aliasing and leakage: Impulses that are close neighbours in the continuous-world spectrum (such as $\mathbf{f}_1$ and $\mathbf{f}_2$ in Fig. 8.16(b)) may still appear in the corresponding DFT spectrum far away from each other, in opposite ends of the spectrum. And indeed, as we have seen in the examples above, a glance at the continuous-world spectrum may often shed new light into the DFT spectrum, and help in its correct interpretation. ■

Remark 8.4: Another pitfall relating to DFT spectra originates from a possible ambiguity in the use of the terms "wraparound" and "foldover". See Problem 8-17 and Remark C.10 at the end of Appendix C. ■

In conclusion, the surprising fact that low-frequency artifacts may exist in a sampled signal even when the sampling rate is located within the frequency range in which sampling is supposed to be safe by virtue of the sampling theorem, is based, in fact, on a misconception. "Safe" in terms of the sampling theorem does not mean that under the specified conditions (i.e. when the sampling rate is at least twice the highest frequency present in the signal) the sampled signal is perfectly faithful to the original continuous-world signal, and does not show new parasite structures due to sampling. The sampling theorem simply states that under the specified conditions there is no aliasing, and the original continuous signal can be perfectly reconstructed from its sampled version by convolution with a specified sinc function. But no guarantee is provided that the sampled signal looks to us exactly the same as the original one, and that it has no new apparent artifacts. This is true, of course, for signals of all dimensions, 1D, 2D or MD.

8.7 Displaying considerations

We have seen throughout the previous chapters several displaying considerations, that are helpful and sometimes even crucial for the correct interpretation of the DFT results. These include, among others, the reorganization issues (Chapter 3); the use of true units along the axes (Chapter 4); the optimal choice of the DFT parameters N, Δx, Δu, etc. (Chapter 7); the size of the final drawing, which may also affect the visibility of detail in the displayed data (Sec. 7.2); the possibility of plotting only the most interesting part of the data (Sec. 7.2); the choice of the number of periods to be taken into consideration, in case of a periodic function (Remark 7.1); and the reconstruction issues related to the plotting method being used to display the data (Sec. 8.5).

Another important displaying consideration concerns the amplitude of the DFT spectrum (or of its magnitude). Plots of the DFT spectrum or magnitude tend to be visually dominated by very low frequencies and in particular by the zero frequency (DC); these spectral components are often so strong that the entire DFT plot (after reorganization) seems to consist of a single spot in the center, surrounded by weak, negligible data. This is, of course, highly undesirable, since most of the interesting information usually resides at frequencies farther away from the center. One possible way to deal with this problem (that we have indeed used in some of our figures) is to truncate the exceeding values of the DC and of the surrounding low frequencies, in order to adapt the scaling of the vertical axis to the values of the farther and more interesting parts of the spectrum. Another possible solution is to use a non-linear scale, such as a logarithmic scale, for displaying the DFT values along the vertical axis (or even along other axes).

When choosing the most judicious plotting method for displaying the given data and the DFT results, one should also consider the various printer-dependent printing artifacts that may occur. Typical examples are irregularities that may appear when printing large uniform gray level spans, or moiré effects which may occur when printing a periodic structure (due to an interference with the halftone screen or the underlying periodic grid used by the printing device [Amidror09 p. 432]). Note that moiré artifacts are not only limited to printed plots, and they may also occur when viewing periodic structures on computer displays (again, due to their discrete nature).

A more detailed discussion on the graphic presentation of the DFT input and output signals in one, two or more dimensions is provided in Sec. 1.5.

8.8 Numeric precision considerations

As we already know, a multitude of errors of various origins may occur when using the DFT to approximate the continuous Fourier transform. Of most significant importance are the errors due to the aliasing and leakage artifacts, as explained in detail in Chapters 5 and 6. What happens, however, when the required conditions are met and no leakage and aliasing occur? Does the DFT in such cases perfectly approximate the continuous Fourier transform of the function underlying the data samples? The answer is, of course, no: The DFT can only be an approximation, since it is only based on a finite number N of input values (samples of the original function), and it yields the same number N of output values (discrete frequencies). But are these discrete values themselves precise samples of the continuous Fourier transform? The answer is, again, no, even if no aliasing and leakage artifacts occur. The reason can be best understood when we investigate the mechanism in which the DFT approximates the continuous Fourier transform. As already mentioned in Remark 2.2, the DFT can be derived from the continuous Fourier transform as a numerical approximation to the integrals of Eqs. (2.1)–(2.2), using the rectangular (or the trapezoidal) rules. These rules are simple, straightforward methods for numerically calculating integrals; they approximate the area below the function curve by a sum of $N-1$ narrow rectangles (or trapezoids) of width Δx (see, for example, [Harris98 pp. 567–568] or [Spiegel68 p. 95]). This approximation process unavoidably contains certain discretization errors that arise because the step Δx is small but not zero. Letting $N \rightarrow \infty$ and $\Delta x \rightarrow 0$ reduces this error, but in any realizable case some discretization error will subsist. A more detailed discussion on the inherent errors in the approximation of the Fourier transform by DFT can be found in [Briggs95 pp. 53–55], and for the case of the inverse DFT in [Briggs95 pp. 56–59 and Sec. 6.8].

Another potential source of error in the computation of the DFT is due to the precision of the hardware and software being used; for example, one may consider the effects of word length and of round-off or quantization errors on the precision of the DFT results. But when modern computers are being used, this source of error can be considered as marginal and it can be usually neglected (except, perhaps, when considering very unstable situations like the phase of the DFT spectrum in areas where the absolute value of the DFT is almost zero (see Sec. 8.3)). A discussion on such numeric precision considerations can be found, for example, in [Oppenheim75, Sec. 9.5].

PROBLEMS

8-1. *Representation of a rect function in the discrete world.* As we have seen in Sec. 8.2, the correct discrete representation of the rect function should take into account the midvalue rule, as shown in Fig. 8.2(a). Without taking this subtlety into account, the resulting DFT will not give the expected sinc function but rather a somewhat corrupted version thereof (see Figs. 8.2(b),(c)). However, even if we do take into account the midvalue rule, the resulting DFT does not perfectly fit the expected sinc function, because of some unavoidable aliasing effects (remember that the rect function is not band limited). Suppose that you insist on getting in your DFT output a good match with the expected continuous-world sinc function, i.e. a faithful sampled version of the sinc function which is not flawed by aliasing or other artifacts. How should the rect function be represented in the input signal in order to obtain this goal? *Hint*: Have a look at Fig. 8.1.

8-2. *Representation of a rect function in the discrete world (continued).* Figs. 8.1 and 8.2 show several different discrete representations of the rect function. Make a list of the various representations, explain each of them and discuss the resulting DFT spectra. Can you think of other possible representations and their impact on the resulting DFT? Which of the different representations are most used in practice? You may have a look at the figures in various books or internet resources.

8-3. *The discrete counterpart of the midvalue rule.* We have seen in Sec. 8.2 that if a discontinuity of a continuous-world function falls *between* the discrete elements n and $n+1$, then these discrete elements take intermediate values which depend on their relative distance from the true discontinuity point between them. This rule is applied automatically by the DFT in the output array whenever a discontinuity occurs in the continuous-world spectrum. But when it is *our* role to sample a continuous-world function having a discontinuity, the application of this midvalue rule during the sampling process is clearly less obvious. The best we can do in this case is to reasonably estimate (interpolate) the intermediate values at the two input array elements n and $n+1$ in accordance with their respective distance from the discontinuity point between them. However, interpolation can be done in many different ways, each giving slightly different results. What kind of interpolation should we apply here to get the most accurate results?

8-4. The original continuous-world function being sampled in Fig. 8.2(a) is a rect function whose width is 1. However, the sampled signal we obtain in this case could be also considered as the sampled version of a continuous-world trapezoid (to see this, connect each of the midvalue points to its two direct neighbouring points by a straight line). It is obvious that this trapezoid and the rect function have different continuous Fourier transforms, but since they are represented in the discrete world by the same discrete signal, they will have identical DFT spectra. How do you explain this contradiction?

8-5. We have seen in Sec. 8.4 that a symmetric rect function in the discrete world must have an odd number of 1-valued elements across its width, namely, the center of the rectangle (which corresponds to $x = 0$), and to each of its sides an equal number of neighbouring 1-valued elements (usually followed by an element with the midvalue $\frac{1}{2}$, as we have seen in Sec. 8.2). Any other possibility will result in a non purely-real DFT. Interestingly, this is true independently of the array length N, be it an even or an odd number. For example, using the Mathematica® software package, which is capable of computing the DFT of an input sequence of any given length N [Wolfram96 p. 1093], we find that for $N = 4$ the DFT of the sequence $\{1,1,0,1\}$ is purely real-valued, $\{1.5,0.5,-0.5,0.5\}$, while the DFT of the sequence $\{1,0,0,1\}$ is complex. Similarly, for $N = 5$, the DFT of the sequence $\{1,1,0,0,1\}$ is purely real, $\{1.34,0.72,-0.28,-0.28,0.72\}$, while the DFT of the sequence $\{1,1,0,1,1\}$ is complex. As we can see, although the input sequences $\{1,0,0,1\}$ and $\{1,1,0,1,1\}$ look symmetric, in terms of the DFT they are not considered as even. Explain these results. Note that all the sequences are given here as they are actually fed to or received from the DFT, i.e. with no reorganizations or amplitude corrections; in particular, their first element corresponds to $x = 0$ (in the signal domain) or to the DC (in the spectral domain). Obviously, before plotting these sequences or interpreting them as CFT approximations, reorganization and amplitude corrections should be done in order to avoid confusion.

8-6. As mentioned at the end of Sec. 8.4, in order to avoid abnormal situations where 2D continuous-world Fourier symmetry rules are violated in the 2D discrete case, we have intentionally zeroed the last row and/or the leftmost column in the 2D input array whenever required. This was done, for example, in Figs. 5.13, 5.14, 5.15 and 8.8(a), but not in other cases such as Figs. 5.16, 5.17 and 8.7(a). Discuss each of these cases and explain why a correction was or was not required there.

8-7. Can you guess how we have generated the relatively complicated yet perfectly symmetric pattern shown in row (d) of Fig. 8.4? *Hint*: This is simply a lower-resolution counterpart of Fig. 5.18(e), where the circular nature of the original continuous-world function being sampled is rendered unrecognizable due to undersampling (aliasing).

8-8. *Jaggies and aliasing*. Consider Fig. 8.7(a). The continuous-world spectrum of a rotated grating consists of an infinite impulse comb passing through the origin, which is rotated by the same angle as the grating. But as we can see in this figure, the DFT counterpart of this spectrum contains also many impulses which are not located along this comb. Student A says that from the point of view of aliasing, these impulses are simply folded-over continuations of the main impulse comb which exceed from one end or the other of the DFT spectrum (due to the insufficient sampling rate) and re-enter from the opposite end. Student B, however, says that from the point of view of the jaggies, the impulses in question simply represent the jaggies which are present in the sampled grating of Fig. 8.7(a), and which do not exist in the continuous-world grating. Student C claims that both of these interpretations are correct, and that this is indeed the reason why such jaggies are considered in the literature as aliasing effects (both jaggies and aliasing being side effects of the same sampling process). What is your opinion? Can you extend your reasoning to Fig. 8.8? And to Figs. 8.5(a),(b)?

8-9. Suppose that the grating of Fig. 8.7(a) was not rotated, so that its lines were perfectly vertical. In this case, no jaggies are visible in the discrete, sampled image (see Fig. 8.6(a) and Fig. 8.9). Can you explain this case, too, using the same reasoning?

8-10. The spectra shown in Figs. 8.6–8.9 were all obtained by DFT, and therefore they only include the frequency range going between $-\frac{1}{2}f_s$ and $\frac{1}{2}f_s$ along each axis, where f_s is the sampling frequency in both directions. However, the spectral domain explanation of 2D sampling is easier to understand in the continuous-world spectrum, which extends throughout the u,v plane and clearly shows the replicas due to the convolution with the 2D nailbed corresponding to the 2D sampling frequencies (see Fig. 5.12). Draw the continuous-world counterpart of each of the cases shown in Figs. 8.6–8.9 and explain what you see in the signal and spectral domains in each of these cases.

8-11. In particular, consider the continuous-world spectrum that corresponds to the unrotated vertical grating shown in Fig. 8.6(a) (the spectral domain shown in this figure is the DFT counterpart of this continuous-world spectrum). As we already know, the continuous-world spectrum of an unrotated vertical grating consists of a comb of impulses along the u axis, whose amplitudes are modulated by a sinc function. Due to the 2D sampling, this continuous-world spectrum is convolved with a 2D nailbed whose impulse distance to both directions equals the sampling frequency f_s. This convolution generates replicas of our comb throughout the u,v plane, away from the u axis, too. A student asks the following question: Since in this case there are no jaggies, the sampled grating (Fig. 8.6(a)) looks exactly the same as the original unsampled grating. How can it be that their continuous-world spectra are nevertheless different? *Hint*: Although no jaggies appear in this case, the sampled and unsampled vertical gratings are not really identical: the former only consists of impulses, while in the latter the lines are really continuous. Note that what we see in the signal domain of Fig. 8.6(a) is not really the sampled grating, but rather its reconstructed version, where each impulse is convolved with a square pixel. This reconstructed signal looks, indeed, very similar to the original unsampled grating (up to a small possible difference in the line widths, due to their rounding to the closest multiple of full square pixels).

8-12. Fig. 8.10 shows a slightly rotated low-frequency cosinusoidal grating, which is clearly band limited and has no sampling aliasing. If reconstruction is done correctly, as stipulated by the sampling theorem, no jaggies should appear in the reconstructed signal; and if jaggies do appear — as in Fig. 8.10 — this is only due to poor reconstruction, which causes reconstruction aliasing. And indeed, a student says that jaggies are not aliasing artifacts, as often claimed in the literature, but rather reconstruction artifacts (see also [Gomes97 p. 197; Wolberg90 pp. 107–108]). What is your opinion? *Hint*: Note that in cases like Figs. 8.7 and 8.8 ideal reconstruction alone will not eliminate the jaggies. To see this, remember that ideal reconstruction is equivalent to multiplying the spectrum of the sampled signal with a rect function that cuts off all the frequencies beyond half of the sampling frequency. But looking at the spectrum in Figs. 8.7 and 8.8 we see that this multiplication will not eliminate the stray impulses that are present *within* this range and the artifacts they represent in the signal domain.

8-13. *Sub-Nyquist artifacts.* Although in Sec. 8.6 we insist on the fact that sub-Nyquist artifacts may occur at frequencies *below* half of the sampling frequency, so they are certainly not caused by aliasing, Fig. 8.13 shows that a similar artifact may also occur *above* half of the sampling frequency, where aliasing does occur (see row (h) in the figure, and compare it with row (f)). Note that in both cases no impulses belonging to the new, visible low frequency in question is present in the spectrum. In Sec. 8.6 we concentrated on the explanation of the case shown in row (f) of the figure, but the explanation of the case shown in row (h) is quite similar. Can you give the detailed explanation for this case?

8-14. *Sub-Nyquist artifacts.* As we saw in the previous problem, sub-Nyquist artifacts may also occur at frequencies *above* half of the sampling frequency, so that the term "sub-Nyquist artifacts" is not really optimal. Can you think of a better term, for example a term reflecting the main property of these artifacts, the fact that they do not possess in the spectrum impulses that correspond to their frequencies?

8-15. *Sub-Nyquist artifacts.* After carefully studying Sec. 8.6, a student points out that the sub-Nyquist artifact is not *generated* by reconstruction errors, as sometimes suggested in the literature. This beating effect already exists in the sampled signal itself, much before the reconstruction stage. Simply, while ideal reconstruction would eliminate this beating effect from the reconstructed signal, non-ideal reconstructions fail to eliminate it (or even may further amplify it), so that this sampling artifact remains visible in the reconstructed continuous signal, too. In other words, reconstruction errors are not the *cause* of this beating effect, but they only allow it to subsist after the reconstruction step. What is your opinion?

8-16. Looking at Fig. 8.15(d), student A says that the frequency of the given 2D cosine (which is represented by the two impulses at the corners of the DFT sprctrum) exceeds beyond half of the sampling frequency, $\frac{1}{2}f_s = 2$ Hz, since the distance of each of these impulses from the origin is almost $\frac{1}{2}\sqrt{2}f_s$, which is higher than $\frac{1}{2}f_s$. Student B, however, says that our 2D cosine does not exceed half of the sampling frequency, since in both of its impulses, located at $\pm(u,v)$, each of the Cartesian components u and v is below $\frac{1}{2}f_s$. What is your opinion?

8-17. *Foldover vs. wraparound.* We have seen in Remark 8.4 that the terms *foldover* and *wraparound* should be used with care in order to avoid confusion. Explain the exact meaning of each of these terms, the difference between them, and the precise circumstances in which they can be used without leading to confusion.

Hint: We have seen in Sec. 2.5 and in Chapter 5 that the sampling of an original continuous signal generates in the continuous-world spectrum new frequencies due to the replication of the original spectrum. When aliasing occurs, frequencies from the new replicas penetrate additively into their neighbouring replicas, and in particular into the main replica about the origin (which corresponds also to the territory of the DFT spectrum). However, as we have seen in Chapter 5, this additive intrusion into neighbouring replicas can be also considered from the point of view of the DFT spectrum as *folding over*, namely: "impulses (or other spectral elements) that exceed from one end of the DFT spectrum re-enter from the opposite end". This can be seen, for example, in Figs. 8.18(b), 5.13, etc. This interpretation as folding over is, indeed, convenient and helpful for the explanation of aliasing in the DFT spectrum, but it should not be confused with wraparound. Consider Fig. 8.16(b), where no aliasing occurs, and no new impulses penetrate into the territory of the main replica (and hence into the DFT spectrum). Because of the periodicity of the DFT spectrum along both axes, one can say that each of the identical replicas in the continuous-world spectrum (due to sampling) *wraps around* and coincides with the main replica (which represents the DFT spectrum). However, in this case the term folding over should be avoided since no addition is involved here.

Let us illustrate the difference using Figs. 8.16 and 8.18. Consider first Fig. 8.18(b), and suppose that the original continuous-world spectrum already contained impulses of its own at the locations $\pm\mathbf{f}_M$. After sampling, because of the aliasing effect, replicas of the impulses $\pm\mathbf{f}_1$ penetrate into the main replica, and fall there on top of the already existing impulses $\pm\mathbf{f}_M$. As already mentioned, this can be also considered as *folding over*. Since the replicas due to sampling are generated by a convolution operation (more precisely, the convolution of the original spectrum with a 2D nailbed containing all the integer combinations of the sampling frequency vectors), our impulse superposition due to folding over implies that *the amplitudes of the coinciding impulses be summed up*. Let us now return to Fig. 8.16(b), where no aliasing occurs and no impulses penetrate into neighbouring replicas. In this case, too, it is legitimate to say that due to the periodicity of the DFT spectrum along both axes all the other replicas *wrap around* and coincide with the main replica (the DFT territory). But of course, an existing impulse and a wrapped-around impulse which coincide within the main replica *are not summed up*: They are simply identified with each other (or construed as congruent modulo the DFT spectrum periodicity[21]), meaning that each of them represents in fact the same entity in the DFT. In this case the use of the term *wraparound* is legitimate, but not the use of the term *foldover* (which implies the addition of amplitudes, and is only to be used in the case of aliasing). See also Remark C.10 at the end of Appendix C.

8-18. We saw in Sec. 8.4 that in the continuous world, when $g(x)$ is real valued, $G(0)$ is real valued, too (since the Hermitian nature of $G(u)$ implies that $\text{Im}[G(0)] = -\text{Im}[G(0)]$, which means in turn that $\text{Im}[G(0)] = 0$). However, in the discrete world it turns out that when the input signal of the DFT is real valued, not only the output element corresponding to the zero frequency is real valued, but also the output element representing the highest (or lowest) frequency in the spectrum, which is located $N/2$ elements away from the zero-frequency element [Briggs95 pp. 77, 120]; see, for example, the leftmost spectral element in Figs. 3.1(d) and 3.3(f). How can you demonstrate this? *Hint*: See point (1) in Sec. D.2 of Appendix D.

---

[21] A number a is said to be congruent to b modulo p if p divides the difference $a - b$ [Weisstein99 p. 303], i.e. if $a - b$ is an integer multiple of p (which means that $a - b = kp$ for some integer k). Thus, if F is the frequency range length of the DFT spectrum, then the continuous-world frequencies f_2 and f_1 are congruent mofulo F if $f_2 = f_1 + kF$ for some integer k. Congruence of frequency vectors $\mathbf{f}_2 = (u_2, v_2)$ and $\mathbf{f}_1 = (u_1, v_1)$ modulo the 2D DFT spectrum range (F, F) is interpreted component-wise. The extension to the MD case is straightforward, too.

Appendix A

Impulses in the continuous and discrete worlds

A.1 Introduction

The mathematical concept of an impulse is widely utilized in many disciplines, including physics, electricity, optics, etc. It is used to represent entities that are concentrated in a single point, such as a point mass, a point charge, etc. Although such entities do not exist in the real world, this theoretical concept is very useful since it may considerably simplify derivations [Bracewell86 Chapter 5; Brigham88 Appendix A].

A.2 Continuous-world impulses vs. discrete-world impulses

In the continuous case, an impulse $\delta(u)$ (also known as a *Dirac delta function*) is defined by (see [Bracewell86 p. 70; Jain89 p.12; Wolberg90 p. 15; Chu08 p. 188]):[1]

$$\delta(u) = 0 \qquad u \neq 0$$

$$\text{and:} \qquad \int_{-\infty}^{\infty} \delta(u)\, du = 1 \tag{A.1}$$

This impulse, or more generally its shifted counterpart $\delta(u-a)$ that is located at the point $u = a$, is therefore considered to have a unit integral (or area); but since it is concentrated on a single point $u = a$ and has there zero width it must have an infinite amplitude. Thanks to its sifting property, when this impulse samples a given function $g(u)$, as in $g(u)\delta(u-a)$, its integral (area) takes the value $g(a)$, while its amplitude remains, of course, infinitely high [Bracewell86 pp. 74–77]. By convention, an impulse $\delta(u-a)$ is often represented graphically as a spike of unit height; more generally, impulses are plotted as spikes of height equal to their integral [Bracewell86 p. 75].[2] For this reason the Fourier transform of the constant function $g(x) = 1$, $G(u) = \delta(u)$, is graphically represented by a spike of height 1 at the origin, and the Fourier transform of the periodic function $g(x) = \cos(2\pi f x)$, $G(u) = \frac{1}{2}\delta(u-f) + \frac{1}{2}\delta(u+f)$, is graphically represented by two spikes of height $\frac{1}{2}$ located at the frequencies $\pm f$ [Brigham88 pp. 19–20; Bracewell86 p. 412]. In reality each of these impulses is infinitely high, but according to this graphical convention the spike heights indicate the value of the integral. This convention is very practical, because it really conveys the most important attribute of the impulse, i.e. its *integral* (and not its *height*,

---

[1] Note that $\delta(u)$ is not a function in the usual mathematical sense of this term; it is normally defined as the limit of the integral of a suitable series of functions, as explained, for example, in [Bracewell86, Chapter 5].

[2] Such a spike can be graphically represented by a vertical line of the corresponding length, with or without an arrow at its end (see, for example, Fig. 2.1).

which is always infinite). But this convention may be somewhat dangerous, since it may induce one to erroneously believe that the spike's height represents the amplitude (or height) of the impulse, which is obviously wrong.[3]

These concepts can be also extended without difficulty to the 2D and MD cases (see, for example, [Bracewell86 pp. 84–85]).

In the discrete case, however, the situation is somewhat different. A discrete impulse (also known as a *Kronecker delta function*) is defined by (see, for example, [Jain89 p. 12; Wolberg90 p. 15; Smith07 pp. 122, 184; Chu08 p. 219]):[4]

$$\delta(n) = \begin{cases} 1 & n = 0 \\ 0 & \text{otherwise} \end{cases} \tag{A.2}$$

Therefore, in the output array of the DFT, the discrete counterpart of the shifted continuous-world impulse $\delta(u-a)$ is an array of zero-valued elements, with a single 1-valued element that is located in the array position n which corresponds to the continuous-world frequency $u = a$.[5] Thus, a discrete impulse consists of one non-zero array element. (Concerning the "width" of such an element see Footnote 14 in Sec. 8.5 for the 2D case; its extension to the 1D and MD cases is straightforward.)

By convention, a discrete impulse $c\delta(n)$ is usually represented graphically as a spike of height c. Such a spike can be drawn as a vertical line of the corresponding length, with or without an arrow at its end (see, for example, Fig. B.1). Alternatively, when plotting our discrete data as a *symbol plot* or as a *connected line plot*, the discrete impulse is represented by a single dot or by a narrow isosceles triangle of height c, respectively (see Sec. 1.5.1 and Fig. 1.2). Note that in Appendix A we will only use connected line plots.

A.3 Impulses in the spectral domain

Having seen the definition and the graphical representation of impulses in the discrete world, what can be said now about the height of a discrete impulse in the DFT spectrum? Consider, for example, the continuous function $g(x) = 1$, whose spectrum is $G(u) = \delta(u)$. The discrete counterpart of $g(x)$ consists of a sequence of N samples all having the value of 1; if we take its DFT we obtain a sequence consisting of zeroes except for the element

---

[3] Sometimes the term "impulse amplitude" is used, by abuse of language, referring to the spike's height in the plot (i.e. the value of the integral). This practice can be tolerated as long as it does not lead to confusion. A better choice would be "impulse strength", as in [Bracewell86 p. 76; Castleman79 p. 228].

[4] In fact, in the context of the DFT, where both the input and the output arrays are considered to be periodic with period N (the array length), the discrete impulse, too, may be considered as periodic with period N. In this case, the condition in the first row becomes: "$n = 0$ or a multiple of N" like in [Briggs95 p. 28].

[5] It is assumed here that this frequency falls exactly on an output array element of the DFT and not between two elements (i.e. that the value of n which corresponds to the frequency $u = a_{Hz}$ according to Eq. (4.14) is precisely integer), in order to avoid leakage artifacts (see Sec. 6.5). Otherwise, the correct discrete counterpart of the impulse is a sampled, narrow sinc function (see Remark 6.5).

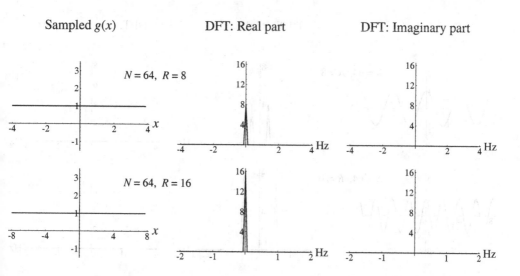

Figure A.1: The discrete counterpart of the function $g(x) = 1$ and its DFT spectrum (after the corrections of Chapters 3 and 4). The array length (number of samples) being used is $N = 64$. Top row: with a sampling range length of $R = 8$ (i.e. sampling interval $\Delta x = 8/64 = 0.125$); bottom row: with a sampling range length of $R = 16$ (i.e. sampling interval $\Delta x = 16/64 = 0.25$). Note the difference in the impulse heights between the two DFT spectra. In the continuous case this impulse is represented in the spectrum by a spike of height 1.

representing the DC, whose height has a finite positive value that depends on the DFT definition being used (see, for example, [Bracewell86 p. 364]; [Weaver89 p. 250]). After applying the standard amplitude scaling correction (see Sec. 4.4) one could expect the height of this DC impulse to be 1, in agreement with the conventional graphic representation of its continuous counterpart as a spike of height 1. However, as shown in Fig. A.1, it turns out that the height of the discrete DC impulse we obtain by the DFT (after the standard amplitude scaling correction) is not 1 but rather $R = N\Delta x$, the length of the range within which the original function has been sampled to form our N-elements input array (i.e. $x_{max} - x_{min}$).

As a second example, consider the continuous function $g(x) = \cos(2\pi f x)$, whose spectrum is $G(u) = \frac{1}{2}\delta(u{-}f) + \frac{1}{2}\delta(u{+}f)$. Even if this function is chosen and sampled so as to avoid DFT artifacts (aliasing and leakage), as explained in Chapters 5 and 6, the impulse heights obtained by the DFT after performing the standard amplitude scaling correction are not $\frac{1}{2}$, as could be expected from the conventional graphic representation used in the continuous case. Rather, just as in our first example, the height of each impulse is further multiplied by $R = N\Delta x$, the length of the range within which the original function has been sampled to form our N-elements input array (see Fig. A.2).

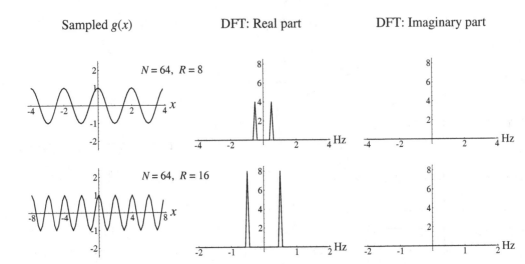

Figure A.2: The discrete counterpart of the function $g(x) = \cos(2\pi fx)$ (with $f = \frac{1}{2}$) and its DFT spectrum (after the corrections of Chapters 3 and 4). The array length being used is $N = 64$. Top row: with a sampling range length of $R = 8$ (i.e. sampling interval $\Delta x = 8/64 = 0.125$); bottom row: with a sampling range length of $R = 16$ (i.e. sampling interval $\Delta x = 16/64 = 0.25$). Note the difference in the impulse heights between the two DFT spectra. In the continuous case these impulses are represented in the spectrum by spikes of height $\frac{1}{2}$.

This rule applies, indeed, to the DFT of all functions having impulsive spectra. Similarly, in the 2D case any impulse amplitude in the DFT output is multiplied by $R^2 = N^2\Delta x^2$, and in the general MD case the factor is $R^M = N^M\Delta x^M$.

It follows, therefore, that the height of the discrete impulse in the DFT spectrum is not fixed, since it varies according to the choice of R. This may seem quite surprising, because in functions having non-impulsive spectra the height of the DFT spectrum (after the standard amplitude scaling correction) corresponds to the true height of the original underlying continuous spectrum (except for possible aliasing or leakage distortions, if any), and it does not depend on the choice of R. For example, as shown in Fig. A.3, the height of the DFT spectrum of the function $\text{sinc}(2x)$, after the standard amplitude scaling correction, is $\frac{1}{2}$ independently of the choice of R.

In order to understand this seemingly abnormal behaviour of the DFT in cases involving impulsive spectra, let us recall the following Fourier-theory results. In the case of the continuous Fourier transform we have [Bracewell86 p. 136]:

$$G(0) = \int_{-\infty}^{\infty} g(x)\, dx \tag{A.3}$$

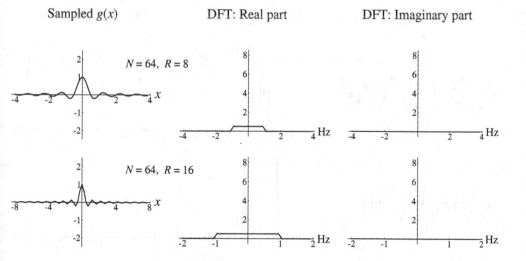

Figure A.3: The discrete counterpart of the function $g(x) = \text{sinc}(2x)$ and its non-impulsive DFT spectrum (after the corrections of Chapters 3 and 4). The array length being used is $N = 64$. Top row: with a sampling range length of $R = 8$ (i.e. sampling interval $\Delta x = 8/64 = 0.125$); bottom row: with a sampling range length of $R = 16$ (i.e. sampling interval $\Delta x = 16/64 = 0.25$). Note that the height of the two DFT spectra is $\frac{1}{2}$, just as in the continuous spectrum, and it does not depend on R.

This result is obtained from Eq. (2.1) for the frequency $u = 0$. (Note, however, that if we define the continuous Fourier transform using other conventions, such as Eq. (2.5), we may have here an additional constant factor preceding the integral.)

The discrete counterpart of Eq. (A.3) can be easily derived by inserting $n = 0$ in Eq. (2.13):

$$G(0) = \sum_{k=0}^{N-1} g(k) \qquad (A.4)$$

(Note that here, too, a constant factor may be needed if we use a different definition of the DFT; for example, with the definition used in the Mathematica® software package [Wolfram96 p. 868] the required constant factor is $\frac{1}{\sqrt{N}}$, and with the definition used in [Bracewell86 p. 358] the constant factor is $\frac{1}{N}$.)

Eq. (A.3) means that in the continuous case the value of the spectrum at the DC frequency $u = 0$ equals the area under the original signal-domain function. Its discrete counterpart, Eq. (A.4), means that the value of the DFT spectrum at the discrete DC frequency (i.e. at the array element representing the frequency 0) simply equals the sum of all the samples of the signal-domain function in the input array.

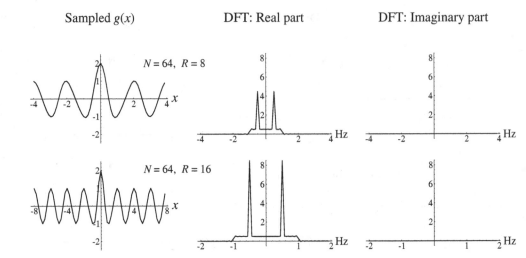

Figure A.4: The discrete counterpart of the function $g(x) = \mathrm{sinc}(2x) + \cos(2\pi fx)$ (with $f = \frac{1}{2}$) and its DFT spectrum (after the corrections of Chapters 3 and 4). The array length being used is $N = 64$. Top row: with a sampling range length of $R = 8$ (i.e. sampling interval $\Delta x = 8/64 = 0.125$); bottom row: with a sampling range length of $R = 16$ (i.e. sampling interval $\Delta x = 16/64 = 0.25$). Note that the height of the square pulse in both DFT spectra is $\frac{1}{2}$, just as in the continuous spectrum, and it does not depend on R, while the height of the two impulses does depend on R. Compare with Figs. A.2 and A.3.

Returning now to our first example above, the function $g(x) = 1$, we see by Eq. (A.4) that the value of the DC in the output array of the DFT[6] is $\sum_{k=0}^{N-1} 1 = N$ (or with the DFT definition used by Mathematica®: $\frac{1}{\sqrt{N}}\sum_{k=0}^{N-1} 1 = \frac{1}{\sqrt{N}}N = \sqrt{N}$; or with the DFT definition used in [Bracewell86 p. 358]: $\frac{1}{N}\sum_{k=0}^{N-1} 1 = 1$). And indeed, after applying the standard amplitude scaling correction of Chapter 4 (which corresponds, in the cited cases, to a multiplication by Δx, $\sqrt{N}\Delta x$ or $N\Delta x$, respectively), the value of the DC becomes, as we have seen above, $N\Delta x = R$ (rather than 1, the height of the DC spike in the continuous case according to the graphic convention of Sec. A.2).

Proceeding to our second example, $g(x) = \cos(2\pi fx)$, let us recall the following theorem [Weaver89 p. 248]: If the DFT of the N-element sequence $g(k)$ is the sequence $G(m)$, then the DFT of the sequence $g(k)\cos(2\pi kn/N)$ is the sequence $\frac{1}{2}G(m+n) + \frac{1}{2}G(m-n)$. This is,

---

[6] Remember that the DC is provided by the DFT in the first element ($n = 0$) of the output array; if required, a reorganization of the output array can be done later for display purposes (see Chapter 3).

Sampled $g(x)$ DFT: Real part   DFT: Imaginary part

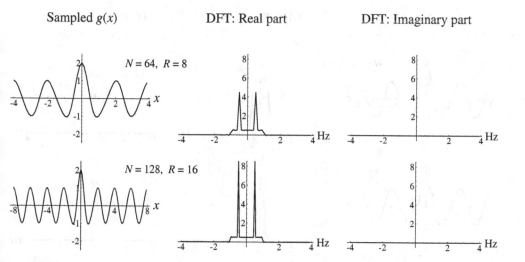

Figure A.5: Same as Fig. A.4, but in the bottom row the sampling range length has been doubled to $R = 16$ without modifying the sampling interval Δx, so that N has been doubled to $N = 128$.

in fact, the discrete counterpart of the modulation theorem [Bracewell86 p. 108]. By applying this theorem to the constant function $g(k) = 1$ of our first example it follows that the DFT of the function $\cos(2\pi kn/N)$, $k = 0,...,N-1$ consists of two impulses that are located $\pm n$ elements away from the zero frequency, each having half the value of the DC impulse of the constant function $g(k) = 1$, i.e., depending on the DFT definition being used, $N/2$, $\sqrt{N}/2$ or $\frac{1}{2}$ (see, for example, [Brigham88 pp. 176–177]; [Weaver89 pp. 250–251]). This means that in all cases, after applying the standard amplitude scaling correction of Chapter 4, the height of the two impulses in the DFT spectrum is $\frac{1}{2}R$ (rather than $\frac{1}{2}$, the height of the two spikes in the spectrum of the continuous case). A similar reasoning is also true for the sine function $g(x) = \sin(2\pi fx)$.[7]

Now, in order to understand what happens in the general impulsive case, we recall that any function having an impulsive spectrum (i.e. any periodic, almost periodic or constant function) can be represented as a Fourier series. This Fourier series is, in fact, a sum of cosine or sine functions (that contribute to the spectrum the impulses of the corresponding frequencies) and possibly a constant function (that contributes to the spectrum the DC impulse). This explains, indeed, based on our previous results for the constant, cosine and sine functions, why in the general impulsive case, too, the impulses in the DFT spectrum (after the standard amplitude scaling correction) are R times higher than in the corresponding CFT plot.

---

[7] This time using the sine counterpart of the above mentioned theorem, according to which the DFT of the sequence $g(k) \sin(2\pi kn/N)$ is the sequence $\frac{i}{2}G(m+n) - \frac{i}{2}G(m-n)$ [Weaver89 p. 248].

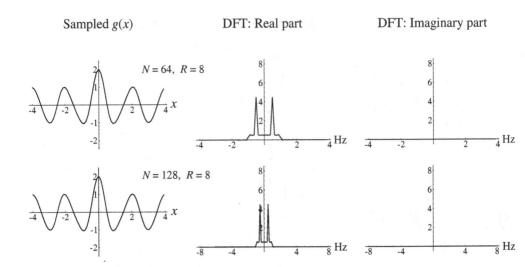

Figure A.6: Same as Fig. A.4, but in the bottom row N has been doubled to $N = 128$ without modifying the sampling range length $R = 8$, so that the sampling interval has been halved to $\Delta x = 8/128 = 0.0625$.

In contrast, in functions whose spectra are non-impulsive, an increase in the range length R causes an equivalent increase in the *width* of the DFT spectrum,[8] and therefore it does not affect the *height* of the DFT spectrum (see, for example, Fig. A.3).

The different behaviour of impulsive and non-impulsive spectra obtained by DFT is best illustrated by a combination of both cases in a single function. Consider, for example, the continuous-world function $g(x) = \text{sinc}(2x) + \cos(2\pi f x)$. Thanks to the addition theorem the continuous-world spectrum of this function is the sum of the individual spectra of its two terms, namely, a square pulse of height $\frac{1}{2}$ within the frequency range $-1 \le u \le 1$ plus a pair of impulses at the frequencies $u = \pm f$ which are represented by spikes of height $\frac{1}{2}$ on top of the square pulse, and whose heads reach therefore the total height 1: $G(u) = \frac{1}{2}\text{rect}(u/2) + [\frac{1}{2}\delta(u-f) + \frac{1}{2}\delta(u+f)]$. When the function $g(x)$ is sampled and fed to the DFT, the result (after reorganization and application of the standard amplitude scaling correction) shows, indeed, a square pulse of height $\frac{1}{2}$ within the range $-1 \le u \le 1$ plus an impulse pair to both sides of the origin (see Fig. A.4). But as shown in this figure, while the height of the square pulse remains $\frac{1}{2}$ independently of the range length R being used in the signal domain, the total height of each of the two impulses is $\frac{1}{2}R + \frac{1}{2}$, which *does* depend on the choice of R.

---

[8] Not in terms of true frequencies in Hz, but in terms of the index n of the output array, i.e. in terms of array elements (since by increasing R and hence Δx, for a given array length N, the frequency step in the output array is reduced in accordance with Eq. (4.4)). See the influence of modifying R on the output of the DFT in Sec. 7.4.

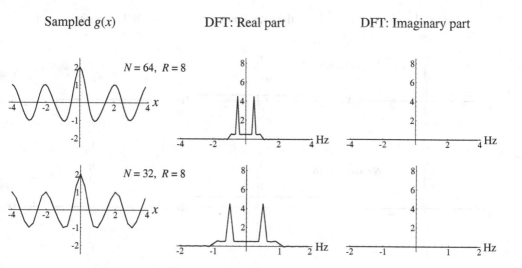

Figure A.7: Same as Fig. A.4, but in the bottom row the the sampling interval has been doubled to $\Delta x = 0.25$ without modifying the sampling range length $R = 8$, so that N has been halved to $N = 32$.

In fact, this behaviour is quite logical: as we can see, doubling R without changing N results in doubling the width of the square pulse in the DFT spectrum (in terms of the array index n, but not in terms of the real frequency in Hz); but when the spectrum consists of impulses, whose width cannot be increased, it is clear that the *height* of the impulses must be doubled.

Finally, Figs. A.5–A.7 show the behavoiur of the DFT of the same function as in Fig. A.4, $g(x) = \mathrm{sinc}(2x) + \cos(2\pi f x)$, under other possible parameter changes based on Eq. (4.2), $R = N\Delta x$ (see Table 7.1): While Fig. A.4 shows what happens when R is doubled but N remains unchanged (so that Δx is doubled), Fig. A.5 shows what happens when R is doubled but Δx remains unchanged (so that N is doubled). Similarly, Fig. A.6 shows what happens when N is doubled but R remains unchanged (so that Δx is halved), and Fig. A.7 shows what happens when Δx is doubled but R remains unchanged (so that N is halved). Note that in the two last cases the impulse height is not affected, since the value of R remains unchanged.

Remark A.1: (Impulses in the spectral domain): The fact that the impulse amplitudes in the DFT spectrum (after the application of the standard amplitude scaling correction of Sec. 4.4) are R times higher than in the corresponding CFT means that in order to obtain the expected results in the DFT spectrum, we need to divide the amplitude of each of the impulses by R. However, there exists an important difference between this amplitude correction and the standard amplitude scaling correction of Chapter 4: While the standard

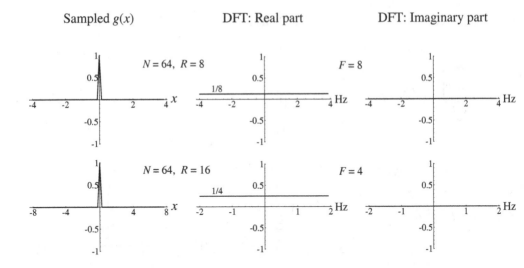

Figure A.8: Example with an impulse in the *signal domain* (see Sec. A.4): The discrete counterpart of $g(x) = \delta(x)$ and its DFT spectrum (after the corrections of Chapters 3 and 4). The array length being used is $N = 64$. Top row: with a sampling range length of $R = 8$ (i.e. sampling interval $\Delta x = 8/64 = 0.125$); bottom row: with a sampling range length of $R = 16$ (i.e. sampling interval $\Delta x = 16/64 = 0.25$). Note the amplitude difference between the two DFT spectra. In the continuous case the corresponding spectrum amplitude is 1.

amplitude scaling correction should be applied in all cases to the entire DFT output, the present impulse amplitude correction only applies to *impulses* in the DFT output (and their leaked tails, if any; see Remark A.2 below). Thus, when the DFT output is hybrid and contains both continuous and impulsive elements, like in Figs. A.4–A.7, the present amplitude correction should be only applied to the impulses themselves (and their leaked tails, if any), but not to the continuous spectral elements. For example, in the above mentioned figures, the division by R should only be applied to the two elements of the output array that contain impulses, and even there it should be done with care, since this division only concerns the amplitude contribution of the impulse itself (in the present example, $\frac{1}{2}R$), but not that of the underlying continuous element (in the present example, $\frac{1}{2}$).

If this selective amplitude correction within the DFT output array is not feasible, no major harm will be caused — one should simply remember that the resulting impulse amplitudes in the DFT output will be R times higher than in the corresponding CFT.[9] ■

---

[9] In simple hybrid cases where the input signal is known to be the sum of a periodic function and an aperiodic function, it would be easier to apply the DFT to each of the two functions separately and then sum up the two resulting DFT spectra (thanks to the addition theorem). Here, the division by R can be applied to the *entire* DFT spectrum of the periodic function, before adding up the two DFT spectra.

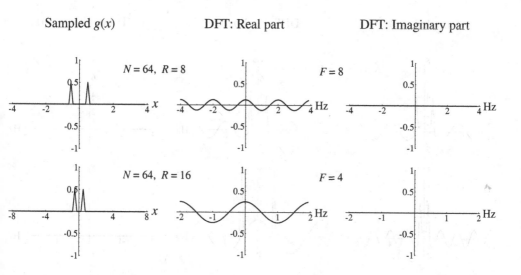

Figure A.9: Example with impulses in the *signal domain* (see Sec. A.4): The discrete counterpart of $g(x) = \frac{1}{2}\delta(x-a) + \frac{1}{2}\delta(x+a)$ (with $a = \frac{1}{2}$) and its DFT spectrum (after the corrections of Chapters 3 and 4). The array length being used is $N = 64$. Top row: with a sampling range length of $R = 8$ (i.e. sampling interval $\Delta x = 8/64 = 0.125$); bottom row: with a sampling range length of $R = 16$ (i.e. sampling interval $\Delta x = 16/64 = 0.25$). Note the amplitude difference between the two DFT spectra. In the continuous case the corresponding spectrum amplitude is 1.

Remark A.2: The particular behaviour of impulses in the DFT spectrum can be also explained from a different point of view. As we have seen in Chapter 6, unlike the continuous Fourier transform, the DFT can only take into account a finite range of the original, underlying function $g(x)$, meaning that $g(x)$ must be truncated before the application of the DFT. This truncation operation can be viewed, still in the continuous world, as a multiplication of the underlying function $g(x)$ by a rectangular "truncation window" $w(x)$ whose length equals the sampling range $R = N\Delta x$, namely, $w(x) = \text{rect}(x/R)$. The spectral-domain counterpart of this truncation is that the DFT does not approximate the original continuous-world Fourier transform $G(u)$, but rather the convolution of $G(u)$ with the Fourier transform of the truncation window $w(x)$, $W(u) = R\text{sinc}(Ru)$. As long as $G(u)$ is non-impulsive, the convolution with $W(u)$ will only smooth out $G(u)$ and add a ripple effect (see Fig. 6.1, and a worked-out example in Sec. D.5). However, when $G(u)$ includes impulses, the convolution with $W(u)$ will replace each impulse $c_i\delta(u-a_i)$ by an additive replica of $R\text{sinc}(Ru)$ that is scaled by the impulse amplitude c_i and centered about the impulse location a_i (Fig. 6.2). As shown in Sec. 6.5, if the continuous-world impulse is located *exactly* on an element of the DFT output array, giving a sharp discrete-world impulse in the DFT spectrum, this replica of $R\text{sinc}(Ru)$ will simply multiply the discrete impulse amplitude c_i by R, but have no effect on the other elements of the output array.

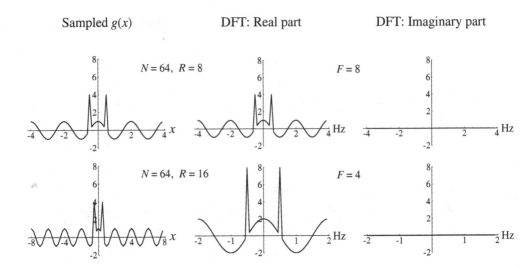

Figure A.10: The discrete counterpart of the signal $g(x) = c[\frac{1}{2}\delta(x-a) + \frac{1}{2}\delta(x+a)] + \cos(2\pi f x)$ (with $a = \frac{1}{2}$, $c = 8$ and $f = \frac{1}{2}$) and its DFT spectrum (after the corrections of Chapters 3 and 4). The array length (number of samples) being used is $N = 64$. Top row: with a sampling range length of $R = 8$ (i.e. sampling interval $\Delta x = 8/64 = 0.125$); bottom row: with a sampling range length of $R = 16$ (i.e. sampling interval $\Delta x = 16/64 = 0.25$). Note that in the DFT spectrum the two impulses are R times higher than in the continuous-world spectrum, whereas the amplitude of the cosine (the Fourier transform of the impulse pair in the signal domain) is $1/F$ times that of the continuous-world spectrum. Compare with Figs. A.2 and A.9. This figure is based on the same concept as Fig. A.4 (the combination of impulsive and non-impulsive signals in the same example), but in the present case both the signal and the spectral domains contain such a combination. See Sec. A.4.

And if the continuous-world impulse is located *between* elements of the DFT output array, so that the resulting discrete impulse is smoothed out and smeared due to leakage, the multiplication by R (and hence the amplitude dependence on the choice of the sampling range length R) will affect both the smoothed out impulse and its leaked tails, but not the other components of the DFT output array. This explains, indeed, why the impulse amplitude in the DFT spectrum is multiplied by R with respect to the original continuous-world spectrum $G(u)$, and why this dependence of the DFT on the choice of R affects only impulses (and their leaked tails, if any) but not other components of $G(u)$, as indeed illustrated in Fig. A.4. More details on this subject are provided in Sec. D.6. ■

The particular behaviour of impulses in the DFT spectrum is explained from yet another point of view in Sec. D.9 of Appendix D, and several further examples are given there in Sec. D.9.1.

Sampled $g(x)$	DFT: Real part	DFT: Imaginary part

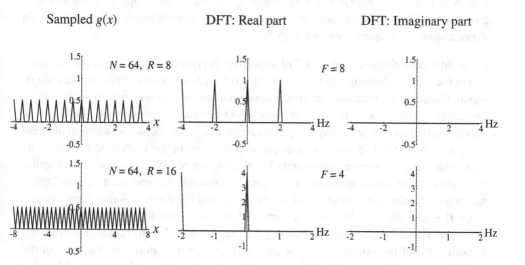

Figure A.11: Example where both the signal domain and the spectral domain are purely impulsive (see Sec. A.4): The discrete counterpart of the comb signal $g(x) = III(2x)$ and of its continuous-world spectrum $G(u) = \frac{1}{2}III(u/2)$ as obtained by DFT (after the corrections of Chapters 3, 4). The array length being used is $N = 64$. Top: with a sampling range length of $R = 8$ (i.e. sampling interval $\Delta x = 8/64 = 0.125$); bottom: with a sampling range length of $R = 16$ (i.e. sampling interval $\Delta x = 16/64 = 0.25$). Note that in the DFT spectrum the impulses are R/F times higher than in the continuous-world spectrum $G(u) = \frac{1}{2}III(u/2)$, in which the impulse heights are 1 [Bracewell86 p. 414].

A.4 Impulses in the signal domain

So far, we have discussed the behaviour of impulses in the spectral domain. What happens when impulses appear in the signal domain?

Consider, for example, the continuous-world case of $g(x) = \delta(x)$, whose spectrum is the constant function $G(u) = 1$. The discrete counterpart of $g(x)$ is a sequence consisting of zeroes except for the element at the origin, whose height is 1. As shown in Fig. A.8, if we take the DFT of this sequence we obtain (after applying the standard amplitude scaling correction) a constant sequence whose height is $1/F$, where $F = N\Delta u$ is the frequency range length (the range length of the *non-impulsive domain*).

As a second example, consider the continuous-world case of $g(x) = \frac{1}{2}\delta(x-a) + \frac{1}{2}\delta(x+a)$, whose spectrum is $G(u) = \cos(2\pi au)$. The discrete counterpart of $g(x)$ is a sequence consisting of zeroes except for two $\frac{1}{2}$-valued elements that are located at the same distance a to both sides of the origin (we assume here that a is an integer multiple of the sampling interval Δx, so that the two impulses fall exactly on a sampling point). Now, as shown in

Fig. A.9, if we take the DFT of this sequence we obtain (after applying the standard amplitude scaling correction) a cosinusoidal signal whose amplitude is $1/F$, which is again dependent on the frequency range length F.

The different behaviour under DFT of impulsive and non-impulsive signals in the signal and in the spectral domains can be best illustrated by a combination of all cases in a single signal. Consider, for example, the continuous-world signal $g(x) = c[\frac{1}{2}\delta(x-a) + \frac{1}{2}\delta(x+a)] + \cos(2\pi f x)$, which consists of a pair of impulses of height $\frac{1}{2}c$ located at the points $x = \pm a$ plus a continuous cosine of frequency f and amplitude 1. Thanks to the addition theorem the continuous-world spectrum of this signal is the sum of the individual spectra of its two terms, namely, a cosine with amplitude c and frequency a (the spectrum of the impulse pair) plus a pair of impulses at the frequencies $u = \pm f$ (the spectrum of $\cos(2\pi f x)$), which is represented by two spikes of height $\frac{1}{2}$ on top of the cosine: $G(u) = c\cos(2\pi a u) + [\frac{1}{2}\delta(u-f) + \frac{1}{2}\delta(u+f)]$. The discrete counterpart of $g(x)$ is a sequence consisting of samples of the signal $\cos(2\pi f x)$ plus two c-valued spikes that are added to the cosine at a distance a to both sides of the origin (again, we assume here that a is an integer multiple of the sampling interval Δx, so that the two impulses fall exactly on a sampling point). When this input signal is fed to the DFT, the result (after reorganization and application of the standard amplitude scaling correction) shows indeed a cosine signal plus an impulse pair to both sides of the origin, as expected (see Fig. A.10). But as shown in this figure, the amplitude of the cosine in the DFT spectrum is multiplied by $1/F$ with respect to the continuous-world spectrum (like in Fig. A.9), while the height of the impulse pair is multiplied by R with respect to the continuous-world spectrum (just as in Fig. A.2). This confirms, indeed, the behaviour under DFT of impulsive signals in both domains, as explained above. And it also confirms that when two signals are summed together in the signal domain each of them still behaves in the DFT of the sum exactly as it would when it undergoes DFT alone; this is most spectacular if the individual behaviour under DFT of each of the summed-up signals is different.

In conclusion, the behaviour under DFT of discrete impulses in the signal domain is dual to their behaviour in the spectral domain: When impulses appear in the spectral domain, their heights in the DFT spectrum are multiplied by R with respect to the continuous-world spectrum (as shown in Figs. A.1 and A.2); but when impulses appear in the signal domain, the height of their individual contribution to the DFT spectrum is multiplied by $1/F$ with respect to the continuous-world spectrum (see Figs. A.8 and A.9).

Remark A.3: (Impulses in the signal domain — signal-domain counterpart of Remark A.1): The fact that the DFT spectrum of an impulsive signal (after the application of the standard amplitude scaling correction of Sec. 4.4) is F times smaller than the corresponding CFT means that in order to obtain the expected DFT spectrum, we need to perform a multiplication by F. Once again, this treatment should only be applied to the spectral components that belong to the impulsive input in question (note that this treatment should be applied even if these input impulses suffer from signal-domain leakage). If such a selective correction is not feasible, the respective elements in the DFT output will simply have a different amplitude than in the CFT. ■

Remark A.4: (Signal-domain counterpart of Remark A.2): The particular behaviour of impulses in the DFT input can be also explained from a different point of view. As we already know, the DFT can only take into account a finite range of frequencies. Therefore, if the original, underlying continuous-world spectrum $G(u)$ does not die out at the borders of this range, $G(u)$ should be truncated before the transition to the discrete world (in order to avoid frequency foldover due to aliasing). This truncation operation can be viewed, still in the continuous world, as a multiplication of the underlying spectrum $G(u)$ by a rectangular "truncation window" $W(u)$ whose length equals the DFT frequency range length $F = N\Delta u$, namely, $W(u) = \text{rect}(u/F)$. The signal-domain counterpart of this truncation is that the input signal of the DFT should not be $g(x)$ itself, but rather the convolution of $g(x)$ with the inverse Fourier transform of the truncation window $W(u)$, $w(x) = F\text{sinc}(Fx)$. As long as $g(x)$ is non-impulsive, the convolution with $w(x)$ will only smooth out $g(x)$ and add a ripple effect (see a worked-out example in Sec. D.5 of Appendix D). However, when $g(x)$ includes impulses, the convolution with $w(x)$ will replace each impulse $c_i\delta(x-a_i)$ by an additive replica of $F\text{sinc}(Fx)$ that is vertically scaled by the impulse amplitude c_i and centered about the impulse location a_i. But unlike in the non-impulsive case, this tacitly implies a vertical scaling by F. If the original continuous-world impulse is located *exactly* on an element of the DFT input array, this convolution will simply multiply the discrete impulse amplitude c_i by F, but have no effect on the other elements of the input array. And if the continuous-world impulse is located *between* elements of the DFT input array, the convolution will scale up the impulse by F, but also force it to leak out. In both cases, the multiplication by F (and hence the amplitude dependence on the choice of the DFT frequency range length F) will only affect the impulses (and their leaked tails, if any), but not the other components of the DFT input array. It follows, therefore, that all impulses in the DFT input, either leaked out or not, should be scaled up by F with respect to the original continuous-world function $g(x)$ (see, for example, the left-hand column in Fig. 6.11). This multiplication by F of the impulse heights in the DFT input explains, indeed, thanks to the scaling theorem (see rule 1a in Sec. 2.4.4) why the corresponding spectral components in the DFT output should be multiplied by F in order to correctly represent the CFT $G(u)$, as we have seen above. ■

The particular behaviour of impulses in the DFT input can be explained from yet another point of view, which is analogous to the explanation provided in Sec. D.9 of Appendix D.

We have already seen above what happens when both the signal and the spectral domains are hybrid, i.e. contain a sum of impulsive and non-impulsive signals. What happens, however, when both domains are purely impulsive? This happens in the case of the continuous-world comb signal $g(x) = \text{III}(x)$, for which we have the following continuous-world Fourier pair [Bracewell86 pp. 214, 414]:

$$\text{III}(x) \leftrightarrow \text{III}(u)$$

or more generally, by virtue of the dilation theorem [Bracewell86 p. 122]:

$$III(ax) \leftrightarrow \tfrac{1}{|a|}III(u/a)$$

Note, however, that the spike heights in $III(ax)$ are $\tfrac{1}{|a|}$, and the spike heights in $III(u/a)$ are $|a|$ (see, for example, [Gaskill78 pp. 60–61] or [Bracewell86 p. 414]). So what happens to such continuous-world signals under DFT?

The discrete counterpart of the signal $g(x) = III(ax)$ is a sequence consisting of zeroes except for all the elements that are located at integer multiples of x/a, which contain the value $\tfrac{1}{|a|}$ (again, we assume here that these locations are multiples of the sampling interval Δx, so that all impulses fall exactly on sampling points). Now, as shown in Fig. A.11, if we take the DFT of this sequence we obtain (after applying the standard amplitude scaling correction) a signal corresponding to the continuous-world spectrum $\tfrac{1}{|a|}III(u/a)$, whose amplitude is further multiplied by R/F. This means that in this case the DFT spectrum is multiplied both by R and by $1/F$ with respect to the continuous-world spectrum.

Figure A.12 shows this doubly-impulsive case, in which both the signal and the spectral domains are purely impulsive, as the limit of a gradual sequence of impulsive signals with non-impulsive spectra. (For the sake of comparison, a similar continuous-world counterpart of this figure can be found in [Brigham88 p. 21].)

Interestingly, the "abnormality" in the behaviour of impulses in the discrete case can be partially resolved if we modify our definition of a discrete impulse (Eq. (A.2)) by multiplying the Kronecker delta function by a factor S, which stands for *the length of the other domain*, i.e. $R = N\Delta x$ or $F = N\Delta u$, depending on the case:[10]

$$\delta(n) = \begin{cases} S & n = 0 \\ 0 & \text{otherwise} \end{cases} \qquad\qquad (A.5)$$

This way the factor S is already incorporated in the discrete impulse definition, and therefore the Fourier pairs obtained by DFT (after applying the standard amplitude scaling correction of Chapter 4) will be the same as in the continuous case; for example, the DFT of the constant sequence consisting of 1-valued elements will be $\delta(n)$, and vice versa. Note that this new definition of the discrete impulse is closer to the definition of the continuous-world impulse (Eq. (A.1)), in which the height of the impulse at the point $u = 0$ is infinite, so that it is equal, indeed, to the infinite range of definition (the continuous counterpart of R or F). However, although this definition resolves our *notational* problem (in the sense that the continuous and discrete Fourier pairs involving impulses are expressed now exactly in the same way), it does not eliminate the actual dependence on the value of R or F in DFTs involving impulses, as shown in our figures throughout this appendix.

---

[10] This multiplication of the impulse amplitude by R or by F corresponds, indeed, to the effect of convolving the impulse with the function $R\,\text{sinc}(Ru)$ or $F\,\text{sinc}(Fu)$, respectively, due to the truncation in the other domain (see Remarks A.2 and A.4). As shown in Fig. 6.4(a), when the impulse falls exactly on an array element, this convolution only affects the impulse height.

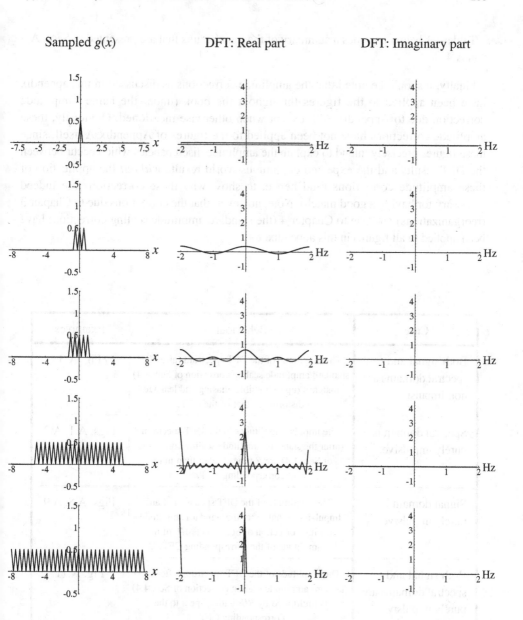

Figure A.12: A sequence showing how the discrete counterpart of the comb signal $g(x) = \text{III}(2x)$ is obtained by adding more and more impulse pairs. The DFT of each intermediate case is still non-impulsive (a sum of cosines), but it gradually converges to an impulsive comb, as shown in the last row. In each intermediate row the amplitude of the DFT spectrum (after the corrections of Chapters 3, 4) is $1/F$ times that of the continuous-world spectrum, due to the impulsive nature of the signal domain. But in the last row, where both the signal domain and the spectral domain are impulsive, the DFT spectrum is $R/F = 4$ times that of its continuous-world counterpart $G(u) = \frac{1}{2}\text{III}(u/2)$, in which the impulse heights are 1 [Bracewell86 p. 414].

Table A.1 below provides a summary of the main results that are presented in Secs. A.3 and A.4.

Finally, it should be noted that the amplitude corrections as discussed in this appendix have been applied to the figures throughout the book (under the name "amplitude correction due to Appendix A"), except when otherwise mentioned. Obviously, these amplitude corrections have not been applied to the figures of Appendix A itself, since these figures precisely intend to explain the amplitude discrepancies which occur between the DFT results and the expected continuous-world results *without* the application of these amplitude corrections (and hence, to show why these corrections are indeed necessary for having a good match). Note, however, that the corrections due to Chapter 3 (reorganizations) and due to Chapter 4 (the standard amplitude scaling correction) have been applied to all figures in this appendix, too.

Case	Behaviour	Examples
Both signal and spectral domains are non-impulsive:	The amplitude of the DFT spectrum (after the standard amplitude scaling correction of Sec. 4.4) matches (up to possible aliasing and leakage distortions) that of the CFT.	Fig. A.3
Spectral domain is purely impulsive:	The impulse amplitudes in the DFT spectrum (after the standard amplitude scaling correction of Sec. 4.4) are R times higher than in the corresponding CFT.	Figs. A.1, A.2
Signal domain is purely impulsive:	The amplitude of the DFT spectrum of an impulsive signal (after the standard amplitude scaling correction of Sec. 4.4) is $1/F$ of the amplitude of the corresponding CFT.	Figs. A.8, A.9
Both signal and spectral domains are purely impulsive:	The amplitude of the DFT spectrum (after the standard amplitude scaling correction of Sec. 4.4) is multiplied by R/F with respect to the corresponding CFT.	Fig. A.11
Hybrid cases:	The DFT of a sum of various components is the sum of the DFTs of the individual components, by virtue of the addition theorem.	Figs. A.4–A.7, A.10

Table A.1: Synoptic overview of the cases discussed in Secs. A.3 and A.4. Impulse amplitudes in the continuous world are to be understood according to the graphic convention of Sec. A.2.

Appendix B

Data extensions and their effects on the DFT results

B.1 Introduction

1D DFT algorithms receive as input a 1D array of N complex numbers (say, samples of a given continuous complex- or real-valued function $y = g(x)$), and return as their output a 1D array of N complex numbers which contains the discrete frequency spectrum. However, when the result obtained by DFT with N data values is not yet satisfactory, one may wonder if a larger number of data values could improve the results. In fact, there exist several possible ways to extend the input data for the DFT, each of which yields quite different results. In this appendix we examine 6 of the most simple ways of extending the input data, and explain how they influence the DFT results (including the possible effects on the DFT artifacts). As we will see, some of these 6 methods are indeed useful and advantageous, depending on one's aims and also on the nature of the data, while others are rather misleading and may introduce significant artifacts. Note that although the discussion in this appendix is presented for the 1D case, its generalization to 2D or MD DFT is straightforward.

B.2 Method 1: Extending the input data by adding new values beyond the original range

Assume that we need to find the DFT of a function $g(x)$ which extends to infinity, such as $g(x) = e^{-x^2}$. It is clear that any number N of samples limits us to a finite sampling range; but if we maintain the sampling interval Δx fixed and take more samples of $g(x)$ beyond the last original sample, a larger part of $g(x)$ will be sampled and submitted to DFT. Suppose that we take KN samples rather than N; how will it affect the DFT results?

It follows from Eq. (4.5) that increasing N by a factor of K while Δx remains unchanged decreases the frequency interval Δu by the same factor. This means that the resolution of the new DFT spectrum resulting from this extention is densified K times. In fact, all values of the original DFT spectrum will still appear in the new transformed signal as before,[1] but each original value will be followed by $K-1$ new values, giving more frequency details (a higher spectral resolution) in the new DFT spectrum. However, due to Eq. (4.9), the frequency range covered by the extended transformed signal does not change, and the maximum frequency of the transformed signal remains the same as before. This is schematically illustrated in Fig. B.1(b).

---

[1] Up to a possible constant scaling factor, if the DFT definition being used contains a scaling factor such as $1/N$ or $1/\sqrt{N}$ (see Chapter 2).

This extention method may be useful when a higher resolution of the spectrum is desired, for instance in order to better distinguish between some closely spaced spectral peaks. It is interesting to note how the DFT artifacts in the original transformed signal are affected by this extention: leakage (see Chapter 6) will be somewhat reduced (since the truncation window has been enlarged and therefore its sinc-shaped transform has been narrowed; see, for example, Fig. 2.1(d) in Chapter 2); but aliasing (see Chapter 5) will be carried over to the new transformed signal and will not be modified (except for manifesting itself at a higher resolution).

B.3 Method 2: Extending the input data by denser sampling within the original range

A second way to increase the size of the DFT consists of decreasing the sampling interval Δx, and taking more samples of the original function within the same range length R as before. If the original signal is extended from N to KN values by such an increase of the sampling rate, the transformed signal will also be extended K times, but this time by increasing K times the frequency range length F (and the maximum frequency $f_{\max}$ obtained by the DFT), while keeping the spectral resolution Δu unchanged (see Eqs. (7.4), (4.10) and (7.3), respectively). In other words, the N values of the original DFT spectrum will continue to represent the same frequencies as before, but new values will appear beyond them, which correspond to higher frequencies that did not appear in the original DFT spectrum (see Fig. B.1(c)).

This does not mean, however, that the N original values in the DFT spectrum will remain exactly as before. First of all, they may be affected by a possible scaling factor, depending on the DFT definition being used. But even more importantly, any aliasing artifacts which occurred in the DFT spectrum before the extention, due to insufficient frequency range

Figure B.1: Schematic illustration of the 6 extension methods discussed in Appendix B. Left column: the given input data sequence; right column: its reorganized DFT (absolute value only). (a) The original N-points input data sequence and its reorganized DFT. (b)–(g) The extended data sequences of KN points according to methods 1–6, and their respective reorganized DFT. Note that if one uses a DFT definition which contains a scaling factor (such as $1/N$ or $1/\sqrt{N}$; see Sec. 2.3 in Chapter 2), appropriate amplitude scaling should be applied to the resulting DFT spectra. All cases are drawn with respect to the true continuous-world signal-domain and frequency-domain axes, which are shown at the bottom of the figure. (This figure could be alternatively drawn with respect to the discrete axes k and n, i.e. with the same element intervals in all cases, so that all rows would have the same geometric length. But then the connection with the real-world interpretation of the results would be lost.)

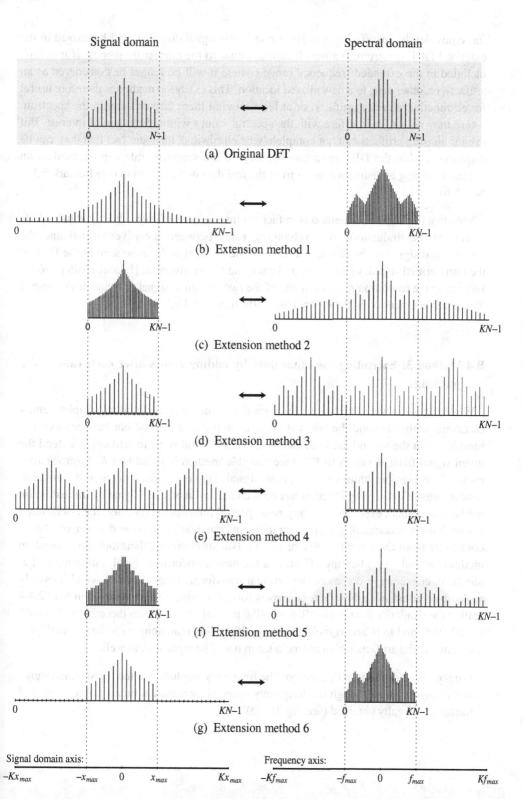

(a) Original DFT

(b) Extension method 1

(c) Extension method 2

(d) Extension method 3

(e) Extension method 4

(f) Extension method 5

(g) Extension method 6

(or, equivalently, insufficient sampling rate in the signal domain), will be moved in the extended DFT spectrum to a new location: either to the correct frequency, if it is now included in the extended frequency range (where it will no longer be considered as an artifact), or, otherwise, to a new aliased location. This extention method is therefore useful for eliminating aliasing artifacts, or at least moving them farther away in the spectrum where they no longer interfere with the spectral values within the range of interest. But even if aliasing artifacts can not completely be eliminated, the mere fact that they can be displaced within the DFT spectrum by varying the sampling rate can be used as an indicator helping to distinguish them from the real data in the spectrum (see Remark 5.3 in Sec. 5.6).

Note that this extention method is in fact the inverse of the first method; and indeed, its effect can be understood by exchanging roles between the given signal and the transformed signal in the first method (since the given signal can be seen as the DFT of the transformed signal, up to a scaling factor and a sign inversion [Bracewell86 p. 364]). Increasing the resolution K times in one of the two domains (signal or frequency) causes a K-fold range extension in the other. This is illustrated in Fig. B.1 (a)–(c).

B.4 Method 3: Extending the input data by adding zeroes after each value (zero packing)

When the underlying continuous function is not known and additional samples cannot be taken, neither beyond the original range as in the first method nor between existing samples as in the second method, there still exist several ways to artificially extend the given signal from N values to KN. One possible method is by adding $K-1$ zeroes after each of the original values in the given signal. However, since no new information concerning the underlying function has been added to the original signal by this extention method, we cannot expect to gain any new spectral information (because the information in one domain is completely equivalent to the information in the other domain, one being obtainable from the other by DFT or IDFT). Not surprisingly, therefore, this extention method has only a replicating effect, and the new transformed signal will consist of K identical, consecutive copies of the original transformed signal (all of which possibly scaled by a constant scaling factor). This is formally expressed by point 11 in Sec. 2.4.4 (called the similarity theorem in [Bracewell86 p. 370], or the stretch theorem in [Smith07 p. 157–158] and in [Cartwright90 pp. 204–205]). Note that along with the original DFT spectrum, all the artifacts which are included in it will be replicated, as well.

We see, therefore, that in this method the frequency resolution in the transformed signal is not changed, and although the frequency range is increased K times, no new spectral information is really obtained (see Fig. B.1(d)).

B.5 Method 4: Extending the input data by replicating it

Another possible method to extend the given signal from N to KN values is by simply replicating it $K-1$ more times. Obviously, in this case, too, no gain of information can be expected in the spectral domain. In fact, this extending method is the inverse of the previous method, and its effect on the transformed signal, within a possible constant scaling, is merely the insertion of $K-1$ zeroes after each original value (this can be formally obtained by an inverse application of the similarity theorem in [Bracewell86 p. 370] or the stretch theorem in [Smith07 pp. 157–158]; see point 11a in Sec. 2.4.4). The frequency range of the transformed signal is not changed by this operation, nor does the frequency represented by each of its original N elements, now being separated by zeroes in the transformed signal. Although the DFT spectrum contains now more values per frequency unit, no new spectral information has been obtained. The benefits of such an extention of the DFT are therefore rather limited. This method is illustrated in Fig. B.1(e). A similar case of data replication and its spectral consequences (albeit within the original array length N) is presented in Remark 7.1 and Fig. 7.1.

B.6 Method 5: Extending the input data by replicating each of its elements

Still another way to extend the given signal from N to KN data values is obtained by replicating each of its values $K-1$ more times. Here, too, no gain of information can be expected in the spectral domain, since no new information has been added to the original signal concerning the underlying function. As illustrated in Fig. B.1(f), the result of this extention method in the spectral domain is a K-fold replication of the original transformed signal, which is then modulated (i.e. multiplied) by a sinc-shaped envelope. This multiplication reflects the fact that the value replications in the signal domain are in fact a convolution with a K-values wide rect function; according to the convolution theorem, this implies in the spectral domain a multiplication with the Fourier transform of this function, which is the sinc function in question. Note that the zeroes of the sinc fall exactly on the central frequency (the replicated DC) in all the new duplicates, reducing them to zero.

If the underlying function $g(x)$ is a piecewise constant function such as a train of pulses or a histogram, then at first sight this method and method 2 (see Sec. B.3) seem to be identical; however, they conceptually differ since in the present method all the new in-between values are just identical copies of a neighboring value, while in method 2 they more finely approximate the up and down transitions of $g(x)$. The use of the present method should normally be avoided, due to the significant distortions it introduces to the resulting DFT spectrum.

B.7 Method 6: Extending the input data by adding zeroes beyond its original range (zero padding)

This is a variant of extention method 1 (see Sec. B.2), which obviously coincides with it if the underlying function $g(x)$ has non-zero values only within a limited range which is already included in the original sampled signal. If, however, $g(x)$ has non-zero values beyond the range covered by the original signal, too, then although the effect of extending the signal K times by adding extra zeroes beyond the original N samples may look similar to the effect of method 1, the new intermediate values obtained in the transformed signal will not give any new informative values (see Fig. B.1(g)).

In fact, appending zeroes to the original signal rather than appending new samples from the underlying function $g(x)$ can be considered as a multiplication of the periodic extension of the signal by a rectangular wave with a period of KN elements and pulse width of N elements (the original signal length). This multiplication in the signal domain implies in the transformed signal a convolution with the transform of the rectangular wave, which is a sampled sinc function whose frequency step is K times smaller than in the original signal. In other words, what we obtain in the transformed signal is an addition of $K-1$ intermediate values after each value of the original transformed signal, but unlike in method 1, where the intermediate values really give finer frequency values, here the intermediate values simply interpolate the values of the original transformed signal with the sinc function mentioned above [Brigham88 pp. 170–172; Smith07 pp. 133–134, 159–160; Bracewell86 p. 369, the Packing Theorem].

In conclusion, although the smooth results obtained by the present method give an illusion of an increased spectral resolution, in fact no additional resolution has been gained (unless the underlying function $g(x)$ is really zero in the interval where the new zeroes are appended [Brigham88 p. 172]). Note, however, that padding with zeroes is not always useless. For example, this method can be used advantageously as a convenient tool for *interpolating* a given sampled function by means of DFT [Brigham88 pp. 198–199]: We first compute the DFT of the given sampled function, and then add padding zeroes beyond its original frequency range (i.e. after the index $n = N/2$ of the non-organized output array). And indeed, the inverse DFT of this extended spectrum gives, back in the signal domain, a more densely interpolated version of the original signal. Padding the original signal with zeroes is also necessary when we wish to find the convolution of two finite-duration signals using DFT, by virtue of the convolution theorem. When convolution is computed with the aid of DFT, both of the given signals are treated as being periodic (recall that DFT considers both its input and its output data as an N-element period of a periodic function), and therefore the resulting convolution is *cyclical*. This means that the outer ends of both input signals inevitably penetrate (additively) into the other side of the resulting convolution. This generates a cyclical wrap-around effect which corrupts the resulting convolution with respect to the expected non-cyclical convolution. This wrap-around effect can be avoided by padding both of the signals to be convolved by a

Method number:	Operations on the DFT input:	Results in the DFT output:**
1	K-fold extension by adding new values beyond the original range.	K-fold increase in the spectral resolution; frequency range unchanged.
2	K-fold extension by performing a K times denser sampling within the original range.	K-fold increase in the frequency range; spectral resolution unchanged.
3*	K-fold extension by adding $K-1$ zeroes after each value.	K-fold replication of the transformed signal.
4*	K-fold extension by replicating the entire given signal $K-1$ more times.	Insertion of $K-1$ zeroes after each original value in the transformed signal.
5*	K-fold extension by replicating each value of the given signal $K-1$ more times.	K-fold replication of the transformed signal, and modulation by a sinc function.
6*	K-fold extension by padding zeroes beyond the original range.	Insertion of $K-1$ sinc-interpolation values after each original value in the transformed signal.

* These cases are "dummy" extension methods which do not really add new spectral information in the transformed signal.

** An appropriate scaling factor may be also applied in each case, depending on the DFT definition being used.

Table B.1: Summary of the 6 extension methods.

sufficiently long span of zeroes before applying the DFT convolution. A detailed and illustrated explanation of this subject can be found, for example, in [Bergland69 p. 48] or in [Brigham88, Chapter 10] for the 1D case, and in [Brigham88 Sec. 11.4] for the 2D case.

It may be noted in this context that in cases where the original data must be extended in order to fit into an input array length N which is a power of 2 (for the sake of using FFT), adding zeroes might not always be the best method. If the addition of trailing zeroes causes jumps in the original data (including between its last and first elements, considering the data as cyclical due to the periodic nature of DFT), a large spectral range of high frequencies will be introduced into the transformed signal, which may cause in the resulting DFT spectrum an oscillatory behaviour due to the leakage artifact (see Chapter 6). Aliasing artifacts (see Chapter 5) may be also caused, since the sampling frequency of the input signal may not be sufficient to capture the high frequencies due to the jumps. Padding with an appropriate non-zero value in order to prevent significant jumps in the signal could be more judicious in such cases [Bracewell86 pp. 375–376].

B.8 Conclusions

A summary of the different extension methods which have been discussed above is given in Table B.1 and in Fig. B.1 for quick reference. Understanding the advantages and shortcomings of each method may help the DFT user to select the most appropriate method for his needs, and to avoid pitfalls and false interpretations of the DFT results.

It should be noted that although only the one-dimensional case has been discussed here, similar results can be also obtained for two or higher dimensional DFT, by applying the same considerations to each dimension separately.

Appendix C

The roles of p and q and their interconnections

C.1 Introduction

As we have seen throughout this book, and in particular in Chapters 5 and 6, cases in which the original input function is periodic are particularly interesting, among other reasons due to the impulsive nature of their spectra. In this appendix we study in more detail some properties of such periodic cases in the discrete world, and provide their multidimensional generalization.

C.2 The one dimensional case

Suppose we are given a continuous periodic function $g(x)$ whose frequency is f, so that its period is $P = 1/f$. As we already know, in order to apply DFT to the function $g(x)$ we first need to truncate it into a finite range of length R and sample it within this range. The resulting N values serve as the input array of the DFT, and the DFT spectrum is obtained in the N values of the output array. The number of periods of $g(x)$ within the sampling range of length R is, of course:[1]

$$q = R/P = Rf \qquad (C.1)$$

The same number of periods will also appear within the range $0,...,N-1$ of the discrete (sampled) version of our function, in the input array of the DFT. This means that in the discrete world the length of each period of our function is, in terms of array elements:

$$p = N/q \qquad (C.2)$$

This is, indeed, the discrete counterpart of the continuous-world relation $P = 1/f$. Formulated in a more symmetric way, the discrete counterpart of $Pf = 1$ is, therefore:

$$pq = N \qquad (C.3)$$

We dwell here on the symmetric form because, as we will see later in Sec. C.3, this is the form that lends itself best to multidimensional generalization.

Remark C.1: Note the dual role of q in the signal and spectral domains: In the signal domain, q is the number of periods of $g(x)$ within the sampling range length R (or within the input array length N). But in terms of the DFT spectrum, q is the distance of the first

---

[1] On the advantages and shortcomings of the inclusion of more than one period of $g(x)$ within the sampling range (the input array of the DFT), see Remark 7.1.

harmonic impulse from the origin (i.e. the frequency of $g(x)$ in terms of array elements), as well as the distance between successive harmonic impulses (again, in terms of array elements). Note that q needs not necessarily be integer; as explained in Chapter 6, when q is non-integer the DFT spectrum suffers from a leakage artifact. Similarly, p, too, plays a double role: In the signal domain p is the length of one period of $g(x)$ in terms of array elements; but in the DFT spectrum p is the number of q-element intervals, i.e. the number of harmonic impulses for which there is room (without exceeding and folding over) in the DFT spectrum of the periodic function $g(x)$.[2] Note that p, too, is not necessarily integer (see below); in fact, the only restriction on p and q is that they must satisfy $pq = N$ (Eq. (C.3)), where N is obviously an integer number. See Example C.1 and Figs. C.2, C.3. ∎

As we see, p and q are given in terms of continuous-world parameters by the relations:

$$p = F/f, \qquad q = R/P \qquad\qquad (C.4)$$

whose respective discrete-world counterparts are simply (see Eq. (C.3)):

$$p = N/q, \qquad q = N/p$$

Furthermore, the connection between the frequency f of our periodic function $g(x)$, in terms of Hz, and its discrete counterpart q, in terms of array elements, is given by:

$$q = f/\Delta u \qquad\qquad (C.5)$$

(this can be also obtained from Eqs. (C.1) and (7.3)). Similarly, we have also the following connection between the period P in terms of continuous-world units and its discrete-world counterpart p in terms of array elements:

$$p = P/\Delta x \qquad\qquad (C.6)$$

The nature of p (integer or non-integer) has particular significance if high harmonic impulses of the periodic function $g(x)$ fall beyond the borders of the DFT spectrum, i.e. if our sampled version of $g(x)$ suffers from aliasing. Using our present terminology we can now reformulate Remark 5.1 as follows:

Remark C.2: There exists a particular case in which no "stray impulses" appear in the DFT spectrum of a periodic function $g(x)$ even if strong aliasing occurs. This happens if p is integer, i.e. if the distance q between the origin and the first harmonic impulse of the periodic function $g(x)$ in the DFT spectrum (in terms of output array elements) satisfies $N = pq$ for an integer p (see Figs. C.2(a) and (c)). In this case, higher harmonic impulses which exceed from one end of the DFT spectrum due to aliasing (foldover) will fall exactly on top of true impulse locations (harmonic frequencies)[3] in the opposite end of the DFT spectrum; their heights will be simply summed up. ∎

---

[2] We count here *all* the harmonic impulses of $g(x)$, even those whose amplitude is zero. See Footnote 3.
[3] To be more precise, we speak here of impulse locations in terms of the *algebraic support* of the spectrum of our periodic function, which consists of *all* the integer multiples of the basic frequency f of $g(x)$, even if the corresponding impulse amplitudes happen to be zero [Amidror09 p. 110].

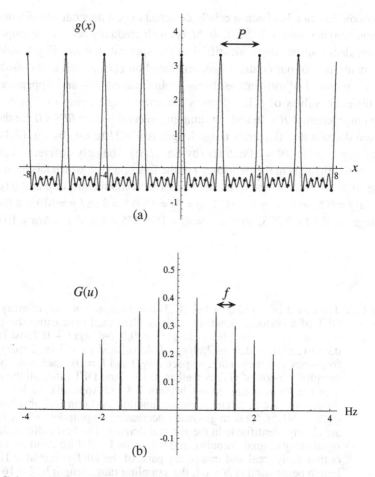

(a)

(b)

Figure C.1: (a) The continuous-world periodic function $g(x) = 0.8\cos(2\pi fx) +$
0.7cos(2π2fx) + 0.6cos(2π3fx) + 0.5cos(2π4fx) + 0.4cos(2π5fx) +
0.3cos(2π6fx) (see Example C.1). In the particular case plotted here
the frequency and period of the function are $f = 0.5$ and $P = 1/f = 2$.
(b) The continuous-world spectrum of $g(x)$, $G(u)$. The black dots in
(a) correspond to the sampled counterpart of $g(x)$ with a sampling
step of $\Delta x = 0.25$. Note that this sampling rate is not sufficient to
capture the detail of the oscillations of $g(x)$ (see below the *x* axis), and
it causes visible aliasing; the same sampled signal and its aliased
DFT spectrum are shown in row (a) of Fig. C.2.

Example C.1: The effect of *p* and *q* on the input and output arrays of the DFT of a
periodic function:

Consider the following periodic function with frequency *f* and period $P = 1/f$ (Fig. C.1):

$$g(x) = 0.8\cos(2\pi fx) + 0.7\cos(2\pi 2fx) + 0.6\cos(2\pi 3fx)$$

$$+ 0.5\cos(2\pi 4fx) + 0.4\cos(2\pi 5fx) + 0.3\cos(2\pi 6fx)$$

This periodic function has been specially designed to best illustrate our discussion here: It has 6 non-zero harmonics (f, $2f$, $3f$, $4f$, $5f$, $6f$) with gradually decreasing amplitudes, so that its impulses can be easily identified in the spectral domain. Fig. C.2 shows the discrete counterpart of our continuous-world function $g(x)$ as well as its resulting DFT (after all the required adjustments as discussed in Chapters 3, 4 and Appendix A), for 4 slightly different values of f. In all cases the array length being used is $N = 64$, the sampling range length is $R = 16$ and the sampling interval is $\Delta x = R/N = 0.25$; therefore in the spectral domain the frequency range length is $F = 1/\Delta x = 4$ (by Eq. (7.4)) and the frequency step is $\Delta u = 1/R = 0.0625$ Hz (by Eq. (7.3)). The only difference between the four rows of the figure is in the frequency f of the function $g(x)$, which has been chosen in each case so as to illustrate one of the four possible combinations of p and q (given that $pq = N$): (a) $f = 0.5$, so that by Eq. (C.1) $q = Rf = 16 \cdot 0.5 = 8$ and $p = N/q = 8$ (both p and q are integers); (b) $f = 0.375$, so that $q = Rf = 16 \cdot 0.375 = 6$ and $p = N/q = 10.666$ (q is

Figure C.2: The effect of p and q of Eq. (C.3) on the input and output arrays of the DFT of a periodic function. This is illustrated here using the periodic function shown in Fig. C.1, $g(x) = 0.8\cos(2\pi fx) + 0.7\cos(2\pi 2fx) + 0.6\cos(2\pi 3fx) + 0.5\cos(2\pi 4fx) + 0.4\cos(2\pi 5fx) + 0.3\cos(2\pi 6fx)$, whose frequency and period are, respectively, f and $P = 1/f$. Each row shows the sampled version of $g(x)$ as well as its resulting DFT (after all the required adjustments as discussed in Chapters 3, 4 and Appendix A), for a slightly different value of f. The periodic function $g(x)$ consists of 6 harmonics (f, $2f$, $3f$, $4f$, $5f$, $6f$) with gradually decreasing amplitudes, whose impulses are clearly identifiable in the spectral domain. The first column shows the input array (signal domain), and the second and third columns show the output array (real and imaginary parts of the DFT spectrum). The array length being used is $N = 64$, the sampling range length is $R = 16$ and the sampling interval is $\Delta x = R/N = 0.25$; therefore in the spectral domain the frequency range length is $F = 1/\Delta x = 4$ and the frequency step is $\Delta u = 1/R = 0.0625$ Hz. The only difference between the four rows is in the frequency f of the function $g(x)$, which has been chosen in each case so as to illustrate one of the four possible combinations of p and q (given that $pq = N$): (a) $f = 0.5$, so that by Eq. (C.4) $p = F/f = 4/0.5 = 8$ and $q = N/p = 8$ (both p and q are integers); (b) $f = 0.375$, so that $p = F/f = 4/0.375 = 10.666$ and $q = N/p = 6$ (q is integer but not p); (c) $f = 0.666$, so that $p = F/f = 4/0.666 = 6$ and $q = N/p = 10.666$ (p is integer but not q); (d) $f = 0.471$, so that $p = F/f = 4/0.471 = 8.5$ and $q = N/p = 7.53$ (neither p nor q are integers). Whenever q is integer (rows (a) and (b)) the spectral impulses are sharp and no leakage occurs. In row (b) all the 6 harmonic impulses are clearly visible; note that the 6-th harmonic impulse has been folded over into the other end of the spectrum due to aliasing. In row (a) the 3 highest harmonic impulses have been folded over due to aliasing, but unlike in row (b) they fall exactly on top of existing low-frequency impulses, and their amplitudes are summed up. In rows (c) and (d) q is non-integer, so that leakage occurs in the DFT spectrum. Just like in the first two rows, in row (d) all the 6 harmonic impulses are clearly visible (the last two being folded-over due to

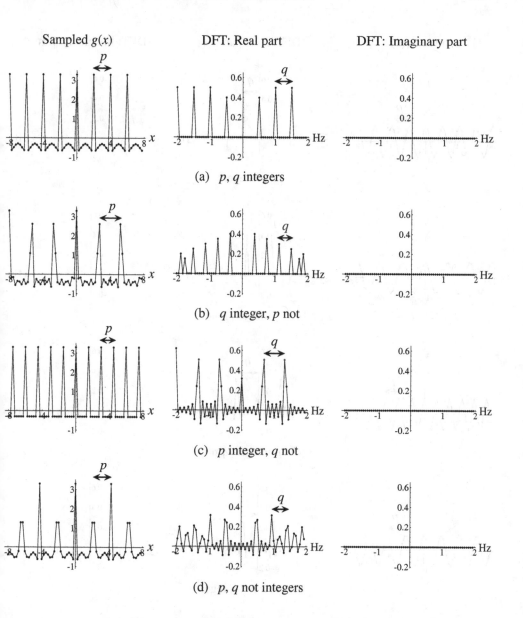

Sampled *g(x)*   DFT: Real part   DFT: Imaginary part

(a) *p, q* integers

(b) *q* integer, *p* not

(c) *p* integer, *q* not

(d) *p, q* not integers

aliasing), while in (c) the folded-over impulses fall exactly on top of existing low-frequency impulses, and their amplitudes are summed up. Note that the high-frequency "grass" at the bottom of the original continuous-world signal *g(x)* (see Fig. C.1(a)) is mimicked here in all rows by a lower frequency grass, due to the insufficient sampling rate and the resulting aliasing effect. The shape of the grass differs from row to row because the aliasing is different in each case; the irregular shape of the grass in rows (b) and (d) is due to the non-integer value of *p*. All the discrete signals and DFT spectra are drawn here as connected symbol plots (see Sec. 1.5.1).

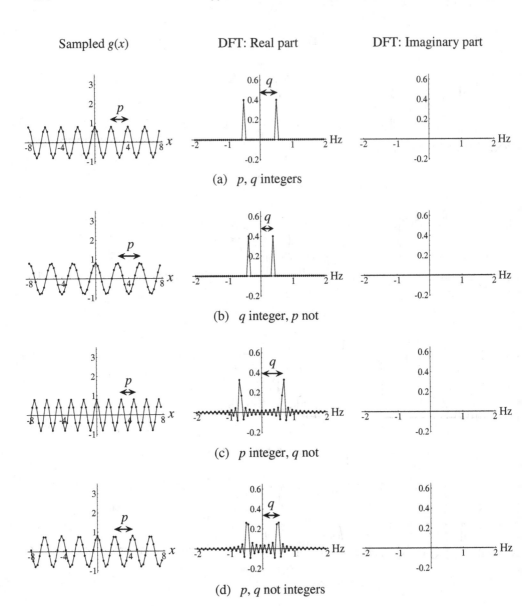

Sampled g(x)   DFT: Real part   DFT: Imaginary part

(a) *p, q* integers

(b) *q* integer, *p* not

(c) *p* integer, *q* not

(d) *p, q* not integers

Figure C.3: Same as Fig. C.2, except that the function $g(x)$ consists here of the first harmonic alone: $g(x) = 0.8\cos(2\pi f x)$. The roles of p and q in terms of the signal domain are clearly visible: p is the length of one period of $g(x)$ in terms of array elements, while q represents the number of periods of $g(x)$ within the input array. In rows (a) and (b) q is exactly integer, so that the input array is perfectly cyclical (seamlessly wraparound); but this is not the case in rows (c) and (d). In terms of the spectral domain, the impulses in rows (a) and (b) are located exactly on an output array element because q is integer; but in rows (c) and (d) the impulses are located *between* output array elements, causing a significant leakage effect.

integer but not p); (c) $f = 0.666$, so that $q = Rf = 16 \cdot 0.666 = 10.666$ and $p = N/q = 6$ (p is integer but not q); (d) $f = 0.471$, so that $q = Rf = 16 \cdot 0.471 = 7.53$ and $p = N/q = 8.5$ (neither p nor q are integers). As we can see in the figure, whenever q is integer (cases (a) and (b)) the spectral impulses are sharp and no leakage occurs in the DFT spectrum. But when q is non-integer (cases (c) and (d)) the spectral impulses fall between output-array elements, so that leakage does occur. On the other hand, whenever p is integer (cases (a) and (c)), there is room in the DFT spectrum exactly for an integer number of harmonic impulses (this is also true in cases where some or even most of the harmonic impulses have zero amplitude, as in the case of the function $g(x) = \cos(2\pi f x)$; see Fig. C.3). This is particularly significant when aliasing occurs, since in this case all the folded-over impulses due to aliasing fall exactly on top of existing lower-frequency impulse locations (see Remark C.2). But when p is non-integer (cases (b) and (d)), there is room in the DFT spectrum for a non-integer number of harmonic impulses. In this case, if aliasing occurs, the impulses that are folded-over due to aliasing fall *between* existing low-frequency impulse locations, and they become visible in the DFT spectrum as stray impulses.

Note that our function $g(x)$ has been intentionally designed to have its highest harmonic impulses beyond the range of the DFT spectrum (which means that $g(x)$ suffers here from aliasing, i.e. insufficient sampling rate in the signal domain): This allows us to show how the exceeding harmonic impulses behave in each of the four cases. A best grasp of the intended continuous-world spectrum (Fig. C.1(b)) is obtained in case (b), where all the impulses are sharp (no leakage occurs), and the folded-over (aliased) impulses do not overlap with existing impulses, so they are clearly visible. In case (d), too, the folded-over impulses do not overlap with existing impulses, but they are more difficult to identify due to the leakage artifact. In cases (a) and (c), however, the folded-over impulses overlap (additively) with existing lower-frequency impulses in the other end of the DFT spectrum, so that no intruding stray impulses are visible in spite of the folding over. In such cases the only effect of aliasing on the DFT spectrum is in modifying the amplitudes of the affected low-frequency impulses. ■

For the sake of comparison we also provide here Fig. C.3, which is identical to Fig. C.2 except that the periodic function it shows contains only the first harmonic of $g(x)$, so that no aliasing occurs.

We terminate with a consequence of Remark C.1, which is, in fact, a reformulation of Remarks 6.3 and 6.4 (as we will see soon, this result is more interesting in the MD case):

Remark C.3: A periodic function $g(x)$ is perfectly cyclical (seamlessly wraparound) within the input array, meaning that it has no discontinuities on the array boundaries when the input array is repeated periodically to pave the entire 1D space, *if and only if* its impulses in the DFT spectrum are located exactly on output array elements (so there is no leakage), i.e. *if and only if q is integer.* ■

Table C.1 provides a synoptic review of the roles of p and q in the 1D case, both in the signal and in the spectral domains.

	Effect on the input array (signal domain)	Effect on the output array (DFT spectrum)	Examples
q is integer	Integer number of periods of $g(x)$ within the input array (sampling range); input array is perfectly cyclical (seamlessly wraparound).	Impulse intervals consist of an integer number of array elements; impulses are located exactly on array elements; no leakage.	Figs. 6.3(a),(e),(f), C.2(a),(b) and C.3(a),(b)
q is non-integer	Non-integer number of periods of $g(x)$ within the input array (sampling range); input array is not perfectly cyclical.	Impulse intervals consist of a non-integer number of array elements; impulses are located between array elements; leakage.	Figs. 6.3(b),(c),(d), C.2(c),(d) and C.3(c),(d)
p is integer	Period length of $g(x)$ is exactly an integer number of array elements.	Integer number of harmonic impulse locations within the output array (DFT spectrum). In particular, if aliasing occurs, folded-over impulses fall exactly on top of lower-harmonic impulse locations, and their amplitudes are summed up.	Figs. C.2(a),(c) and C.3(a),(c)
p is non-integer	Period length of $g(x)$ is not an integer number of array elements.	Non-integer number of harmonic impulse locations within the output array (DFT spectrum). In particular, if aliasing occurs, folded-over impulses fall between lower-harmonic impulse locations, giving visible stray impulses.	Figs. C.2(b),(d) and C.3(b),(d)

Table C.1: The effects of *p* and *q* of Eq. (C.3) on the input and output arrays of the DFT of a periodic function $g(x)$. Note that the effects of *p* and *q* are orthogonal (i.e. independent of each other), and all the 4 combinations of *p* and *q* may exist, as shown in Fig. C.2.

C.3 Generalization of *p* and *q* to the multidimensional case

As it could be expected, the situation in the 2D case is more complex and more interesting than in the 1D case. This is already true in the continuous world: First of all, the period and the frequency of a 2D periodic function $g(x,y)$ are no longer scalars but rather vectors in the x,y space and in the spectral u,v space, respectively. Second, a 2D periodic function $g(x,y)$ may be either 1-fold periodic or 2-fold periodic, so that it may have either one or two period-vectors in the x,y space, and hence either one or two frequency-vectors in the u,v space. Furthermore, it turns out that the relationship between the period-vectors and the frequency-vectors in the 2D case is no longer as simple as the relationship between the period P and the frequency f in the 1D case. For example, if $g(x,y)$ is 2-fold periodic, then its two period-vectors $\mathbf{P}_1$ and $\mathbf{P}_2$ do not have the same orientation as the respective frequency-vectors $\mathbf{f}_1$ and $\mathbf{f}_2$ in the spectral domain; rather, $\mathbf{P}_1$ is orthogonal to $\mathbf{f}_2$, and $\mathbf{P}_2$ is orthogonal to $\mathbf{f}_1$. It can be expected, therefore, that the situation in the discrete world, too, will not be as simple as in the 1D case. So how can we generalize the scalars p and q and the relationship $pq = N$ between them to the multidimensional case?

As we already know, $pq = N$ is the discrete counterpart of the continuous-world relationship $Pf = 1$ between the period P and the frequency f of a 1D periodic function $g(x)$. Let us start with the multidimensional generalization of the continuous-world relationship, $Pf = 1$. Then, based on the experience we gain from this case, we will proceed to the analogous generalization in the discrete world.

C.3.1 Multidimensional generalization in the continuous world

The multidimensional generalization of the continuous-world relationship $Pf = 1$ can be found, for example, in [Rosenfeld82 p. 75]; a more detailed review is also provided in Secs. A4–A.5 of Appendix A in [Amidror09]. Let us briefly reformulate here the main continuous-world results.

Definition C.1: If $\mathbf{P}_1,...,\mathbf{P}_n \in \mathbb{R}^M$ are n fundamental period-vectors of the n-fold periodic function $g(x_1,...,x_M)$ in the MD case (where obviously $1 \le n \le M$), then the corresponding frequency-vectors $\mathbf{f}_1,...,\mathbf{f}_n \in \mathbb{R}^M$ in the spectral domain are defined by the following relationship:[4]

$$\mathbf{P}_i \cdot \mathbf{f}_j = \begin{cases} 1 & \text{if } i = j \\ 0 & \text{if } i \neq j \end{cases} \tag{C.7}$$

where $i,j = 1,...,n$. ∎

Note that because Eq. (C.7) is symmetric, it can also be used the other way around to determine the period-vectors $\mathbf{P}_1,...,\mathbf{P}_n$ given the frequency-vectors $\mathbf{f}_1,...,\mathbf{f}_n$. The only requirement in both cases is that the vectors $\mathbf{P}_1,...,\mathbf{P}_n$ as well as the vectors $\mathbf{f}_1,...,\mathbf{f}_n$ be independent; the existence of such sets of n independent vectors is indeed guaranteed because $g(x_1,...,x_M)$ is n-fold periodic.

---

[4] As illustrated in Figs. C.5 and C.6 (and explained in greater detail in Sec. A.3.4 of Appendix A in [Amidror09]), the period-vectors $\mathbf{P}_1,...,\mathbf{P}_n$ as well as the frequency-vectors $\mathbf{f}_1,...,\mathbf{f}_n$ are not unique. For example, every 2D 2-fold periodic function has infinitely many sets of period-vectors (and of frequency-vectors), all of which are equivalent and define the same periodicity. But each set of period-vectors $\mathbf{P}_1,...,\mathbf{P}_n$ defines by Eq. (C.7) its own unique set of frequency-vectors $\mathbf{f}_1,...,\mathbf{f}_n$. This set of frequency vectors is said to be *reciprocal* to its corresponding set of period-vectors.

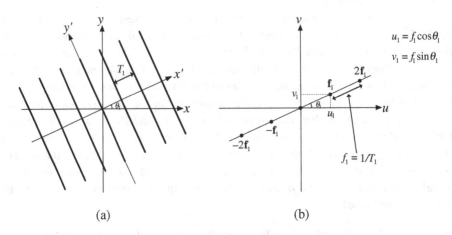

Figure C.4: Schematic plot of a 1-fold periodic function $g(x,y)$ in the signal domain (a), and of the corresponding spectrum (impulse-comb) $G(u,v)$ in the spectral domain (b).

The geometric interpretation of this definition is as follows:

(a) The second condition in Eq. (C.7) means that the vector $\mathbf{f}_j$ is perpendicular to all the vectors $\mathbf{P}_i$ with $i \neq j$; this determines the direction of the line through the origin of $\mathbb{R}^M$ on which $\mathbf{f}_j$ is situated. (For example, in the 3D case $\mathbf{f}_1$ is situated on the line emanating from the origin of $\mathbb{R}^3$ perpendicularly to the plane spanned by $\mathbf{P}_2$ and $\mathbf{P}_3$).

(b) The first condition in Eq. (C.7) determines the precise length and direction of the vector $\mathbf{f}_j$ on the line defined in (a). $\mathbf{P}_j \cdot \mathbf{f}_j = 1$ means that the length of $\mathbf{f}_j$ on this line is reciprocal to the length of the projection of $\mathbf{P}_j$ on the same line: Since by [Vygodski73 p. 142] $\mathbf{P} \cdot \mathbf{f} = |\mathbf{f}| \, |\text{proj}(\mathbf{P})_\mathbf{f}|$ (where $\text{proj}(\mathbf{P})_\mathbf{f}$ denotes the projection of $\mathbf{P}$ on $\mathbf{f}$), and since we have here $\mathbf{P} \cdot \mathbf{f} = 1$, it follows that $|\mathbf{f}| = 1/|\text{proj}(\mathbf{P})_\mathbf{f}|$. Furthermore, since we have $\mathbf{P}_j \cdot \mathbf{f}_j = 1 > 0$ the direction of the vector $\mathbf{f}_j$ on the same line is determined such that the angle between $\mathbf{f}_j$ and $\mathbf{P}_j$ is sharp [*ibid.*].

Hence, the two conditions of Eq. (C.7) fully determine each of the n vectors $\mathbf{f}_j$.

Note that when $M = 1$ and $n = 1$ Eq. (C.7) simply gives the 1D relationship $Pf = 1$. Another simple particular case of Eq. (C.7) is that of a 2D 1-fold periodic function $g(x,y)$, such as a line grating on the x,y plane that is possibly rotated by angle θ (see Fig. C.4); in this case $M = 2$ and $n = 1$, so that the function $g(x,y)$ has a single period-vector $\mathbf{P} = (x,y)$ and a single frequency-vector $\mathbf{f} = (u,v)$, both having the same direction θ and reciprocal lengths along this direction. This case is discussed in more detail in the beginning of Sec. 6.6.1. Yet another particular case of Eq. (C.7) is that of a 2D 2-fold periodic function $g(x,y)$, in which $M = 2$ and $n = 2$; we will return to this case in Example C.2 below.

Based on Eq. (C.7) the reciprocity between the period-vectors $\mathbf{P}_i$ and the frequency-vectors $\mathbf{f}_i$ can be also expressed in matrix notation:[5]

$$\begin{pmatrix} \mathbf{P}_1 \\ \vdots \\ \mathbf{P}_n \end{pmatrix} (\mathbf{f}_1,...,\mathbf{f}_n) = \begin{pmatrix} \mathbf{P}_1 \cdot \mathbf{f}_1 & \cdots & \mathbf{P}_1 \cdot \mathbf{f}_n \\ \cdots & \cdots & \cdots \\ \mathbf{P}_n \cdot \mathbf{f}_1 & \cdots & \mathbf{P}_n \cdot \mathbf{f}_n \end{pmatrix} = \begin{pmatrix} 1 & & 0 \\ & \ddots & \\ 0 & & 1 \end{pmatrix}$$

Therefore, if the matrix $(\mathbf{f}_1,...,\mathbf{f}_n)$ is invertible, i.e., non-singular (which is true *iff* the vectors $\mathbf{f}_1,...,\mathbf{f}_n$ are linearly independent over $\mathbb{R}$ [Birkhoff77 pp. 237–238], a condition that is guaranteed in our case because $g(x_1,...,x_M)$ is *n*-fold periodic), then:

$$\begin{pmatrix} \mathbf{P}_1 \\ \vdots \\ \mathbf{P}_n \end{pmatrix} = (\mathbf{f}_1,...,\mathbf{f}_n)^{-1}$$

and by writing both sides as columns we obtain:

$$\begin{pmatrix} \mathbf{P}_1 \\ \vdots \\ \mathbf{P}_n \end{pmatrix} = \begin{pmatrix} \mathbf{f}_1 \\ \vdots \\ \mathbf{f}_n \end{pmatrix}^{-T} \quad \text{or:} \quad \begin{pmatrix} \mathbf{f}_1 \\ \vdots \\ \mathbf{f}_n \end{pmatrix} = \begin{pmatrix} \mathbf{P}_1 \\ \vdots \\ \mathbf{P}_n \end{pmatrix}^{-T} \tag{C.8}$$

where "–T" means the transpose of the inverse matrix. If we denote these two matrices in short by P and F, then Eq. (C.8) becomes $P = F^{-T}$ or $F = P^{-T}$.

Example C.2: Let us consider the 2D 2-fold periodic case ($M = 2$, $n = 2$); see Fig. C.5. If the two period-vectors are $\mathbf{P}_1 = (x_1,y_1)$ and $\mathbf{P}_2 = (x_2,y_2)$, then:

$$\begin{pmatrix} \mathbf{f}_1 \\ \mathbf{f}_2 \end{pmatrix} = \begin{pmatrix} \mathbf{P}_1 \\ \mathbf{P}_2 \end{pmatrix}^{-T} = \begin{pmatrix} x_1 & y_1 \\ x_2 & y_2 \end{pmatrix}^{-T} = \frac{1}{x_1 y_2 - y_1 x_2} \begin{pmatrix} y_2 & -x_2 \\ -y_1 & x_1 \end{pmatrix} \tag{C.9}$$

where $\bar{A} = x_1 y_2 - y_1 x_2$ is the area of the parallelogram A defined by the two vectors $\mathbf{P}_1$ and $\mathbf{P}_2$: $\bar{A} = \mathbf{P}_1 \times \mathbf{P}_2 = x_1 y_2 - y_1 x_2$. Note that $\bar{A}$ is, in fact, the determinant of the matrix $P = \begin{pmatrix} \mathbf{P}_1 \\ \mathbf{P}_2 \end{pmatrix} = \begin{pmatrix} x_1 & y_1 \\ x_2 & y_2 \end{pmatrix}$.[6]

The two frequency-vectors $\mathbf{f}_1 = (u_1,v_1)$ and $\mathbf{f}_2 = (u_2,v_2)$ are given, therefore, by:

$$\mathbf{f}_1 = \tfrac{1}{A}(y_2,-x_2)$$
$$\mathbf{f}_2 = \tfrac{1}{A}(-y_1,x_1) \tag{C.10}$$

Consequently, we obtain here the following interesting properties [Amidror09 p. 388]:

---

[5] Note that each vector in the following expressions represents an *M*-tuple of coordinates, and therefore each entity in parentheses is, in fact, a *matrix*. If $n < M$ this matrix is not square, but we can always choose a new basis in $\mathbb{R}^M$ along the *n* axes of periodicity. When expressed using this basis each of the vectors will have *n* non-zero coordinates, so if we omit the trailing $M - n$ zeros each vector becomes an *n*-tuple and all the matrices are therefore square with $n \times n$ elements. For example, in the 2D 1-fold periodic case (see Fig. C.4) the original 2D vectors become 1D along the axis of the 1-fold periodicity.

[6] It is interesting to note that in linear algebra determinants play the role of area (or volume) functions [Lay03 pp. 204–209]. For example, the determinant $\bar{A}$ of the matrix P gives the area of the parallelogram that is determined by the rows of the matrix, i.e. by the vectors $\mathbf{P}_1 = (x_1,y_1)$ and $\mathbf{P}_2 = (x_2,y_2)$.

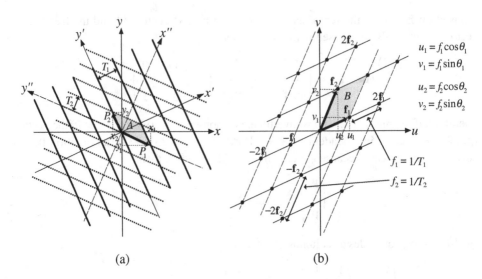

(a)   (b)

Figure C.5: Schematic plot of a 2-fold periodic (skew-periodic) function $g(x,y)$ in the signal domain (a), and its skewed impulse-nailbed $G(u,v)$ in the spectral domain (b). The gray parallelogram A in the signal domain represents a one-period element (tile) of $g(x,y)$. P_1 and P_2 are segments of this parallelogram that coincide with the period-vectors $\mathbf{P}_1$ and $\mathbf{P}_2$.

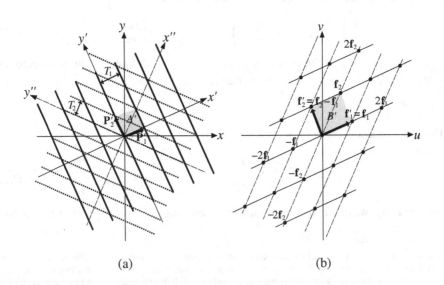

(a)   (b)

Figure C.6: A different set of fundamental period-vectors $\mathbf{P}'_1$, $\mathbf{P}'_2$ and fundamental frequency vectors $\mathbf{f}'_1$, $\mathbf{f}'_2$ for the same function $g(x,y)$ as in Fig. C.5. The parallelograms A' and B' they define have the same respective areas as the parallelograms A and B in Fig. C.5, but they are almost square.

(1) Orthogonality: $\mathbf{f}_1 \perp \mathbf{P}_2$ and $\mathbf{f}_2 \perp \mathbf{P}_1$.

(2) Vector lengths: $|\mathbf{P}_1| = \sqrt{x_1^2 + y_1^2}$

$$|\mathbf{P}_2| = \sqrt{x_2^2 + y_2^2}$$

and from Eq. (C.10): $|\mathbf{f}_1| = \frac{1}{A}\sqrt{x_2^2 + y_2^2} = \frac{1}{A}|\mathbf{P}_2|$

$$|\mathbf{f}_2| = \frac{1}{A}\sqrt{x_1^2 + y_1^2} = \frac{1}{A}|\mathbf{P}_1|$$

This means that the length ratio between the two pairs of fundamental vectors is preserved reciprocally:

$$\frac{|\mathbf{f}_1|}{|\mathbf{f}_2|} = \frac{|\mathbf{P}_2|}{|\mathbf{P}_1|}$$

For example, if in the signal domain the length of $\mathbf{P}_1$ is twice the length of $\mathbf{P}_2$, then in the spectrum the length of $\mathbf{f}_1$ is half the length of $\mathbf{f}_2$.

(3) Another interesting result concerns the areas of the period-parallelogram A that is defined by the vectors $\mathbf{P}_1$ and $\mathbf{P}_2$ in the signal domain and the frequency-parallelogram B that is defined by the vectors $\mathbf{f}_1$ and $\mathbf{f}_2$ in the spectrum (see Fig. C.5):

$$\bar{A} = \mathbf{P}_1 \times \mathbf{P}_2 = x_1 y_2 - y_1 x_2$$

$$\bar{B} = \mathbf{f}_1 \times \mathbf{f}_2 = u_1 v_2 - v_1 u_2 = \frac{1}{A^2}(x_1 y_2 - y_1 x_2) = \frac{1}{A}$$

This means that the areas of the parallelograms A and B are, indeed, reciprocal:[7]

$$\bar{A}\bar{B} = 1$$

Due to these reciprocity relations which prevail between the period-vectors and the frequency-vectors these two sets of vectors are said to be *reciprocal*.

Finally, it is easy to verify that the mixed scalar products of the period and frequency vectors give, indeed:

$$\mathbf{P}_1 \cdot \mathbf{f}_1 = x_1\frac{y_2}{A} - y_1\frac{x_2}{A} = \frac{1}{A}\bar{A} = 1 \qquad \mathbf{P}_1 \cdot \mathbf{f}_2 = -x_1\frac{y_1}{A} + y_1\frac{x_1}{A} = 0$$

$$\mathbf{P}_2 \cdot \mathbf{f}_1 = x_2\frac{y_2}{A} - y_2\frac{x_2}{A} = 0 \qquad \mathbf{P}_2 \cdot \mathbf{f}_2 = -x_2\frac{y_1}{A} + y_2\frac{x_1}{A} = \frac{1}{A}\bar{A} = 1$$

in accordance with Eq. (C.7).   ∎

C.3.2 Multidimensional generalization in the discrete world

Having generalized in the continuous world the 1D relationship $Pf = 1$ to the MD case, we are ready to proceed now to the discrete world. The approach we take in the discrete world is very similar, except that this time we start from the 1D relationship $pq = N$, which

---

[7] The generalization of this result to the MD case is based on determinants: As suggested by the previous footnote, the volume $\bar{A}$ defined by the vectors $\mathbf{P}_1,...,\mathbf{P}_n$ is the determinant of the matrix P of Eq. (C.8) (see, for example, under "parallelepiped" in [Weisstein99 p. 1314]). Similarly, the volume $\bar{B}$ defined by the vectors $\mathbf{f}_1,...,\mathbf{f}_n$ in the spectral domain is the determinant of the matrix $F = P^{-T}$, which equals, indeed, $1/\bar{A}$. We therefore have in the general MD case, too, $\bar{A}\bar{B} = 1$. This result is already found (in a slightly different form due to the use of different Fourier conventions) in Appendix B of [Petersen62].

is the particular case that we expect to get when $M = 1$ and $n = 1$. Let us denote by $\mathbf{p}_1,...,\mathbf{p}_n$ the counterparts in the MD input array of the continuous-world period vectors $\mathbf{P}_1,...,\mathbf{P}_n$, and by $\mathbf{q}_1,...,\mathbf{q}_n$ the counterparts in the MD output array (DFT spectrum) of the continuous-world frequency vectors $\mathbf{f}_1,...,\mathbf{f}_n$ (see Fig. C.7). We will henceforth call these discrete-world vectors *p-vectors* and *q-vectors*, respectively. Note that the coordinates of all the *p*- and *q*-vectors are specified in terms of array elements. We therefore obtain:

Definition C.2: If $\mathbf{p}_1,...,\mathbf{p}_n \in \mathbb{R}^M$ are n fundamental *p*-vectors of the discrete (sampled) version of the n-fold periodic function $g(x_1,...,x_M)$ in the MD case (where obviously $1 \le n \le M$), then the corresponding *q*-vectors $\mathbf{q}_1,...,\mathbf{q}_n \in \mathbb{R}^M$ in the spectral domain are defined by the following relationship:[8]

$$\mathbf{p}_i \cdot \mathbf{q}_j = \begin{cases} N & \text{if } i = j \\ 0 & \text{if } i \neq j \end{cases} \tag{C.11}$$

where $i,j = 1,...,n$. ∎

Because Eq. (C.11) is symmetric, it can also be used the other way around to determine the *p*-vectors $\mathbf{p}_1,...,\mathbf{p}_n$ given the *q*-vectors $\mathbf{q}_1,...,\mathbf{q}_n$. The only requirement in both cases is, again, that the vectors $\mathbf{p}_1,...,\mathbf{p}_n$ as well as the vectors $\mathbf{q}_1,...,\mathbf{q}_n$ be independent; the existence of such sets of n independent vectors is guaranteed since our function is n-fold periodic.

This definition is coherent with its continuous-world counterpart, Definition C.1, since just like in the 1D case (see Eqs. (C.5) and (C.6)), in the MD case, too, the connection between the continuous-world vectors and their discrete-world counterparts is given by:

$$\mathbf{p}_i = \mathbf{P}_i / \Delta x \tag{C.12}$$

$$\mathbf{q}_j = \mathbf{f}_j / \Delta u \tag{C.13}$$

so that by using Eq. (4.4) we obtain, indeed, Eq. (C.11):

$$\mathbf{p}_i \cdot \mathbf{q}_j = (\mathbf{P}_i \cdot \mathbf{f}_j) / \Delta x \Delta u = N \mathbf{P}_i \cdot \mathbf{f}_j$$

Furtheremore, we also have here the following MD generalization of Eq. (C.4):

$$\mathbf{p}_i = F \mathbf{P}_i = \frac{N}{R} \mathbf{P}_i \tag{C.14}$$

$$\mathbf{q}_i = R \mathbf{f}_i = \frac{N}{F} \mathbf{f}_i \tag{C.15}$$

Note that in definition C.2 we say that $\mathbf{p}_1,...,\mathbf{p}_n$ and $\mathbf{q}_1,...,\mathbf{q}_n$ are vectors in $\mathbb{R}^M$ rather than in $\mathbb{Z}^M$, although we are dealing here with the discrete world. The reason is that, just as in the 1D case, both the *p*-vectors and the *q*-vectors need not necessarily have integer coordinates; the only restriction which applies is that they must satisfy Eq. (C.11) where obviously N is integer. We will return to this point in Remark C.4.

---

[8] Just as in the continuous case (Definition C.1), the *p*-vectors $\mathbf{p}_1,...,\mathbf{p}_n$ as well as the *q*-vectors $\mathbf{q}_1,...,\mathbf{q}_n$ are not unique; but each set of *p*-vectors $\mathbf{p}_1,...,\mathbf{p}_n$ defines by Eq. (C.11) its own unique set of *q*-vectors $\mathbf{q}_1,...,\mathbf{q}_n$. This set of *q*-vectors is said to be *reciprocal* to its corresponding set of *p*-vectors.

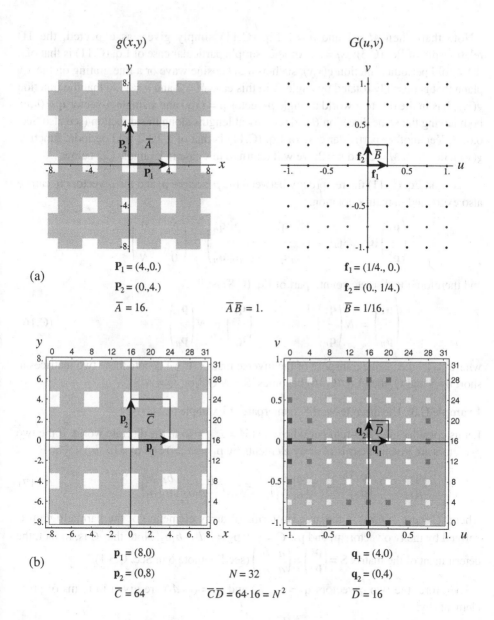

Figure C.7: (a) A 2D 2-fold periodic function $g(x,y)$ and its spectrum $G(u,v)$ in the continuous world. (b) Their discrete-world counterparts using 2D DFT. Compare the period vectors $\mathbf{P}_1$, $\mathbf{P}_2$ and the frequency vectors $\mathbf{f}_1$, $\mathbf{f}_2$ in (a) with their respective discrete-world counterparts in (b), $\mathbf{p}_1$, $\mathbf{p}_2$ and $\mathbf{q}_1$, $\mathbf{q}_2$, whose coordinates are specified in terms of array elements. We use in (b) $N = 32$ in order that the individual array elements be sufficiently large and their locations easily identified. The 2D input array corresponds to the continuous-world range $-8...8$ along both axes, and the 2D DFT spectrum covers the frequency range $-1...1$ Hz along both axes, so that $R = 16$ and $F = 2$.

Note that when $M = 1$ and $n = 1$ Eq. (C.11) simply gives, as expected, the 1D relationship of Eq. (C.3): $pq = N$. Another simple particular case of Eq. (C.11) is that of a 2D 1-fold periodic function $g(x,y)$, such as a 2D cosine wave or a line grating on the x,y plane that is possibly rotated by angle θ; in this case $M = 2$ and $n = 1$, so that the function $g(x,y)$ has in the discrete world a single p-vector $\mathbf{p} = (x,y)$ and a single q-vector $\mathbf{q} = (u,v)$, both having the same direction θ and reciprocal lengths along this direction (see also Sec. 6.6.1). Yet another particular case of Eq. (C.11) is that of a 2D 2-fold periodic function $g(x,y)$, in which $M = 2$ and $n = 2$; we will return to this case in Example C.3 below.

Based on Eq. (C.11) the reciprocity between the p-vectors $\mathbf{p}_i$ and the q-vectors $\mathbf{q}_i$ can be also expressed in matrix notation:[9]

$$\begin{pmatrix} \mathbf{p}_1 \\ \vdots \\ \mathbf{p}_n \end{pmatrix} (\mathbf{q}_1,\ldots,\mathbf{q}_n) = \begin{pmatrix} \mathbf{p}_1{\cdot}\mathbf{q}_1 & \cdots & \mathbf{p}_1{\cdot}\mathbf{q}_n \\ \cdots & \cdots & \cdots \\ \mathbf{p}_n{\cdot}\mathbf{q}_1 & \cdots & \mathbf{p}_n{\cdot}\mathbf{q}_n \end{pmatrix} = \begin{pmatrix} N & & 0 \\ & \ddots & \\ 0 & & N \end{pmatrix}$$

and therefore the discrete counterpart of Eq. (C.8) is:[10]

$$\begin{pmatrix} \mathbf{p}_1 \\ \vdots \\ \mathbf{p}_n \end{pmatrix} = N \begin{pmatrix} \mathbf{q}_1 \\ \vdots \\ \mathbf{q}_n \end{pmatrix}^{-T} \quad \text{or:} \quad \begin{pmatrix} \mathbf{q}_1 \\ \vdots \\ \mathbf{q}_n \end{pmatrix} = N \begin{pmatrix} \mathbf{p}_1 \\ \vdots \\ \mathbf{p}_n \end{pmatrix}^{-T} \tag{C.16}$$

where "$-T$" means the transpose of the inverse matrix. If we denote these two matrices in short by S and Q, then Eq. (C.16) becomes $S = N\,Q^{-T}$ or $Q = N\,S^{-T}$.

Example C.3: The discrete-world counterpart of Example C.2 above:

Let us consider the 2D 2-fold periodic case ($M = 2$, $n = 2$) in the discrete world. If the two p-vectors are given in terms of array elements by $\mathbf{p}_1 = (a_1,b_1)$ and $\mathbf{p}_2 = (a_2,b_2)$, then:[11]

$$\begin{pmatrix} \mathbf{q}_1 \\ \mathbf{q}_2 \end{pmatrix} = N \begin{pmatrix} \mathbf{p}_1 \\ \mathbf{p}_2 \end{pmatrix}^{-T} = N \begin{pmatrix} a_1 & b_1 \\ a_2 & b_2 \end{pmatrix}^{-T} = \frac{N}{a_1 b_2 - b_1 a_2} \begin{pmatrix} b_2 & -a_2 \\ -b_1 & a_1 \end{pmatrix} \tag{C.17}$$

where $\overline{C} = a_1 b_2 - b_1 a_2$ is the area (in terms of array elements) of the parallelogram C defined by the two vectors $\mathbf{p}_1$ and $\mathbf{p}_2$: $\overline{C} = \mathbf{p}_1 \times \mathbf{p}_2 = a_1 b_2 - b_1 a_2$. Note that $\overline{C}$ is, in fact, the determinant of the matrix $S = \begin{pmatrix} \mathbf{p}_1 \\ \mathbf{p}_2 \end{pmatrix} = \begin{pmatrix} a_1 & b_1 \\ a_2 & b_2 \end{pmatrix}$ (see Footnote 6 in Sec. C.3.1).

Therefore, the two q-vectors $\mathbf{q}_1 = (c_1,d_1)$ and $\mathbf{q}_2 = (c_2,d_2)$ are given in terms of array elements by:

$$\mathbf{q}_1 = \tfrac{N}{\overline{C}}(b_2,-a_2)$$
$$\mathbf{q}_2 = \tfrac{N}{\overline{C}}(-b_1,a_1) \tag{C.18}$$

Consequently, we obtain here the following properties (compare with Example C.2):

---

[9] Note, again, that each vector in the following expressions represents an M-tuple of coordinates, and therefore each entity in parentheses is, in fact, a *matrix*. And once again, if $n < M$ we can choose a new basis in $\mathbb{R}^M$ along the n axes of periodicity, in which the vectors become n-tuples and the matrices are square with $n{\times}n$ elements. These n axes are also used for sampling the given function.

[10] This is a generalization of the 1D case, in which $pq = N$ (Eq. (C.3)) gives $p = N/q$ or $q = N/p$.

[11] Although a_i,b_i,c_i,d_i are given in terms of array elements, they are not necessarily integer (Remark C.4).

(1) Orthogonality: $\mathbf{q}_1 \perp \mathbf{p}_2$ and $\mathbf{q}_2 \perp \mathbf{p}_1$.

(2) Vector lengths (in terms of array elements):

$$|\mathbf{p}_1| = \sqrt{a_1^2 + b_1^2}$$

$$|\mathbf{p}_2| = \sqrt{a_2^2 + b_2^2}$$

and from Eq. (C.18): $|\mathbf{q}_1| = \frac{N}{C}\sqrt{a_2^2 + b_2^2} = \frac{N}{C}|\mathbf{p}_2|$

$$|\mathbf{q}_2| = \frac{N}{C}\sqrt{a_1^2 + b_1^2} = \frac{N}{C}|\mathbf{p}_1|$$

This means that the length ratio between the two pairs of *p*- and *q*-vectors is preserved reciprocally:

$$\frac{|\mathbf{q}_1|}{|\mathbf{q}_2|} = \frac{|\mathbf{p}_2|}{|\mathbf{p}_1|}$$

For example, if in the input array the length of $\mathbf{p}_1$ is twice the length of $\mathbf{p}_2$, then in the DFT spectrum the length of $\mathbf{q}_1$ is half the length of $\mathbf{q}_2$.

(3) The area of the period-parallelogram *C* that is defined by the vectors $\mathbf{p}_1$ and $\mathbf{p}_2$ in the input array and the area of the frequency-parallelogram *D* that is defined by the vectors $\mathbf{q}_1$ and $\mathbf{q}_2$ in the DFT spectrum are given, in terms of array elements, by:

$$\bar{C} = \mathbf{p}_1 \times \mathbf{p}_2 = a_1 b_2 - b_1 a_2$$

$$\bar{D} = \mathbf{q}_1 \times \mathbf{q}_2 = c_1 d_2 - d_1 c_2 = \left(\frac{N}{C}\right)^2 (a_1 b_2 - b_1 a_2) = \frac{N^2}{C} \qquad (\text{C.19})$$

This means that the areas of the parallelograms *C* and *D* in terms of array elements are reciprocal, in the sense that:[12]

$$\bar{C}\bar{D} = N^2 \qquad\qquad (\text{C.20})$$

This is, indeed, a generalization of the discrete 1D relationship $pq = N$.

Due to these reciprocity relations which prevail between the *p*-vectors and the *q*-vectors these two sets of vectors are said to be *reciprocal*.

Finally, it is easy to verify that the mixed scalar products of the *p*- and *q*-vectors give, indeed:

$$\mathbf{p}_1 \cdot \mathbf{q}_1 = \frac{N}{C}(a_1 b_2 - b_1 a_2) = \frac{N}{C}\bar{C} = N \quad\quad \mathbf{p}_1 \cdot \mathbf{q}_2 = \frac{N}{C}(-a_1 b_1 + b_1 a_1) = 0$$

$$\mathbf{p}_2 \cdot \mathbf{q}_1 = \frac{N}{C}(a_2 b_2 - b_2 a_2) = 0 \quad\quad \mathbf{p}_2 \cdot \mathbf{q}_2 = \frac{N}{C}(-a_2 b_1 + b_2 a_1) = \frac{N}{C}\bar{C} = N$$

in accordance with Eq. (C.11).

The connections between $\bar{C}$ and $\bar{D}$ and their continuous-world counterparts $\bar{A}$ and $\bar{B}$ can be obtained using Eqs. (C.14) and (C.15). Thus, for the case of $n = 2$ we get:

$$\bar{C} = \mathbf{p}_1 \times \mathbf{p}_2 = F^2 \mathbf{P}_1 \times \mathbf{P}_2 = F^2 \bar{A} = F^2 / \bar{B} \qquad (\text{C.21})$$

$$\bar{D} = \mathbf{q}_1 \times \mathbf{q}_2 = R^2 \mathbf{f}_1 \times \mathbf{f}_2 = R^2 \bar{B} = R^2 / \bar{A} \qquad (\text{C.22})$$

---

[12] The generalization of this discrete-world result to the MD case is $\bar{C}\bar{D} = N^n$ (where $1 \leq n \leq M$ is the number of *p*-vectors $\mathbf{p}_1,...,\mathbf{p}_n$), as opposed to $\bar{A}\bar{B} = 1$ in the continuous-world case of Example C.2.

Note also that the surface areas of the input array and of the output array are N^2 array elements, which is obviously an integer number. However, the area of each period-parallelogram in the input array, as well as the number of such period-parallelograms that pave the entire input array, may be either integer or not, depending on the case. The same is also true for the frequency-parallelograms in the output array. For example, in Fig. C.8(b) we have $N = 32$, $\overline{C} = 9 \cdot 4 = 36$ and $\overline{D} = 3.555 \cdot 8 = 28.444$; and indeed, $\overline{CD} = 36 \cdot 28.444 = 32^2$. Note that if the area of a single period-parallelogram in the input array is $\overline{C}$ array elements, then the number of period-parallelograms $\overline{C}$ within the $N \times N$-element input array is precisely $N^2 / \overline{C} = \overline{D}$. And similarly, if the area of a single frequency-parallelogram in the output array is $\overline{D}$ array elements, then the number of frequency-parallelograms $\overline{D}$ within the $N \times N$-element output array is precisely $N^2 / \overline{D} = \overline{C}$. This is, indeed, a 2D generalization of the roles of p and q in the 1D case (see Remark C.1). ∎

It should be mentioned that the generalization of Eqs. (C.21) and (C.22) to the n-fold periodic MD case is straightforward, using determinants rather than vector products:[9]

$$\overline{C} = \det(\mathbf{p}_1,...,\mathbf{p}_n) = F^n \det(\mathbf{P}_1,...,\mathbf{P}_n) = F^n \overline{A} = F^n / \overline{B} \qquad (C.23)$$

$$\overline{D} = \det(\mathbf{q}_1,...,\mathbf{q}_n) = R^n \det(\mathbf{f}_1,...,\mathbf{f}_n) = R^n \overline{B} = R^n / \overline{A} \qquad (C.24)$$

Based on Definition C.2 we can now formulate the generalization of Remark C.1 to the MD n-fold periodic case with $1 \leq n \leq M$ (as a simple visual illustration consider, for example, the 3D 2-fold periodic case or the 3D 3-fold periodic case):[13]

Remark C.4: Note the dual role of the vectors $\mathbf{q}_1,...,\mathbf{q}_n$ in the signal and spectral domains: In the signal domain, the vectors $\mathbf{q}_1,...,\mathbf{q}_n$ determine (via their determinant $\overline{D}$; see Footnotes 6 and 7 in Sec. C.3.1) the number of period-parallelepipeds of the MD n-fold periodic function $g(x_1,...,x_M)$, $1 \leq n \leq M$, that:

(a) Fill the periodic n-dimensional sub-volume R^n of the sampling range volume R^M (since by Eq. (C.24) we have $\overline{D} = R^n / \overline{A}$); or, equivalently,

(b) Fill the periodic n-dimensional sub-volume N^n of the input array volume N^M (since by the general version of Eq. (C.20) we have $\overline{D} = N^n / \overline{C}$).

But in terms of the DFT spectrum, the vectors $\mathbf{q}_1,...,\mathbf{q}_n$ give the locations of the n fundamental first-harmonic impulses (i.e. the n frequency-vectors of $g(x_1,...,x_M)$ in terms of array elements), as well as the distances between successive harmonic impulses (again, in terms of array elements). Note that $\mathbf{q}_1,...,\mathbf{q}_n$ need not necessarily be integer; when $\mathbf{q}_j$ is non-integer in its i-th coordinate the DFT spectrum suffers from a leakage artifact along its i-th coordinate.

Similarly, the vectors $\mathbf{p}_1,...,\mathbf{p}_n$, too, play a double role: In the signal domain they determine the n-dimensional period-parallelepiped of $g(x_1,...,x_M)$ in terms of array

---

[13] Once again, if $n < M$ we can choose a new basis in $\mathbb{R}^M$ along the n axes of periodicity, in which the vectors become n-tuples. Thus, if $n < M$ we consider the restriction of $\mathbb{R}^M$ to the n-dimensional subspace $\mathbb{R}^n$ in which the given n-fold periodic function is indeed periodic. To avoid complications, we assume that these n axes are also used for sampling the given function. Note however that this assumption on the sampling axes excludes here discrete cases that are not sampled along the axes of periodicity, like the 2D 1-fold periodic cases shown in Fig. C.9, although $p = N/q$ is still valid there (see Sec. 6.6.1).

elements. But in the DFT spectrum the vectors $\mathbf{p}_1,...,\mathbf{p}_n$ determine (via their determinant $\overline{C}$; see Footnotes 6 and 7 in Sec. C.3.1) the number of frequency-parallelepipeds that:

(a) Fill the *n*-dimensional sub-volume F^n of the DFT frequency range volume F^M (since by Eq. (C.23) we have $\overline{C} = F^n / \overline{B}$); or equivalently

(b) Fill the *n*-dimensional sub-volume N^n of the output array volume N^M (since by the general version of Eq. (C.20) we have $\overline{C} = N^n / \overline{D}$).

Figure C.8: The effect of the *p*-vectors and the *q*-vectors of Eq. (C.11) on the input and output arrays of the DFT of a periodic function in the multidimensional case. Each row shows a 2D 2-fold periodic function (dot lattice made of 1-valued impulses) as well as its DFT spectrum (real and imaginary parts), after all the required adjustments as discussed in Chapters 3, 4 and Appendix A. We use here $N = 32$, so that the individual array elements are sufficiently large and their locations can be easily identified. Each 2D input array corresponds to the continuous-world range −8...8 along both axes, and each 2D DFT spectrum covers the frequency range −1...1 Hz along both axes, so that $R = 16$ and $F = 2$. But all the vectors are given here in terms of array elements ($\mathbf{p}_i$ and $\mathbf{q}_i$ rather than $\mathbf{P}_i$ and $\mathbf{f}_i$). Note that in this figure all the *p*-vectors are purely integer (meaning that all the impulses of the input lattices are exactly located on input-array elements), but the *q*-vectors in the resulting DFT spectra illustrate various integer, non-integer and hybrid cases, as follows (we use italics to highlight the non-integer coordinates):

(a) $\mathbf{p}_1 = (8,0)$, $\mathbf{p}_2 = (0,4)$ so that $\mathbf{q}_1 = (4,0)$, $\mathbf{q}_2 = (0,8)$: both *q*-vectors have purely integer coordinates (so no leakage occurs in the DFT spectrum); the 2D input array is perfectly cyclical along both directions.

(b) $\mathbf{p}_1 = (9,0)$, $\mathbf{p}_2 = (0,4)$ so that $\mathbf{q}_1 = (3.555,0)$, $\mathbf{q}_2 = (0,8)$: $\mathbf{q}_1$ is non-integer in its first coordinate (so leakage occurs in the horizontal direction); the 2D input array is perfectly cyclical (seamlessly wraparound) in the vertical direction but not in the horizontal direction.

(c) $\mathbf{p}_1 = (8,1)$, $\mathbf{p}_2 = (0,8)$ so that $\mathbf{q}_1 = (4,0)$, $\mathbf{q}_2 = (-0.5,4)$: $\mathbf{q}_2$ is non-integer in its first coordinate; the effect is similar to that of case (b).

(d) $\mathbf{p}_1 = (5,1)$, $\mathbf{p}_2 = (0,4)$ so that $\mathbf{q}_1 = (6.4,0)$, $\mathbf{q}_2 = (-1.6,8)$: both $\mathbf{q}_1$ and $\mathbf{q}_2$ are non-integer in their first coordinate; the effect is again as in case (b).

(e) $\mathbf{p}_1 = (5,0)$, $\mathbf{p}_2 = (1,4)$ so that $\mathbf{q}_1 = (6.4,-1.6)$, $\mathbf{q}_2 = (0,8)$: $\mathbf{q}_1$ is non-integer in both coordinates (so that leakage occurs in both directions); the 2D input array is not perfectly cyclical in any of the two directions.

(f) $\mathbf{p}_1 = (9,0)$, $\mathbf{p}_2 = (0,5)$ so that $\mathbf{q}_1 = (3.555,0)$, $\mathbf{q}_2 = (0,6.4)$: $\mathbf{q}_1$ is non-integer in its first coordinate and $\mathbf{q}_2$ is non-integer in its second coordinate; the effect is similar to that of case (e).

(g) $\mathbf{p}_1 = (8,-1)$, $\mathbf{p}_2 = (1,4)$ so that $\mathbf{q}_1 = (3.88,-0.98)$, $\mathbf{q}_2 = (0.98,7.76)$: $\mathbf{q}_1$ and $\mathbf{q}_2$ are non-integer in both of their coordinates; the effect is similar to that of case (e).

Note that in cases (c), (e) and (g) the DFT spectrum has a non-zero imaginary-valued part although in the continuous-world counterparts the spectrum is purely real-valued (as the infinite image-domain signal is centrosymmetric). This artifact of the DFT is explained in Sec. 8.4.

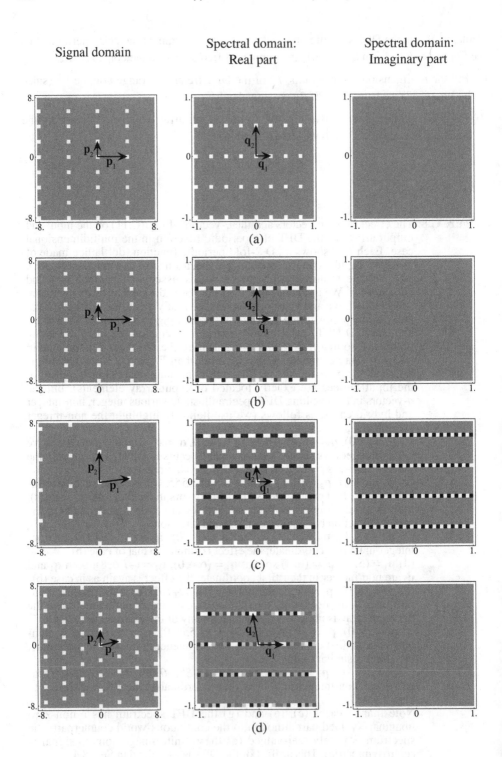

Signal domain

Spectral domain:
Real part

Spectral domain:
Imaginary part

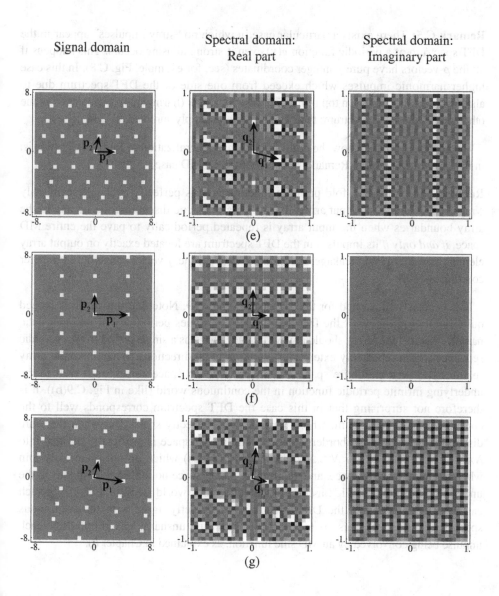

(e)

(f)

(g)

Note that the vectors $\mathbf{p}_1,...,\mathbf{p}_n$, too, need not necessarily be integer; in fact, the only restriction on $\mathbf{p}_1,...,\mathbf{p}_n$ and $\mathbf{q}_1,...,\mathbf{q}_n$ is that they must satisfy Eq. (C.11), where N is obviously an integer number. ■

The nature of the *p*-vectors (integer or non-integer) has particular significance if high harmonic impulses of the periodic function $g(\mathbf{x})$ fall beyond the borders of the DFT spectrum, i.e. if our sampled version of $g(\mathbf{x})$ suffers from aliasing. Using our present terminology we can now formulate the MD generalization of Remark C.2 as follows:[13]

Remark C.5: There exists a particular case in which no "stray impulses" appear in the DFT spectrum of a periodic function $g(\mathbf{x})$ even if strong aliasing occurs. This happens if all the p-vectors have purely integer coordinates (see, for example, Fig. C.8). In this case higher harmonic impulses which exceed from one side of the DFT spectrum due to aliasing will fall exactly on top of true impulse locations (harmonic frequencies)[14] in the other side of the DFT spectrum; their heights will be simply summed up. ∎

We can also formulate now the multidimensional generalization of Remark C.3 (this is, in fact, a reformulation of Remarks 6.3 and 6.4 for the MD case):

Remark C.6: An MD n-fold periodic function $g(\mathbf{x})$ is perfectly cyclical (seamlessly wraparound) within the input array, meaning that it has no discontinuities on any of the array boundaries when the input array is repeated periodically to pave the entire MD space, *if and only if* its impulses in the DFT spectrum are located exactly on output array elements (so there is no leakage), i.e. *if and only if* all the q-vectors have purely integer coordinates. ∎

This is clearly illustrated for the 2D case by Fig. C.9. Note that this result is indeed quite intuitive, in view of the fact that the DFT assumes periodicity of its input data, namely, that its input array should always be understood as a single period from a periodic sequence which repeatedly extends *ad infinitum* in all directions: When the input array seamlessly paves the entire space, this discrete paved space corresponds well to the underlying infinite periodic function in the continuous world (like in Fig. C.9(b)). It is therefore not surprising that in this case the DFT spectrum corresponds well to the continuous-world spectrum, whose impulses are perfectly sharp. But when there are discontinuities along the borders, the resulting paved space is artificially divided into $N{\times}N$-element squares (or $N{\times}...{\times}N$-element MD cubes) which do not seamlessly join when paving the entire space, and the discrete paved space no longer corresponds to the underlying infinite periodic function in the continuous world (see Fig. C.9(a)). In such cases we cannot expect the DFT spectrum to perfectly agree with the continuous spectrum. And indeed, this gives in the DFT spectrum unsharp, leaked impulses (each impulse being convolved by an MD sinc function, as explained in Chapter 6).

Figure C.9: A 2D example illustrating Remark C.6: The 2D input array (containing our sampled periodic function) is perfectly cyclical (seamlessly wraparound) *if and only if* the impulses in the DFT spectrum are located exactly on output array elements (so there is no leakage), i.e. *if and only if* all the q-vectors have purely integer coordinates. The continuous-world function being sampled here is a rotated cosine, whose spectrum

---

[14] To be more precise, we speak here of impulse locations in terms of the *algebraic support* of the spectrum of our periodic function, which consists of *all* the integer linear combinations of the basic frequency vectors $\mathbf{f}_1,...,\mathbf{f}_n$ of $g(\mathbf{x})$, even if the corresponding impulse amplitudes happen to be zero [Amidror09 p. 110].

Signal domain

Spectral domain:
Real part

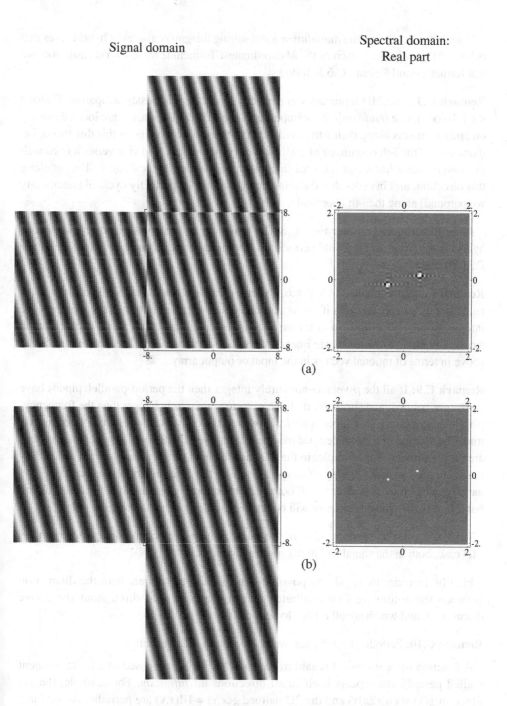

(a)

(b)

consists of an impulse pair centered about the origin. In both rows we use $N = 64$, $R = 16$, $F = 4$, $\Delta x = 16/64 = 0.25$ and $\Delta u = 4/64 = 0.0625$. In the first row the cosine frequency is $f = 0.5$ and its angle is $\theta = 16°$; the second row is a rational approximation of the first row with $\mathbf{q} = (7,2)$, so that $f = 0.455$ and $\theta = \arctan(2/7) = 15.9454°$.

Note, however, that in the multidimensional setting there may also exist hybrid cases that behave differently along each of the M coordinates. To include these hybrid cases, too, we can further extend Remark C.6 as follows:

Remark C.7: The MD input array is perfectly cyclical (seamlessly wraparound) along the i-th coordinate *if and only if* the impulses in the MD DFT spectrum are located exactly on array elements along their i-th coordinate (so there is no leakage in this direction), i.e. *if and only if* the i-th coordinate of all the q-vectors is integer. And vice versa, it is enough that one q-vector has a non-integer i-th coordinate in order that leakage will occur along this direction, and in order that the input array will not be perfectly cyclical (seamlessly wraparound) along the i-th coordinate. ■

As an illustration, consider Figs. C.8–C.9 which show examples belonging to various hybrid cases based on 2D 2-fold periodic or 2D 1-fold periodic functions, along with their DFT spectra.

Remark C.8: As explained in Sec. 6.6, a vector $\mathbf{q}$ in the output array is called *rational* (or equivalently *purely integer*) if all its coordinates in terms of output array elements are integer. Similarly, a vector $\mathbf{p}$ in the input array is called rational if all its coordinates in terms of input array elements are integer. This allows us to reformulate Remarks C.5–C.7 above in terms of rational vectors in the input or output array. ■

Remark C.9: If all the p-vectors are purely integer then the period-parallelepipeds have an integer volume; and similarly, if all the q-vectors are purely integer then the frequency-parallelepipeds have an integer volume. However, the converse claims are not necessarily true: The volume of a parallelepiped may be integer even if not all of its generating vectors are purely integer. For example, in the 2D case the area of the parallelogram defined by the vectors $\mathbf{p}_1 = (a_1, b_1)$ and $\mathbf{p}_2 = (a_2, b_2)$ is given by the determinant of the matrix $\begin{pmatrix} a_1 & b_1 \\ a_2 & b_2 \end{pmatrix}$, namely, $a_1 b_2 - b_1 a_2$. It is clear that if both vectors are purely integer so is the determinant; but if $b_1 = 0$, then the determinant will be integer even if a_2 is not an integer. ■

Table C.2 offers a synoptic view of the roles of the p-vectors and the q-vectors in the MD case, both in the signal and in the spectral domains.

Finally, in order to avoid any possible confusion, let us repeat here the distinction between the following four attributes that are largely used throughout the above discussions, and which are all related to periodicity:

Remark C.10: Periodic, folded-over, wrapped-around and cyclical:

A function (or a structure) is said to be *periodic* if it is composed of a finite segment (called period) that repeats itself in all directions *ad infinitum*. For ecample, the 1D function $g(x) = \cos(2\pi f x)$ and the 2D nailbed $g(x,y) = III(x,y)$ are periodic. As we have seen in Secs. 2.5 and D.1, the input and output arrays of the DFT are periodic: each of them should be understood as a single period from a periodic sequence which repeatedly extends *ad infinitum* in all directions.

	Effect on the input array (signal domain)	Effect on the output array (DFT spectrum)	Examples
All q-vectors are purely integer	Input array is perfectly cyclical (seamlessly wraparound) along all dimensions (see Remarks C.6 and C.7).	Rational angles and frequencies; all impulses are located exactly on array elements; no leakage (see Sec. 6.6).	Figs. C.8(a) and C.9(b)
At least one q-vector $\mathbf{q}_j$ has a non-integer i-th coordinate	Input array is not perfectly cyclical (seamlessly wraparound) along the i-th coordinate (see Remarks C.6 and C.7).	Irrational angle or frequency along the j-th q-vector; impulses are located between array elements along the i-th coordinate; leakage along the i-th coordinate (see Sec. 6.6).	Figs. C.8(b)–(g) and C.9(a)
All p-vectors are purely integer	Rational angles and periods; period area (or volume) of $g(\mathbf{x})$ is exactly an integer number of array elements (see Remark C.9).	If aliasing occurs, folded-over impulses fall exactly on top of lower-harmonic impulse locations, and their amplitudes are summed up (see Remark C.5).	Figs. C.8(a)–(g)
At least one p-vector $\mathbf{p}_j$ has a non-integer i-th coordinate	Irrational angle or period along the j-th p-vector; period area (or volume) of $g(\mathbf{x})$ *may be* a non-integer number of array elements (see Remark C.9).	If aliasing occurs, folded-over impulses fall between lower-harmonic impulse locations, causing visible stray impulses (see Remark C.5).	

Table C.2: Multidimensional generalization of Table C.1: The effects of the p-vectors and of the q-vectors of Eq. (C.11) on the input and output arrays of the MD DFT of an MD n-fold periodic function $g(x_1,...,x_M)$, $1 \leq n \leq M$.[13] Note that the effects of the p- and q-vectors are orthogonal (i.e. independent of each other), and just as in the 1D case each of the 4 combinations may exist.

The term *foldover* or *folding-over* is used when frequencies of the continuous spectrum (or any other components thereof, such as impulses, etc.) fall beyond the boundaries of the DFT spectrum, due to aliasing. Because of the periodicity of the DFT spectrum (the output array), such exceeding entities re-enter from the opposite end of the DFT spectrum and give the impression of folding over (see, for example, row (b) in Fig. C.2). It is important to note that any folded-over entity is additively overlapped (summed up) on top of the spectral components that were originally present at that location. This happens, as we have seen in Chapter 5, due to the additive nature of the convolution with an impulse comb or nailbed.

The term *perfectly cyclical* or *seamlessly wraparound* is used when a periodic function is sampled into our input or output array (1D, 2D or MD), and its periodicity exactly fits into the array, so that when this array is periodically repeated in all directions, no seams appear between its neighbouring copies (see Remarks C.3 and C.6). This is illustrated by the signal domain in Figs. 6.3(a),(e),(f) for the 1D case, or in Fig. C.9(b) for the 2D case. Confusingly, here too one may say that "every element which exceeds from one side of the array re-enters from the opposite side of the array"; however, unlike folding-over, wraparound is not additive, and it should only be understood as a geometric consideration.

When the array in question is not seamlessly wraparound (see, for example, Fig. C.9(a)), it still repeats itself periodically, like any input or output array of the DFT. Simply, it will contain jumps or discontinuities along the seams between neighbouring copies of the array. But because of this periodic repetition, we can still say that all the neighbouring copies of the array wrap around and coincide with the main copy, meaning that each of them represents, in fact, the same entity (see Problem 8-17). In this case, too, it can be said that whatever exceeds from one side of the array re-enters from the opposite side (thanks to the periodicity) — but of course, elements which exceed beyond the boundaries may suffer from jumps or discontinuities.

Note that the term *wraparound* is also used in expressions such as "shift with wraparound" (see, for example, Footnote 2 in Sec. 3.2), referring to the cyclical nature of the operation (meaning, again, that whatever exceeds from one side of the entity in question re-enters from the opposite side). Sometimes the term "additive wraparound" is also used as a synonym to *foldover*.

Finally, the term *cyclical* is more ambiguous, since it can be used as a synonym to periodic or to wraparound, depending on the context. ■

Appendix D

Miscellaneous remarks and derivations

D.1 The periodicity of the input and output arrays of the DFT

As we have seen in Sec. 2.5, the DFT imposes a cyclic behaviour on both of its input and output data sequences, in the sense that each of them should be understood as a single N-element period (or "truncation window") from a periodic sequence which repeatedly extends to both directions *ad infinitum* (even if the underlying continuous data before sampling was not periodic). In the informal graphical development of the DFT, as presented in Sec. 2.5, this periodicity is explained by the fact that sampling (i.e. multiplication with an infinite impulse train) in one domain is equivalent to a convolution with the reciprocal infinite impulse train in the other domain. Thus, sampling in the signal domain results in a periodic behaviour in the frequency domain, and sampling in the frequency domain results in a periodic behaviour in the signal domain (see Fig. 2.1).

More formally, the reason for this periodicity is that the exponential part in the definition of the DFT (for example, in Eq. (2.13)) is periodic in the index n with period N.[1] To see this, let $n = r + N$; we have, therefore:

$$e^{-i2\pi kn/N} = e^{-i2\pi k(r+N)/N} = e^{-i2\pi kr/N} e^{-i2\pi k}$$

$$= e^{-i2\pi kr/N}$$

since $e^{-i2\pi k} = \cos(2\pi k) - i\sin(2\pi k) = 1$ for any integer k [Brigham88 p. 96]. This implies that for any integer j:

$$\widetilde{G}(n + jN) = \widetilde{G}(n) \tag{D.1}$$

Similarly, the periodicity of the exponential part in the definition of the IDFT (for example, in Eq. (2.14)) implies that for any integer j:

$$\widetilde{g}(k + jN) = \widetilde{g}(k) \tag{D.2}$$

Therefore, if we desire to extend the definition of $\widetilde{g}(k)$ and $\widetilde{G}(n)$ beyond the original domains given by $k = 0, \dots, N-1$ and $n = 0, \dots, N-1$, this extension must be governed by Eqs. (D.1) and (D.2). This means that the extensions in the signal and spectral domains are periodic repetitions of the original N-element arrays $\widetilde{g}(k)$, $k = 0, \dots, N-1$ and $\widetilde{G}(n)$, $n = 0, \dots, N-1$, respectively.

The 2D generalization of this reasoning is straightforward, as shown for example in [Rosenfeld82 pp. 22–24]: The two-fold periodicity of the exponential part in the

---

[1] Note that no similar periodicity exists in the exponential part of the various definitions of the CFT (for example, in Eq. (2.1)).

definition of the 2D DFT, for example in Eq. (2.23), implies that for any integers j_1 and j_2:

$$\widetilde{G}(m + j_1 N, \, n + j_2 N) = \widetilde{G}(m,n) \tag{D.3}$$

and similarly, in the 2D IDFT we have for any integers j_1 and j_2:

$$\widetilde{g}(k + j_1 N, \, l + j_2 N) = \widetilde{g}(k,l) \tag{D.4}$$

Therefore, if we desire to extend the definition of $\widetilde{g}(k,l)$ and $\widetilde{G}(m,n)$ beyond the original domains given by $k,l = 0, \dots ,N{-}1$ and $m,n = 0, \dots ,N{-}1$, this extension must be governed by Eqs. (D.3) and (D.4). This means that the extensions in the signal and spectral domains are 2D periodic repetitions of the original $N{\times}N$-element arrays $\widetilde{g}(k,l)$, $k,l = 0, \dots ,N{-}1$ and $\widetilde{G}(m,n)$, $m,n = 0, \dots ,N{-}1$, respectively.

This result can be readily extended to cases having a different number of elements along each dimension, and to cases with any higher number of dimensions.

D.2 Explanation of the element order in the DFT output array

We have seen in Chapter 3 that the elements in the output array of the DFT are not obtained in the usual order, so that a reorganization of the output array is required if we wish to plot the DFT results in the conventional way. In the present section we give the explanation for the 1D case; the extension to the 2D and MD cases can be obtained by applying these considerations to each dimension separately.

As we have seen in Sec. 3.2, 1D DFT algorithms receive as input a 1D array of N complex numbers (say, samples of a given continuous complex- or real-valued function $y = g(x)$), and return as their output a 1D array of N complex numbers, indexed by $n = 0,\dots,N{-}1$, which contains the resulting discrete frequency spectrum. However, the frequency values in the resulting output array are not ordered in the conventional way we use to plot the spectrum: The first element in the array ($n = 0$) corresponds to the frequency 0; the second element ($n = 1$) corresponds to f_1, the smallest frequency above 0 in the discrete spectrum; and the next elements represent the following frequencies of the discrete spectrum, i.e. $2f_1$, $3f_1$, etc. However, this is only true until $n = N/2{-}1$; the output array elements starting from $n = N/2$ and until the last element, $n = N{-}1$, correspond to the negative frequencies of the spectrum. Thus, the last element of the output array, $n = N{-}1$, corresponds to the negative frequency $-f_1$, which should conventionally be plotted just before the frequency 0; its predecessor, the element $n = N{-}2$, corresponds to the frequency $-2f_1$; and finally, the element $n = N/2$ corresponds to $-(N/2)f_1$, the most negative frequency in the spectrum.[2] It follows, therefore, that in order to plot the resulting spectrum in the conventional way, with the zero frequency in the center and the negative frequencies to its

---

[2] In the vast majority of DFT algorithms N is a power of two, or at least an even number. In the rare cases where N is odd the reorganization of the array should be adapted accordingly [Briggs95 Sec. 3.1].

left, we have to reorganize the output array of the DFT by interchanging (swapping) its first half (the segment containing the elements $n = 0,...,N/2–1$) and its second half (the segment containing the elements $n = N/2,...,N–1$).[3] This reorganization of the spectrum is not usually done by the DFT algorithm [Brigham88 p. 169], and it is up to the user to perform it when plotting the DFT results.

Why does the DFT behave this way? First of all, it should be mentioned that this complication only occurs when using a DFT definition with *non-centered* indices, like Eqs. (2.13)–(2.14), in which the indices vary between 0 and $N–1$ rather than between $–N/2$ and $N/2–1$ [Briggs95 p. 67]; but because this is the standard convention used in virtually all software applications of the DFT (see, for example, the partial list given at the end of Sec. 2.3), we have to accept this fact and get used to it. So let us see now in detail why the DFT with non-centered indices behaves this way. This ordering is a consequence of the N-element periodicity of the output array of the DFT (see Sec. D.1), which implies that $\widetilde{G}(N–n) = \widetilde{G}(–n)$. But as shown in Fig. D.1, the periodicity alone does not yet explain this ordering, since it could still allow many other possibilities — for example, the wrong output array ordering shown in Fig. D.1(b), in which the heighest DFT frequency is located in the last element of the output array, $N–1$. In order to obtain the correct ordering, shown in Fig. D.1(c), and to exclude all the other possibilities, we need therefore an additional condition, as explained below.

(1) If the input signal is purely real, which is indeed the case in most situations, then both $G(u)$ and its DFT counterpart are Hermitian (see, for example, [Bracewell86] pp. 15 and 366, respectively). This additional information indeed enforces the correct ordering of the DFT output array shown in Fig. D.1(c): Because we are using the non-centered definition of the DFT (Eq. (2.13)) it follows that the first element in the DFT output array is $n = 0$, which corresponds to the frequency 0 (i.e. to the origin of the CFT). Then, the Hermitian nature of the DFT output implies that for any integer n:

$$\widetilde{G}(n) = \widetilde{G}^*(–n)$$

and due to the periodicity with period N:

$$= \widetilde{G}^*(N–n)$$

which means that $\text{Re}[\widetilde{G}(n)] = \text{Re}[\widetilde{G}(N–n)]$ and $\text{Im}[\widetilde{G}(n)] = –\text{Im}[\widetilde{G}(N–n)]$. In particular we have:

$$\widetilde{G}(1) = \widetilde{G}^*(N–1)$$

$$\widetilde{G}(2) = \widetilde{G}^*(N–2)$$

$$\vdots$$

$$\widetilde{G}(\tfrac{N}{2}) = \widetilde{G}^*(N–\tfrac{N}{2}) = \widetilde{G}^*(\tfrac{N}{2})$$

---

[3] Note that this is also equivalent to a cyclic shift of the array (i.e. a shift with wraparound) through $N/2$ positions in either direction.

(see, for example, Fig. 3.1(b)). This means that the ordering of the DFT output is, indeed, as shown in Fig. D.1(c).

(2) However, even if the input signal is not purely real, meaning that the Hermitian property no longer holds, the DFT output array is still ordered in the same way. To see this, remember that when sampling a continuous function $g(x)$ with a sampling interval Δx, no frequencies beyond $\pm\frac{1}{2\Delta x}$ may appear in the resulting DFT spectrum [Smith07 p. 229; Harris98 p. 713; Briggs95 p. 9; Castleman79 p. 235]. In other words, by virtue of Eq. (4.6), the DFT spectrum can only contain frequencies up to half of the sampling frequency f_s [Smith03 p. 42], i.e. up to $(N/2)f_1$. And indeed, as we have seen in Eq. (4.9), $f_{max} = (N/2-1)f_1$ and $f_{min} = -(N/2)f_1$. Hence, the maximum positive frequency is located in the position $n = N/2-1$ of the output array and the minimum negative frequency is located in the negative position $n = -N/2$ (see Fig. D.1(c)). Since we are using a non-centered definition of the DFT that is formulated for values of n in the interval $0,...,N-1$, negative values of n are not included. But by virtue of the N-element periodicity of the DFT we see that the elements in the positions $N/2, ... ,N-1$ of the output array are indeed a copy of the elements in the negative positions $-N/2, ... ,-1$.

Note that the explanation given in point (2) is more general than that given in point (1) since it covers all cases, including those with purely real input signals. This means that the explanation provided in point (1) is redundant, but we still mention it here because it appears quite frequently in the literature (see, for example, [Gonzalez87 p. 78]).

D.3 The beating effect in a sum of cosines with similar frequencies

Let us consider the sum of two cosines having close frequencies f_1 and f_2. Based on the well-known trigonometric identity [Harris98 p. 284; Spiegel68 p. 17]:

$$\cos\alpha + \cos\beta = 2\cos\frac{\alpha+\beta}{2}\cos\frac{\alpha-\beta}{2}$$

we obtain:

$$\cos(2\pi f_1 x) + \cos(2\pi f_2 x) = 2\cos(2\pi \tfrac{f_1+f_2}{2}x)\cos(2\pi \tfrac{f_1-f_2}{2}x) \qquad (D.5)$$

If the frequencies f_1 and f_2 are similar, this cosine product corresponds to a modulation effect, where the cosine with the higher frequency $\frac{f_1+f_2}{2}$ represents the carrier and the cosine with the low frequency $\frac{f_1-f_2}{2}$ represents the envelope (see Fig. D.2(a)).[4] However, because the envelope periodicity is in fact half of one full cosine period (see Fig. D.2(b)) it turns out that the envelope's frequency is equal to the frequency difference $f_1 - f_2$.

Thus, the sum of two cosines with very close frequencies f_1 and f_2 gives a beating effect (pseudo moiré) whose frequency is $f_1 - f_2$. Note, however, that this beating is not a true

---

[4] If the original frequencies are very different, say $f_1 \ll f_2$, the cosine sum no longer resembles such a modulation, but rather takes the form of a low-frequency cosine wave of frequency f_1 having a high-frequency perturbation of frequency f_2 (or in other words, a high-frequency cosine wave of frequency f_2 riding on a low-frequency cosine wave of frequency f_1).

Spectral domain

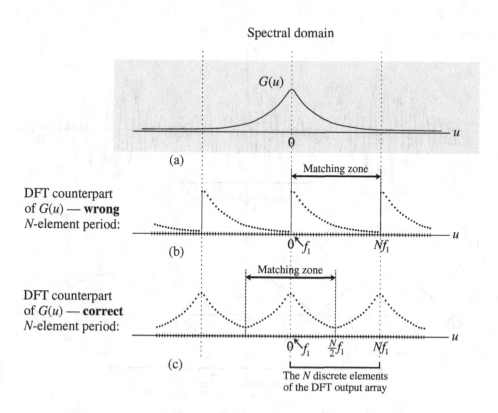

(a)

DFT counterpart
of $G(u)$ — **wrong**
N-element period:

Matching zone

(b)

DFT counterpart
of $G(u)$ — **correct**
N-element period:

Matching zone

(c)

The N discrete elements
of the DFT output array

Figure D.1: The element order in the DFT output array. Suppose we are given a
continuous-world function $g(x)$ whose spectrum (continuous Fourier
transform) is $G(u)$, as shown in (a). The discrete counterpart of $G(u)$
(the output array of the DFT) consists of N complex elements, and it
is periodic with period N. This periodicity means that the DFT output
corresponds to the continuous Fourier transform (a) within only one
sequence of N output elements; but outside this range this sequence
of N elements is repeated *ad infinitum*, and it no longer matches $G(u)$.
In principle, the sequence of N elements that do match $G(u)$ could be
situated in many different ways, for example as shown in (b) or in (c).
However, we show that the only possible ordering of the DFT output
array is indeed the one shown in (c). This means that when using a
non-centered definition of the DFT, indexed by $n = 0, \ldots, N{-}1$, the
elements in the positions $n = N/2, \ldots, N{-}1$ of the DFT output array are
indeed a copy of the elements that are located in the figure in the
negative positions $-N/2, \ldots, -1$. Note that even within the "matching
zone" there still may exist a discrepancy between the DFT results and
$G(u)$ due to aliasing and leakage (see Chapters 5 and 6).

moiré effect, since the corresponding spectrum contains no impulse having the new low
frequency $f_1 - f_2$: According to the addition rule (see point 1b in Sec. 2.4.1), the Fourier
transform of the cosine sum (D.5) only contains the frequencies that are contributed by
the two original cosines themselves, namely, $\pm f_1$ and $\pm f_2$.

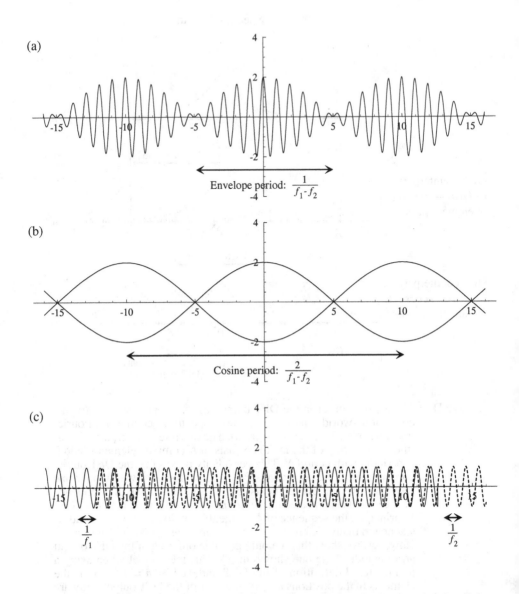

Figure D.2: (a) The sun of two cosinusoidal waves with similar frequencies f_1 and f_2, namely, $\cos(2\pi f_1 x) + \cos(2\pi f_2 x)$. (b) The envelope of the modulated wave (a) is given by the cosine $2\cos(2\pi \frac{f_1 - f_2}{2} x)$, but its period is half of the period of this cosine. The frequency of the envelope is therefore twice $\frac{f_1 - f_2}{2}$, namely $f_1 - f_2$. For the sake of completeness, we show in (c) the two original cosines being used in the present figure: The cosine $\cos(2\pi f_1 x)$ with $f_1 = 1$ is plotted by a continuous line (left), while the cosine $\cos(2\pi f_2 x)$ with $f_2 = 1.1$ is dashed (right); both curves are overprinted in the central part of (c) to allow a better understanding of their sum in (a).

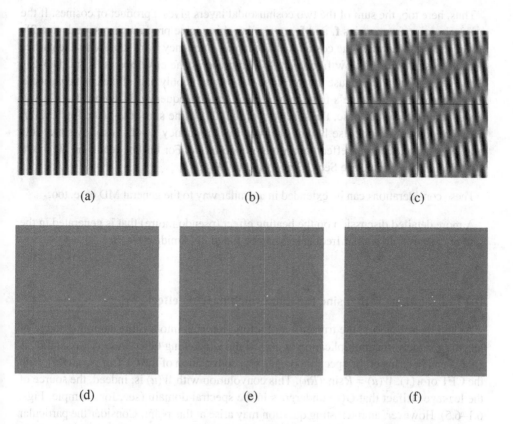

Figure D.3: 2D cosinusoidal signals (a) and (b) having frequencies $\mathbf{f}_1$ and $\mathbf{f}_2$ and their sum (c) in the signal domain; their respective spectra are the impulse pairs shown in (d) and (e) and their sum (f). These spectra were obtained by applying 2D DFT to each of the first-row images. Note the beating effect in the sum (c), whose frequency is $\mathbf{f}_1 - \mathbf{f}_2$. This is a pseudo moiré effect since it does not have corresponding impulses in the spectrum (f).

The situation in the 2D case is similar, but a little more complex since the carrier and the envelope waves in the 2D case may have different orientations (as we can see, for example, in Fig. D.3(c)). To simplify the presentation in the 2D case, we adopt the following vector notation:

$$\cos(2\pi f_i[x\cos\theta_i + y\sin\theta_i]) = \cos(2\pi \mathbf{f}_i \cdot \mathbf{x})$$

where $\mathbf{f}_i = f_i \cdot (\cos\theta_i, \sin\theta_i)$ and $\mathbf{x} = (x,y)$. Using this vector notation the generalization of Eq. (D.5) to the 2D case of Fig. D.3 is:

$$\cos(2\pi \mathbf{f}_1 \cdot \mathbf{x}) + \cos(2\pi \mathbf{f}_2 \cdot \mathbf{x}) = 2\cos(2\pi \tfrac{\mathbf{f}_1 + \mathbf{f}_2}{2} \cdot \mathbf{x})\cos(2\pi \tfrac{\mathbf{f}_1 - \mathbf{f}_2}{2} \cdot \mathbf{x}) \qquad (D.6)$$

Thus, here too, the sum of the two cosinusoidal layers gives a product of cosines. If the original cosine frequencies $\mathbf{f}_1$ and $\mathbf{f}_2$ are similar, this cosine product corresponds to a 2D modulation effect, where the cosine with the higher frequency $\frac{\mathbf{f}_1 + \mathbf{f}_2}{2}$ represents the carrier and the cosine with the low frequency $\frac{\mathbf{f}_1 - \mathbf{f}_2}{2}$ represents the envelope (see Fig. D.3(c)). However, once again, because the envelope periodicity is only half of the cosine period it turns out that the envelope's frequency is equal to the frequency difference $\mathbf{f}_1 - \mathbf{f}_2$. And since, just as in the 1D case, the spectrum of the 2D cosine sum (D.6) (see Fig. D.3(f)) does not contain an impulse having the new low frequency $\mathbf{f}_1 - \mathbf{f}_2$, it follows that this beating is not a true moiré effect, but just a pseudo-moiré. For further details on 2D moiré effects and their spectra, see Sec. 2.3 in [Amidror09].

These considerations can be extended in a similar way to the general MD case, too.

A more detailed discussion on the beating effect (pseudo moiré) that is generated in the sum of cosines with similar frequencies can be found in [Amidror09a, Sec. 3].

D.4 Convolutions with a sinc function which have no effect

As shown in Fig. 6.1, the truncation of a function $g(x)$ into a finite sampling range of length R, namely, the multiplication of $g(x)$ by the windowing function $w(x) = \text{rect}(x/R)$, is equivalent, in terms of the spectral domain, to a convolution of the CFT of $g(x)$, $G(u)$, with the CFT of $w(x)$, $W(u) = R\,\text{sinc}(Ru)$. This convolution with $W(u)$ is, indeed, the source of the leakage artifact that $G(u)$ undergoes in the spectral domain (see, for example, Figs. 6.1–6.5). However, an interesting question may arise at this point: Consider the particular case in which the given function $g(x)$ has a finite duration which is fully included within the sampling range. Obviously, there exist infinitely many different functions $g(x)$ that satisfy this condition. It is clear that in all these cases the truncation into the sampling range will not affect $g(x)$, since $g(x)w(x) = g(x)$. This implies in the spectral domain that $G(u)*W(u) = G(u)$, meaning that in all these cases the convolution of $G(u)$ with $W(u) = R\,\text{sinc}(Ru)$ gives back $G(u)$, so that $G(u)$ is not affected by this convolution and does not suffer from leakage (see case (b) in Remark 6.1). The question is: how can it be that in such cases the convolution with $W(u)$ does not modify (smooth out) $G(u)$? Indeed, it may be surprising to see that in all these different cases the convolution with the sinc function $W(u)$ has no effect whatsoever, not even smoothing out or adding ripple. The situation here is obvious when we consider the product of $g(x)$ with the truncation window $w(x)$; but it remains quite surprising when we consider the dual question in terms of convolution.

To take a concrete example, consider any function $g(x) = \text{rect}(x/a)$ with $a \leq R$. It is clear that in all these cases we have $g(x)w(x) = g(x)$; but the dual result, which says that $a\,\text{sinc}(au)*R\,\text{sinc}(Ru) = a\,\text{sinc}(au)$, is not $a\text{-}priori$ obvious. Similarly, given a finite-duration function $h(x)$, any rectangular windowing function $w_r(x) = \text{rect}(x/r)$ that is larger than $h(x)$ will give exactly the same product $h(x)w_r(x) = h(x)$ (i.e. no effect). But once again, this is not obvious to see when considering the convolution in the other domain.

Similar questions may arise in many different situations and in various disguises (see, for example, case (b) in Remark 6.1, or, with a domain reversal, in Sec. 6.7). This is indeed a good demonstration of the power of the Fourier theory, which may often transform complicated questions in one domain into trivial ones in the dual domain.

D.5 The convolution of a square pulse with a sinc function

In this section we provide the analytic derivation of the convolution of the functions $W(u) = R\,\text{sinc}(Ru)$ and $G(u) = \frac{1}{2a}\text{rect}(\frac{u}{2a})$ (see Example 6.2 in Chapter 6).

$$C(u) = W(u)*G(u) = \int_{-\infty}^{\infty} W(z)G(u-z)\,dz$$

$$= \frac{R}{2a}\int_{-\infty}^{\infty} \text{sinc}(Rz)\,h(u-z)\,dz$$

where the function $h(z) = \text{rect}(\frac{z}{2a})$ has the value 1 within the range $-a...a$, and is zero everywhere else. Because $h(z)$ is symmetric about the origin, $h(-z)$ has the value 1 in the same range $-a...a$, and therefore the function we actually have inside the integral, $h(u-z) = h(-z+u)$, has the value 1 in the range $-a-u...a-u$. We therefore have:

$$= \frac{R}{2a}\int_{-a-u}^{a-u} \frac{\sin(\pi Rz)}{\pi Rz}\,dz$$

This integral can be simplified by using the variable substitution $t = \pi Rz$ which implies also $dz = \frac{1}{\pi R}\,dt$. Following this variable substitution our integration limits:

$$-a-u \le z \le a-u$$

change into:

$$\pi R(-a-u) \le t \le \pi R(a-u)$$

and we therefore obtain:

$$= \frac{R}{2a}\int_{\pi R(-a-u)}^{\pi R(a-u)} \frac{\sin(t)}{t}\frac{1}{\pi R}\,dt$$

$$= \frac{1}{2a\pi}\int_{\pi R(-a-u)}^{0} \frac{\sin(t)}{t}\,dt + \frac{1}{2a\pi}\int_{0}^{\pi R(a-u)} \frac{\sin(t)}{t}\,dt$$

$$= \frac{1}{2a\pi}\int_{0}^{\pi R(u+a)} \frac{\sin(t)}{t}\,dt - \frac{1}{2a\pi}\int_{0}^{\pi R(u-a)} \frac{\sin(t)}{t}\,dt$$

Now, using the definition of the sine integral function [Weisstein99 p. 1646]:

$$\text{Si}(x) = \int_{0}^{x} \frac{\sin(t)}{t}\,dt$$

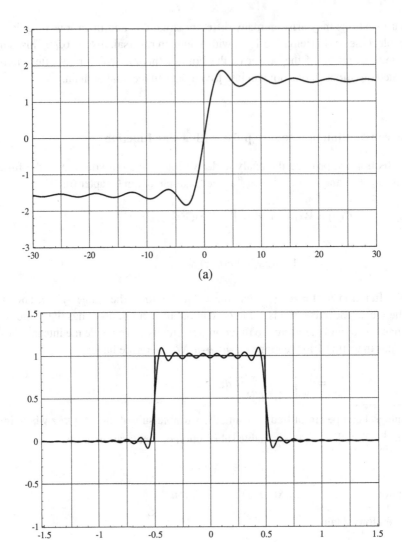

Figure D.4: (a) Graphic representation of the function Si(x). (b) Graphic representation of the functions $G(u) = \frac{1}{2a}\text{rect}(\frac{u}{2a})$ and $G(u)*W(u)$ of Eq. (D.7) for the case of $a = 1/2$.

(see Fig. D.4(a)) we finally get:

$$= \frac{1}{2a\pi}\left[\text{Si}(\pi R(u+a)) - \text{Si}(\pi R(u-a))\right] \tag{D.7}$$

As shown in Fig. D.4(b), this function resembles a smoothed-out version of $G(u)$ (which is a square pulse of height $\frac{1}{2a}$ extending between $-a$ and a), but it has an overshoot to each side of the two transitions, followed by gradually decaying oscillations to both directions (see also the dashed lines in the spectra shown in Fig. 6.4).

Note that the oscillations of the sine integral function are the reason for the overshoot and ripple artifacts which occur in signal processing in various circumstances, such as when using a truncated sinc filter as an approximation to the ideal low-pass filter (see Example 3.3 and Fig. 3.7 in Chapter 3), or the well known Gibbs phenomenon [Kammler07 pp. 44–45; Bracewell86 pp. 209–211].

We terminate this section with an interesting remark that is used later in Sec. D.6:

Remark D.1: Although the function $G(u)*W(u)$ of Eq. (D.7) is just a smoothed-out and slightly oscillating version of the original function $G(u) = \frac{1}{2a}\text{rect}(\frac{u}{2a})$, there still exists between them a major difference in their behaviour when a tends to zero: As $a \to 0$, the value of $G(u)$ at $u = 0$ tends in the limit to infinity, but the value of $G(u)*W(u)$ at $u = 0$ tends in the limit to the finite value R.

The first part of this claim is rather obvious, since the value of $G(u)$ at $u = 0$ is $\frac{1}{2a}$, which tends to infinity when $a \to 0$. To show the second part of the claim, we note from Eq. (D.7) that at $u = 0$ the value of $C(u) = G(u)*W(u)$ is:

$$C(0) = \frac{1}{2a\pi}\left[\text{Si}(\pi Ra) - \text{Si}(-\pi Ra)\right]$$

$$= \frac{1}{2a\pi}\left[\text{Si}(\pi Ra) + \text{Si}(\pi Ra)\right]$$

$$= \frac{1}{\pi a}\text{Si}(\pi Ra)$$

We want to show now that:

$$\lim_{a \to 0} \frac{\text{Si}(\pi Ra)}{\pi a} = R$$

As suggested by the *Wolfram Alpha®* website [WolframAlpha09], we start by factoring out constants:

$$\lim_{a \to 0} \frac{\text{Si}(\pi Ra)}{\pi a} = \frac{1}{\pi}\lim_{a \to 0} \frac{\text{Si}(\pi Ra)}{a}$$

This is a limit of the type 0/0. Applying l'Hopital's rule we have:

$$= \frac{1}{\pi}\lim_{a \to 0} \frac{\frac{d\text{Si}(\pi Ra)}{da}}{\frac{da}{da}}$$

$$= \frac{1}{\pi}\lim_{a \to 0} \pi R\frac{\sin(\pi Ra)}{\pi Ra}$$

$$= \frac{1}{\pi}\lim_{a \to 0} \pi R\,\text{sinc}(Ra) \qquad\qquad (\text{where } \text{sinc}(x) = \frac{\sin(\pi x)}{\pi x})$$

$$= R\lim_{a \to 0} \text{sinc}(Ra)$$

Using the continuity of $\text{sinc}(x)$ at $x = 0$:

$$= R \, \text{sinc}(\lim_{a \to 0} Ra)$$

$$= R \, \text{sinc}(R \lim_{a \to 0} a)$$

$$= R \, \text{sinc}(0)$$

This gives, indeed:

$$= R$$

which was to be demonstrated. ■

The significance of this result is explained in Sec. D.6 below.

D.6 A more detailed discussion on Remark A.2 of Sec. A.3

As we have seen in Chapter 6, unlike the continuous Fourier transform, the DFT can only take into account a finite range of the original, underlying function $g(x)$, meaning that $g(x)$ must be truncated before the application of the DFT. This truncation operation can be viewed as a multiplication of the underlying function $g(x)$ by a rectangular truncation window $w(x)$ whose length equals the sampling range $R = N\Delta x$, namely, $w(x) = \text{rect}(x/R)$.[5] The spectral-domain counterpart of this truncation is that the DFT does not approximate the original Fourier transform $G(u)$, but rather the convolution of $G(u)$ with the Fourier transform of the truncation window $w(x)$, $W(u) = R\,\text{sinc}(Ru)$. As long as $G(u)$ is non-impulsive, the convolution with $W(u)$ will smooth out $G(u)$ and add a ripple effect (see a worked-out example in Sec. D.5). However, when $G(u)$ includes impulses, the convolution with $W(u)$ will replace each impulse $c_i\delta(u-a_i)$ by an additive replica of $R\,\text{sinc}(Ru)$ that is scaled by the impulse amplitude c_i and centered about the impulse location a_i. As we have seen in Sec. 6.5, if the continuous-world impulse is located *exactly* on an element of the DFT output array, giving a sharp discrete-word impulse in the DFT spectrum, this replica of $R\,\text{sinc}(Ru)$ will simply multiply the discrete impulse amplitude c_i by R, but have no effect on the other elements of the output array. And if the continuous-world impulse is located *between* elements of the DFT output array, so that the resulting discrete impulse is smoothed out and smeared due to leakage, the multiplication by R (and hence the dependence on the choice of the sampling range length R) will affect both the smoothed out impulse and its leaked tails, but again it will have no other effect on the DFT output array. This explains, indeed, why the impulse amplitude in the DFT spectrum is multiplied by R with respect to the original continuous-world spectrum $G(u)$, and why this dependence of the DFT on the choice of R affects only impulses (and their leaked tails, if any) but no other components of $G(u)$.

---

[5] This truncation can be done either before or after sampling; see Sec. D.7 below. In practice, this is simply done by sampling the underlying function $g(x)$ within a finite domain.

To better understand this point, it may be instructive to see what happens in the DFT when a non-impulsive continuous spectrum such as $G(u) = \frac{1}{2a}\text{rect}(\frac{u}{2a})$ gradually becomes narrower and taller and approaches an impulse. Figures D.5(a)–(f) on the following pages show a sequence of continuous-world functions $g(x) = \text{sinc}(2ax)$ with gradually decreasing values of the parameter a, going from $a = 1/2$ in (a) to $a = 1/64$ in (f), along with their respective continuous Fourier transforms $G(u) = \frac{1}{2a}\text{rect}(\frac{u}{2a})$ and their DFT counterparts. Each of the cases (a)–(f) is shown twice, with a truncation window of length $R = 8$ (top row) or $R = 16$ (bottom row); the number of sampling points remains $N = 32$ in all cases. As the parameter a tends to zero, the value of $G(u)$ at the point $u = 0$ tends to infinity (since it equals $\frac{1}{2a}$); but the value of the convolution $G(u)*W(u)$ (Eq. (D.7)) at the point $u = 0$ does not tend to infinity but rather to R (see the derivation in Remark D.1 above). And indeed, as we clearly see in the figures, when $a\to0$ the CFT $G(u) = \frac{1}{2a}\text{rect}(\frac{u}{2a})$ (a pulse of height $\frac{1}{2a}$ extending between $-a$ and a, plotted in the spectra by continuous lines) tends in the limit into a continuous-world impulse at $u = 0$ [Bracewell86 pp. 69–72]; but on the other hand, when $a\to0$ the function $G(u)*W(u)$, plotted in the spectra by dashed lines, becomes a sinc-like function whose peak at $u = 0$ tends to R. Since the DFT of the original function consists of samples of $G(u)*W(u)$ rather than samples of $G(u)$, it follows that the discrete-world impulse that is obtained in the DFT when $a\to0$ has the height R (while the corresponding continuous-world impulse is conventionally drawn in the spectrum as a unit-height spike, although its height at $u = 0$ is infinite; see Sec. A.2 of Appendix A). This implies, in turn, that the height of our discrete-world impulse depends on the choice of the truncation range length R. Note that this multiplication of the height by R in the discrete world (and the resulting dependence on the value of R) affect only impulses, but not continuous components in the spectrum: As we can see by comparing the top and bottom rows in each of Figs. D.5(a)–(d), as long as the DFT is not impulsive its height follows the height of the CFT, $G(u) = \frac{1}{2a}\text{rect}(\frac{u}{2a})$, and is therefore independent of R (except for small fluctuations due to the ripple effect). But as we see in Figs. D.5(e)–(f), as soon as a becomes sufficiently small to make the DFT impulsive, or more precisely, as soon as the continuous-world pulse width $2a$ (extending between $-a$ and a) becomes equal or smaller than the DFT frequency step $\Delta u = 1/R$ (see Eq. (7.3)), the height of the resulting DFT impulse at $u = 0$ no longer follows the height of the CFT $G(u)$ but starts to approach R, meaning that now it *does* depend on the choice of R. Note that the discrete impulse height in the DFT spectrum reaches the exact value R only when $a = 0$; in this limit case the original signal-domain function $g(x) = \text{sinc}(2ax)$ becomes a straight horizontal line $g(x) = 1$, and in the DFT spectrum all the samples have the height 0, except for the sample corresponding to $u = 0$ which has the height R. Fig. D.5(f), in which $a = 1/64$, is already quite close to this limit case, both in the signal and in the spectral domains.

Table D.1 summarizes the behaviour when $a\to0$ of the CFT $G(u) = \frac{1}{2a}\text{rect}(\frac{u}{2a})$ and of the corresponding DFT (which follows, as we have seen, the behaviour of $G(u)*W(u)$ rather than that of $G(u)$).

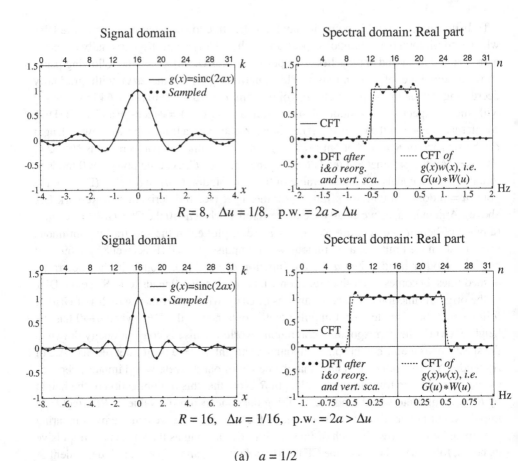

$$R = 8, \quad \Delta u = 1/8, \quad \text{p.w.} = 2a > \Delta u$$

$$R = 16, \quad \Delta u = 1/16, \quad \text{p.w.} = 2a > \Delta u$$

(a) $a = 1/2$

Figure D.5: A gradual transition from a non-impulsive DFT spectrum to an impulsive DFT spectrum, illustrating in particular the transition from R-independent to R-dependent DFT spectra (see Sec. D.6). Parts (a)–(f) of the figure show a series of continuous-world functions $g(x) = \text{sinc}(2ax)$, along with their respective continuous Fourier transforms $G(u) = \frac{1}{2a}\text{rect}(\frac{u}{2a})$, for six gradually decreasing values of a, going from $a = 1/2$ in (a) to $a = 1/64$ in (f). In each case, the continuous lines correspond to the continuous-world function $g(x)$ and to its CFT $G(u)$, and the black dots correspond to the sampled version of $g(x)$ and to its DFT spectrum. The dashed lines in the spectra correspond to the CFT of $g(x)w(x)$ where $w(x) = \text{rect}(x/R)$, i.e. to the continuous-world convolution $G(u) * W(u)$ (see Eq. (D.7)), which is the continuous-world function being sampled by the discrete elements of the DFT spectrum. Each of the cases (a)–(f) is shown twice, with a truncation window of length $R = 8$ (top row) or $R = 16$ (bottom

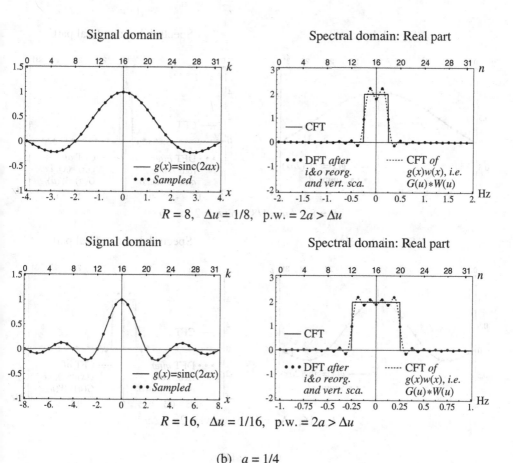

(b) $a = 1/4$

row); the number of sampling points remains $N = 32$ in all cases. In cases (a)–(d) the dashed lines as well as the dots representing the DFT spectrum follow the continuous-world spectrum $G(u)$, up to some minor oscillations (which correspond to the ripple effect due to leakage, as explained in Chapter 6). But when a gets smaller, in cases (e)–(f), the dashed lines and the DFT dots break away from the continuous-world pulse $G(u)$ (whose height continues to grow to infinity), and instead of following it they rather tend to the height R (and hence they become dependent on the choice of the truncation range length R). This break away starts when the continuous-world pulse width (p.w. = $2a$) becomes smaller than the DFT frequency step $\Delta u = 1/R$. And indeed, note that when $R = 8$ (top row) this break away begins in case (e), i.e. starting from $a = 1/32$, but when $R = 16$ (bottom row), where the pulse width is twice wider (in terms of DFT output elements), the break away only starts in case (f), i.e. when $a = 1/64$.

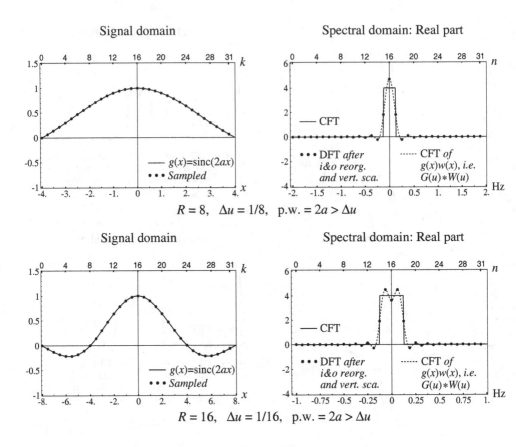

(c) $a = 1/8$

D.7 The effect of the sinc lobes on its convolution with a sharp-edged function

We have seen in the previous sections and in Chapter 6 that the convolution of a sharp-edged function such as a square pulse with the sinc function $R\,\mathrm{sinc}(Rx)$ results in an overshoot to each side of the discontinuity, followed by a gradually decaying ripple effect. It is interesting to see what is the contribution of the main lobe of the sinc function and what is the contribution of its following sidelobes to the shape of the resulting convolution.

As one can see by making a series of convolutions between a step function and a gradually truncated sinc function (see, for example, Fig. D.6), the main lobe of the sinc function only rounds the corners of the step function. The overshoots are obtained when we add the first sidelobe to each side of the main lobe of the sinc function; and the decaying ripple effect beyond the overshoots is obtained when we also add the higher sidelobes of the sinc function. A similar discussion can be found in [Oppenheim99 p. 473] and Fig. 7.23 therein.

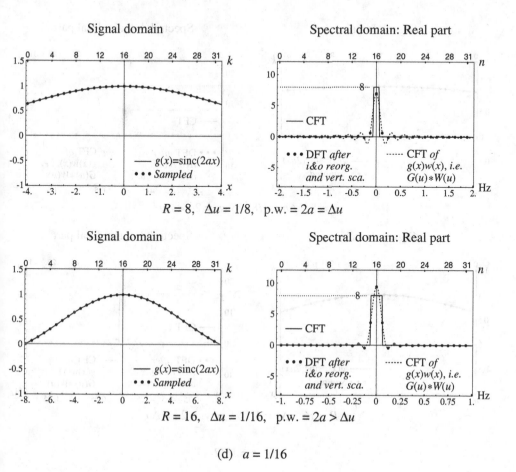

$$R = 8, \quad \Delta u = 1/8, \quad \text{p.w.} = 2a = \Delta u$$

$$R = 16, \quad \Delta u = 1/16, \quad \text{p.w.} = 2a > \Delta u$$

(d) $a = 1/16$

D.8 On the order of applying the sampling and truncation operations

We have seen that in order to apply DFT to an original continuous function $g(x)$, we must first *sample* $g(x)$ and *truncate* it into a finite range. In most texts explaining the relationship between CFT and DFT (see, for example, [Brigham88 pp. 89–97] or [Gonzalez87 pp. 95–97]) these two operations are performed in the above mentioned order: First sampling is applied to $g(x)$, and only then a finite range truncation is applied to the sampled signal. This approach is adopted in the present book, too, as one can see in Sec. 2.5.

It may be useful to note, however, that the order we apply these two operations, sampling and truncation, is immaterial, and it is perfectly admissible to first truncate $g(x)$, giving the truncated, continuous signal $g(x)w(x)$ and its spectral domain counterpart $G(u)*W(u)$, and only then sample $g(x)w(x)$ (which gives in the spectral domain a convolution of $G(u)*W(u)$ with an impulse train, i.e. an infinite number of shifted replicas of

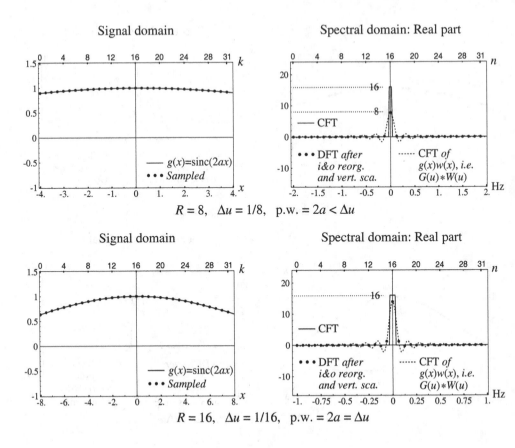

(e) $a = 1/32$

$G(u)*W(u)$). This can be formally explained by the fact that both multiplication and convolution are commutative: in the signal domain we have

$$[g(x)\ \mathrm{III}(x/\Delta x)]\ w(x) = [g(x)\ w(x)]\ \mathrm{III}(x/\Delta x) \tag{D.8}$$

where Δx is the sampling interval, and in the spectral domain we have

$$[G(u) * \Delta x\ \mathrm{III}(\Delta x\ u)] * W(u) = [G(u) * W(u)] * \Delta x\ \mathrm{III}(\Delta x\ u) \tag{D.9}$$

where the spectral-domain impulse train $\Delta x\ \mathrm{III}(\Delta x\ u)$ is the Fourier transform of the sampling impulse train $\mathrm{III}(x/\Delta x)$ (see Eqs. (5.1)–(5.2) in Sec. 5.2).

Hence, whichever order we choose for applying the sampling and truncation operations to the continuous-world signal $g(x)$ will give the same results, and the discrete counterpart of the resulting continuous-world spectrum (D.9), the DFT spectrum, will also be the same in both cases. This result is tacitly used, for example, in the beginning of Sec. 6.2. For more details on the freedom in choosing the order we perform the various operations in Fig. 2.1, see the cubic commutative diagram in [Kammler07 Sec. 1.4].

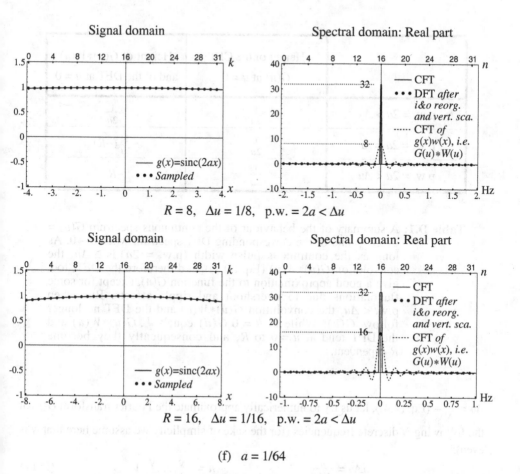

(f) $a = 1/64$

D.9 Relation of the DFT to the CFT and to Fourier series

We have seen in Chapter 2 that although the discrete theory of the DFT stands in its own right, the DFT can be also obtained as a numerical approximation to the CFT (see Remark 2.2). Let us briefly review this result.

Suppose we are given a non-periodic continuous-world function $g(x)$, which satisfies $g(x) \approx 0$ outside a certain finite range $0...R$ along the x axis (the case of periodic functions will be discussed later in this section). We wish to approximate the Fourier transform $G(u)$ from a sequence of N samples of $g(x)$ taken within this finite range with a sampling interval of $\Delta x = R/N$:

$$g(k\Delta x) \qquad\qquad k = 0, ... , N{-}1 \qquad\qquad (D.10)$$

With N input samples of $g(x)$ we can only expect to get N independent output values (frequencies). Therefore, anticipating that the frequency step obtained in the DFT output is

	Height of the CFT $G(u)$ at $u = 0$	Height of $G(u)*W(u)$ and of the DFT at $u = 0$
p.w. $= 2a > \Delta u$	$\frac{1}{2a}$	$\sim\frac{1}{2a}$
p.w. $= 2a = \Delta u$	$\frac{1}{2a}$	$\sim\frac{1}{2a}$ $(\sim R)$
p.w. $= 2a < \Delta u$	$\frac{1}{2a}$	$\sim R$

Table D.1: A summary of the behaviour of the continuous spectrum $G(u) = \frac{1}{2a}\text{rect}(\frac{u}{2a})$ and of the corresponding DFT spectrum when $a \rightarrow 0$. As long as the continuous pulse width (p.w. $= 2a$) is $\geq \Delta u$, the convolution $G(u)*W(u)$ (Eq. (D.7)), and therefore the DFT, too, give a good approximation to the function $G(u)$, except for some fluctuations due to overshoot and ripple effects. But when p.w. $< \Delta u$, the convolution $G(u)*W(u)$ and the DFT no longer follow $G(u)$: While at $u = 0$ $G(u)$ equals $\frac{1}{2a}$, $G(u)*W(u)$ and the DFT tend at $u = 0$ to R, and consequently they become R-dependent.

$\Delta u = \frac{1}{N\Delta x}$ (Eq. (4.4)), let us try to numerically approximate the Fourier transform $G(u)$ at the following N discrete frequencies (for the sake of simplicity we assume here that N is even):[6]

$$n\Delta u = \frac{n}{N\Delta x} \qquad\qquad n = -\frac{N}{2}, \dots, \frac{N}{2}-1 \qquad (D.11)$$

The reasons for choosing this range of n will be explained soon in Remark D.2. For the time being, note that the extreme frequencies obtained with this range of n, namely $-\frac{1}{2\Delta x}$ and $\frac{1}{2\Delta x} - \frac{1}{N\Delta x}$, correspond exactly to the lowest and highest frequencies that can be obtained when sampling $g(x)$ at N points that are spaced by a sampling interval of Δx (see Sec. 4.3).

Now, for any value of n within this range we have from the definition of the CFT (Eq. (2.1)):

$$G(n\Delta u) = \int_{-\infty}^{\infty} g(x)\, e^{-i2\pi n\Delta ux}\, dx$$

Because we assumed that $g(x) \approx 0$ outside our finite sampling range $0...R$, we have:

$$\approx \int_{0}^{R} g(x)\, e^{-i2\pi n\Delta ux}\, dx$$

---

[6] In the vast majority of DFT implementations N is a power of two, or at least an even number. In the rare cases where N is odd the index range should be adapted accordingly [Briggs95 Sec. 3.1].

Given sharp pulse   Truncated 4sinc(4x)   Their convolution

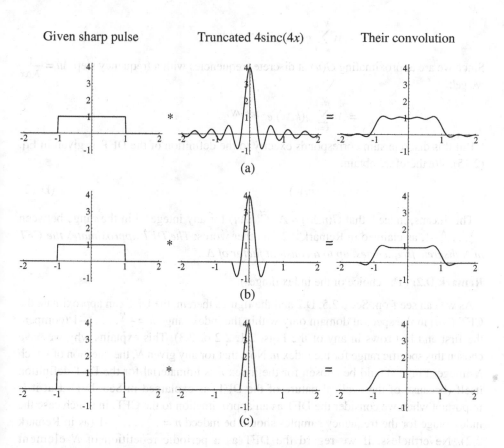

Figure D.6: Convolution of a sharp pulse (consisting of two step functions) with a gradually truncated $R\,\text{sinc}(Rx)$ function, showing the influence of the different sinc lobes on the resulting convolution. All parts of the figure subsist in the signal domain. We have used here the function $4\text{sinc}(4x)$ for the sake of clarity, but higher values of R give the same results. As shown in row (c), the main lobe of the sinc function only rounds the corners of each sharp transition of the pulse. Row (b) shows that the overshoots to both sides of each sharp transition are obtained when we add next to the main lobe of the sinc function the first neighbouring sidelobe to each direction. Row (a) shows that the decaying ripple effect beyond the overshoots is obtained when we also add the higher sidelobes of the sinc function.

Let us approximate this integral by a discrete sum, using the rectangular rule for approximating integrals [Press02 p. 508; Ramirez85 p. 67]:[7]

---

[7] A more detailed and rigorous explanation can be found in [Briggs95, Sec. 2.2], but it is based there on a different variant of the DFT definition, and on the trapezoidal rule rather than the rectangular rule for approximating integrals (these rules of numerical integration can be found, for example, in [Harris98 pp. 567–568], [Kreyszig93 Sec. 18.5] or [Spiegel68 p. 95]).

$$\approx \Delta x \sum_{k=0}^{N-1} g(k\Delta x)\, e^{-i2\pi n\Delta u k\Delta x}$$

Since we are approximating $G(u)$ at discrete frequencies with a frequency step $\Delta u = \dfrac{1}{N\Delta x}$ we get:

$$= \Delta x \sum_{k=0}^{N-1} g(k\Delta x)\, e^{-i2\pi kn/N}$$

But this discrete sum corresponds exactly to the definition of the DFT as given in Eq. (2.15). We therefore obtain:

$$= \Delta x\, \widetilde{G}(n\Delta u) \tag{D.12}$$

This means, indeed, that $G(n\Delta u) \approx \Delta x\, \widetilde{G}(n\Delta u)$ for any integer n in the range between $-\dfrac{N}{2}, \dots, \dfrac{N}{2}-1$, as claimed in Remark 2.2. In other words: *The DFT approximates the CFT at N discrete frequencies, up to a constant factor of Δx.*

Remark D.2: (The choice of the index ranges):

As we can see from Secs. 2.5, D.2 and the figures therein, the DFT can approximate the CFT $G(u)$ in the spectral domain only within the index range $n = -\dfrac{N}{2}, \dots, \dfrac{N}{2}-1$ (compare the first and last rows in any of the Figs. 2.1, 2.2 or 2.3). This explains why we have chosen this specific range for the index n. Note that for any given N, the question of which N-integer range should be chosen for the index n is immaterial for the DFT definition itself (because of the cyclical nature of the DFT, as explained in Sec. D.1). But it *is* important when we consider the DFT as an approximation to the CFT, in which case the index range for the frequency samples should be indeed $n = -\dfrac{N}{2}, \dots, \dfrac{N}{2}-1$ (as in Remark 2.2). Nevertheless, if we regard the DFT as a periodic repetition of N-element replicas, meaning that the index n is considered modulo this range, then any other range of N consecutive integers may also be used for the index n, including the simple range $n = 0, \dots, N-1$ as in Eqs. (2.11)–(2.20). In such cases the resulting N frequency values will still be the same as before, but their order in the output array of the DFT will be cyclically shifted. See also Sec. 3.2 about the output data array of the DFT and its reorganization.

As for the range of the signal-domain index k, its choice basically depends on the nature and the domain of the original continuous-world function $g(x)$, as we can see in Figs. 2.1–2.3. In the general case where the function $g(x)$ is not symmetric (see Figs. 2.2–2.3) it is reasonable to choose $k = 0, \dots, N-1$, as we did in Eq. (D.10). But here, too, due to the cyclical nature of the DFT input (which is periodic with an N-element period), the formal numerical approximation above can be done for any range of the index k consisting of N consecutive integers. For example, in [Briggs95, Chapter 2] the development is done using the range $k = -\dfrac{N}{2}+1, \dots, \dfrac{N}{2}$, while in [Press02 p. 508] it is done using the range $k = 0, \dots, N-1$. See also Sec. 3.3 about the input data array of the DFT and its reorganization. ∎

Let us now proceed to the case of Fourier series, which occurs if the given continuous-world function is periodic. Here, too, a similar result can be formulated to express the relation of the DFT to the spectrum of the periodic function, which consists, in this case, of the Fourier series coefficients of $g(x)$. Let us briefly review this case, too.

Suppose that $p(x)$ is a periodic continuous-world function with period T. Its Fourier series expansion is given by (see, for example, [Briggs95 p. 33], [Kammler07 pp. 5–6]):

$$p(x) \sim \sum_{n=-\infty}^{\infty} c_n\, e^{i2\pi nx/T} \tag{D.13}$$

where the n-th Fourier series coefficients c_n are:

$$c_n = \frac{1}{T}\int_T p(x)\, e^{-i2\pi nx/T}\, dx \tag{D.14}$$

What is the spectrum of the function $p(x)$? The Fourier series expansion of $p(x)$ can serve us here as a link between the original function $p(x)$ in the signal domain and its Fourier transform $P(u)$ in the spectral domain. This is based on the fact that the Fourier transform (the spectrum) of the function $g(x) = e^{i2\pi x/T}$ is $G(u) = \delta(u - \frac{1}{T})$, namely, an impulse at the frequency $f = \frac{1}{T}$ [Bracewell86 p. 101]. Consequently, the spectrum of each term in the Fourier series (D.13) consists of a single impulse at the frequency of n/T, whose amplitude (real or complex) is given by the corresponding Fourier coefficient c_n. It follows therefore (under the appropriate convergence conditions) that the spectrum $P(u)$ of the periodic function $p(x)$ is an impulse-comb with step $f = \frac{1}{T}$, whose n-th impulse is located at the frequency $u = nf$ and has the amplitude c_n. The general form of the spectrum of a periodic function $p(x)$ with period T is, therefore [Papoulis68 p. 107]:

$$P(u) = \sum_{n=-\infty}^{\infty} c_n\, \delta(u - n/T) = \sum_{n=-\infty}^{\infty} c_n\, \delta(u - nf) \tag{D.15}$$

where $\delta(u)$ is the impulse symbol. Note that the range of the index n in this sum may be infinite, depending on the given function $p(x)$.

Now, to find the relation of the DFT to the spectrum of a given periodic function $g(x)$, we proceed in a similar way as above. First of all, we must limit ourselves once again to a finite set of only N discrete frequencies; and indeed, for the same reasons as before, we choose for the index n the range $n = -\frac{N}{2}, \dots, \frac{N}{2} - 1$. For any value of n within this range we have from the definition of the Fourier series coefficient (Eq. (D.14)):

$$c_n = \frac{1}{T}\int_T g(x)\, e^{-i2\pi nx/T}\, dx$$

As before, we approximate this integral over the range T by a discrete sum, using the rectangular rule for approximating integrals:[8]

$$\approx \frac{1}{T}\Delta x \sum_{k=0}^{N-1} g(k\Delta x)\, e^{-i2\pi nk\Delta x/T}$$

Note that in order for this approximation to make sense, our sampling range length, $R = N\Delta x$, should be equal to one period T of our given function. We therefore have:

---

[8] A more detailed and rigorous explanation can be found in [Briggs95, Sec. 2.4], but it is based there on a different variant of the DFT definition, and on the trapezoidal rule rather than the rectangular rule for approximating integrals.

$$= \frac{1}{R} \Delta x \sum_{k=0}^{N-1} g(k\Delta x) \, e^{-i2\pi kn/N}$$

As we can see, this discrete sum corresponds exactly to the definition of the DFT as given in Eq. (2.15). We therefore obtain:

$$= \frac{\Delta x}{R} \, \widetilde{G}(n\Delta u) \qquad\qquad\qquad\qquad\qquad (D.16)$$

This means, indeed, that $c_n \approx \frac{\Delta x}{R} \, \widetilde{G}(n\Delta u)$ for any integer n in the range $-\frac{N}{2}, \ldots, \frac{N}{2}-1$. In other words: *The DFT approximates N Fourier series coefficients of $g(x)$, up to a constant factor of $\frac{\Delta x}{R}$ (which equals $\frac{1}{N}$, since $R = N\Delta x$).*

It is interesting to note at this point that if we take for the discrete approximation of the integral a range length $R = N\Delta x$ that is not equal to the period T of the given function, we cannot expect to obtain a good discrete approximation to c_n. And indeed, in terms of the DFT theory we have seen in Chapter 6, if $R \neq T$ a leakage error may occur and considerably deteriorate the approximation (see Remark 6.1).[9]

In conclusion, we see that the relation between a CFT and its DFT approximation is given by:

$$G(n\Delta u) \approx \Delta x \, \widetilde{G}(n\Delta u), \quad \text{where} \quad \widetilde{G}(n\Delta u) = \sum_{k=0}^{N-1} g(k\Delta x) \, e^{-i2\pi kn/N} \qquad n = -\frac{N}{2}, \ldots, \frac{N}{2}-1$$

And similarly, the relation between a Fourier series and its DFT approximation is:

$$c_n \approx \frac{\Delta x}{R} \, \widetilde{G}(n\Delta u), \qquad \text{where} \quad \widetilde{G}(n\Delta u) = \sum_{k=0}^{N-1} g(k\Delta x) \, e^{-i2\pi kn/N} \qquad n = -\frac{N}{2}, \ldots, \frac{N}{2}-1$$

It is important to note, however, that when using a version of the DFT definition that includes the constant $\frac{1}{N}$ (as is the case, for example, in [Briggs95]), the scaling factors Δx and $\frac{\Delta x}{R}$ that precede the DFT $\widetilde{G}(n\Delta u)$ in the above relations must be multiplied by N (in order to allow for a partial factor of $\frac{1}{N}$ thereof to be taken out and incorporated into the DFT itself without causing any changes). In this case we get the following relations (like in [Briggs95 pp. 22–23 and 40]):

$$G(n\Delta u) \approx R \, \widetilde{G}(n\Delta u), \quad \text{where} \quad \widetilde{G}(n\Delta u) = \frac{1}{N}\sum_{k=0}^{N-1} g(k\Delta x) \, e^{-i2\pi kn/N} \qquad n = -\frac{N}{2}, \ldots, \frac{N}{2}-1$$

and

$$c_n \approx \widetilde{G}(n\Delta u), \qquad \text{where} \quad \widetilde{G}(n\Delta u) = \frac{1}{N}\sum_{k=0}^{N-1} g(k\Delta x) \, e^{-i2\pi kn/N} \qquad n = -\frac{N}{2}, \ldots, \frac{N}{2}-1$$

These results are succinctly summarized in Table D.2. Further insights into the imterconnections between the DFT, the CFT and the Fourier series, and on the ways they approach each other under various limiting considerations, can be found in [Briggs95 pp.

---

[9] In fact, according to Remark 6.1 leakage does not occur whenever the sampling range length R is an exact integer multiple of the period T. But if the integer multiple m in question is bigger than 1, meaning that our sampling range R contains $m > 1$ full periods of $g(x)$ rather than one, then the impulses in the N-element output array of the DFT will be spaced by $m-1$ zeroes, and the number of the approximated coefficients c_n we obtain in the N-element DFT output will be therefore reduced from N to N/m (see Remark 7.1).

	Relation between DFT and CFT	Relation between DFT and Fourier series
Using DFT definition without $\frac{1}{N}$	$G(n\Delta u) \approx \Delta x\, \widetilde{G}(n\Delta u)$	$c_n \approx \frac{\Delta x}{R}\, \widetilde{G}(n\Delta u)$
Using DFT definition with $\frac{1}{N}$	$G(n\Delta u) \approx R\, \widetilde{G}(n\Delta u)$	$c_n \approx \widetilde{G}(n\Delta u)$

Table D.2: The relations between CFT or Fourier series and their DFT approximations.

211–212]. Similar interconnections for the IDFT can be found in [Briggs95 pp. 218–222]. Note, however, that the DFT and IDFT definitions used there are different than ours.

Remark D.3: As suggested by Table D.2, the following relation exists in the continuous world between the CFT and the Fourier series (where Δu stands for $1/R$):

$$G(n\Delta u) = R\, c_n \qquad\qquad (D.17)$$

Note that this underlying relation holds in each of the two rows of Table D.1; and indeed, this relation exists in the continuous world in its own right, independently of the DFT theory. To see this, consider a continuous-world function $g(x)$ that is spatially limited, meaning that $g(x)$ is zero outside a range of length R. The Fourier transform evaluated at the frequency $u = n\Delta u = n/R$ is given by:

$$G(n/R) = \int_R g(x)\, e^{-i2\pi nx/R}\, dx$$

On the other hand, consider the periodic function $p(x)$ that is defined as a periodic extension of $g(x)$ with the period R. The integral for the Fourier coefficients of $p(x)$ on the interval R is given, according to Eq. (D.14), by:

$$c_n = \frac{1}{R} \int_R p(x)\, e^{-i2\pi nx/R}\, dx$$

We see, therefore, that in the case of a function $g(x)$ that is zero outside the interval R, the CFT of $g(x)$ and the Fourier series coefficients of $p(x)$, the periodic extension of $g(x)$, are indeed related by Eq. (D.17) [Brigham88 Sec. 5.2; Briggs95 pp. 40–41, 197; Kammler07 p. 179]. If outside the interval R the original function $g(x)$ only satisfies $g(x) \approx 0$ rather than $g(x) = 0$, then relation (D.17) turns into an approximation:

$$G(n\Delta u) \approx R\, c_n \qquad\qquad (D.18)$$

Note that Eq. (D.17) concerns the relation between the Fourier coefficients c_n of a periodic function $p(x)$ with period R and the Fourier transform $G(u)$ of the spatially-limited function $g(x)$ that consists of *a single period* of $p(x)$ about the origin. The relation between the Fourier coefficients c_n of the periodic function $p(x)$ and the Fourier transform $P(u)$ of $p(x)$ itself is given by Eq. (D.15); in this case no scaling factor is present, and we have, indeed:

$$P(n\Delta u) = c_n \tag{D.19}$$

as explained in the paragraph preceding Eq. (D.15). ∎

Example D.1: To take a concrete example, consider the triangular function tri(x), which extends between –1 and 1; its CFT is sinc$^2(u)$ [Bracewell86 p. 415]. According to the dilation theorem (see point 2 in Sec. 2.4.1), its two-fold narrower counterpart $g(x) =$ tri(2x), which extends between $-\frac{1}{2}$ and $\frac{1}{2}$, has the CFT $G(u) = \frac{1}{2}$sinc$^2(\frac{1}{2}u)$. Now, as shown in Fig. D.7, the periodic triangular wave $p(x) =$ triangwave(2x) that is defined as a periodic extension of $g(x)$ with the period $R = 2$, can be seen as a convolution of the function $g(x) =$ tri(2x) with the infinite impulse train $t(x) = \frac{1}{2}$III($\frac{1}{2}x$) having period $P = 2$ and amplitude 1. The CFT of $t(x)$ is the impulse train $T(u) =$ III(2u), which has impulse intervals of $f = 1/P = \frac{1}{2}$ and amplitude $\frac{1}{2}$ [Bracewell86 p. 414]. Therefore, the CFT $P(u)$ of our periodic triangular wave $p(x) =$ triangwave(2x) consists of the product of $G(u) = \frac{1}{2}$sinc$^2(\frac{1}{2}u)$ and $T(u) =$ III(2u), which is an infinite impulse train that samples the "envelope" $\frac{1}{4}$sinc$^2(\frac{1}{2}u)$ at impulse intervals of $f = \frac{1}{2}$ (see, for example, [Bracewell86 p. 213]). And indeed, we see that the impulse amplitudes in the spectrum of $p(x)$, namely, its Fourier series coefficients c_n, are identical to the corresponding values in the spectrum of $g(x)$ up to a scaling factor of $R = 2$, in accordance with Eq. (D.17). ∎

Eq. (D.17) says, in fact, that it is possible to compute the Fourier series coefficients c_n of $p(x)$ using the CFT of $g(x)$, i.e. by calculating $G(n/R)$ [Kammler07 p. 179]. In this case, the result we obtain is not c_n but rather Rc_n: each impulse in the spectrum is multiplied by R (which is, in fact, the period of the periodic function $p(x)$ whose Fourier series we are computing). This fact sheds, indeed, a new light on our results in Appendix A.3, where we have seen that the impulse amplitudes in the DFT spectrum (after the standard amplitude scaling correction) are R times higher than the true impulse amplitudes c_n in the continuous-world spectrum.

Stating it in other words, remember that the standard amplitude scaling correction that we should always apply to the DFT output according to Sec. 4.4 includes a multiplication by Δx. This is, indeed, in accordance with our result $G(n\Delta u) \approx \Delta x \, \widetilde{G}(n\Delta u)$ from Eq. (D.12). This multiplication by Δx is to be done in all cases, including when the spectrum $G(u)$ contains impulses.[10] We therefore must realize, by virtue of our result $c_n \approx \frac{\Delta x}{R} \widetilde{G}(n\Delta u)$ (see Eq. (D.16)), that if the spectrum $G(u)$ includes impulses, then their amplitudes (which are, according to Eq. (D.15), the Fourier series coefficients c_n) will appear in the DFT output, after the standard amplitude scaling correction, R times higher than their true amplitudes in the continuous-world spectrum $G(u)$.

---

[10] As shown, for example, in Fig. A.4, in certain cases the spectrum $G(u)$ may be hybrid and contain symultaneously both continuous and impulsive components; the multiplication of the DFT output values by Δx will affect both the impulsive and the non-impulsive parts of the DFT spectrum.

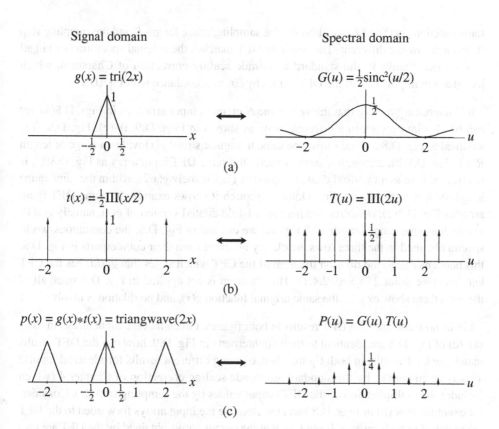

Figure D.7: A concrete continuous-world example illustrating Remark D.3. (a) The continuous function $g(x) = \text{tri}(2x)$, which extends in the range between $-\frac{1}{2}$ and $\frac{1}{2}$, and its CFT $G(u) = \frac{1}{2}\text{sinc}^2(\frac{1}{2}u)$. (b) The impulse train $t(x) = \frac{1}{2}\text{III}(\frac{1}{2}x)$, having period $P = 2$ and amplitude 1, and its CFT $T(u) = \text{III}(2u)$, which is an impulse train with impulse intervals of $f = 1/P = \frac{1}{2}$ and amplitude $\frac{1}{2}$. (c) The periodic triangular wave $p(x) = \text{triangwave}(2x)$ is a convolution of the signals $g(x)$ and $t(x)$ of rows (a) and (b), and therefore its CFT $P(u)$ is the product of the CFTs $G(u)$ and $T(u)$ of rows (a) and (b).

D.9.1 Examples illustrating the relation of the DFT to the CFT and to Fourier series

Let us now consider a few examples that will help us better understand the relationship between the DFT of the function $g(x)$ and the DFT of its periodic extension $p(x)$. These examples illustrate, in particular, the various amplitude corrections that must be applied to the DFT output in order that it may match the corresponding real-world spectra (assuming that no DFT errors such as aliasing or leakage are present).

Fig. D.8 shows the continuous function $g(x) = \text{tri}(2x)$ and its CFT $G(u) = \frac{1}{2}\text{sinc}^2(\frac{1}{2}u)$, as well as their discrete counterparts that are obtained by sampling $g(x)$ and applying DFT (after applying the corrections of Chapters 3 and 4). All the rows of the figure show the same functions $g(x)$ and $G(u)$, although the sampling range length R and the sampling step Δx in each row are different. The resulting DFT matches the original spectrum $G(u)$ in all of the rows, thanks to the standard amplitude scaling correction of Chapter 4, which includes a multiplication of the DFT values by Δx, in accordance with Eq. (D.12).

It is interesting to note that the very same N-element input arrays as in Figs. D.8(a)–(c) can be also interpreted in a different way, as shown in Figs. D.9(a)–(c). Fig. D.9(a) is identical to Fig. D.8(a), and shows the same triangular signal $g(x)$ within a range of length $R = 8$. Fig. D.9(b), although it shows exactly the same DFT input array as Fig. D.8(b), is interpreted here as a two-fold dilated version of $g(x)$, namely $g(x/2)$, within the same range length of $R = 8$. Similarly, Fig. D.9(c), although it shows exactly the same DFT input array as Fig. D.8(c), is interpreted here as a 4-fold dilated version of $g(x)$, namely $g(x/4)$, within the same range length of $R = 8$. As we can see in Fig. D.9, the continuous-world spectra obtained in the three rows are clearly different from their counterparts in Fig. D.8; this happens due to the dilation theorem of the CFT which states that $g(x/a)$ has the CFT $|a|G(au)$ (see point 2 in Sec. 2.4.1). This theorem is not applied in Fig. D.8 since all of the rows there show exactly the same original function $g(x)$, and no dilation is involved.

Let us now consider the DFT results in both figures. Given that the input arrays in rows (a)–(c) of Fig. D.8 are identical to their counterparts in Fig. D.9, how can the DFT results match the CFT results in both figures, hence giving different results for identical inputs? This is taken care of by the standard amplitude scaling correction of Chapter 4 (which includes the multiplication of the DFT output values by the sampling step Δx). Consider, for example, row (b) in Figs. D.8 and D.9. Because the input arrays forwarded to the DFT are identical in both cases, it is obvious that the output arrays obtained by the DFT are also identical. However, in Fig. D.8(b) this raw data is then multiplied by $\Delta x = 4/64 = 1/16$, while in Fig. D.9(b) the same output data is rather multiplied by $\Delta x = 8/64 = 1/8$. Hence the final DFT output amplitudes in Fig. D.9(b) are twice higher than in Fig. D.8(b), so that in both cases the corrected DFT results indeed match the respective CFTs.

Consider now Fig. D.10, which shows the periodic counterpart of Fig. D.8. The periodic function $p(x) = \text{triangwave}(2x)$ is the periodic extension of $g(x) = \text{tri}(2x)$ with the period $P = 2$, and its CFT $P(u)$ is, as we have seen in Example D.1 above, an infinite impulse train that samples the "envelope" $\frac{1}{2}G(u) = \frac{1}{4}\text{sinc}^2(\frac{1}{2}u)$ at impulse intervals of $f = \frac{1}{2}$. Just as in Fig. D.8, all the rows of our figure show the same functions $p(x)$ and $P(u)$, although the sampling range length R and the sampling step Δx in each row are different. The resulting DFT matches the original spectrum $P(u)$ in all of the rows, thanks to the standard amplitude scaling correction of Chapter 4 (which includes a multiplication of the DFT results by Δx) and the impulse amplitude correction of Appendix A (the division of the DFT impulse amplitudes by R), in accordance with Eq. (D.16).

Remark D.4: Note that the input arrays that are forwarded to the DFT in Fig. D.8(c) and in Fig. D.10(c) are strictly identical. The difference between the two figures is only in the interpretation given to this data: While in Fig. D.8(c) the input data is considered to be a sampled version of the aperiodic triangular function $g(x) = \text{tri}(2x)$, in Fig. D.10(c) the same input data is considered to be a sampled version of a single period of the periodic function $p(x) = \text{triangwave}(2x)$, the periodic extension of $g(x)$ with period $P = 2$. Consequently, the very same raw DFT output that is obtained in the two cases is interpreted differently in each case: in Fig. D.8(c) it is considered as an approximation to the continuous CFT of $g(x)$, $G(u) = \frac{1}{2}\text{sinc}^2(\frac{1}{2}u)$, while in Fig. D.10(c) it is interpreted as an approximation to the Fourier series of $p(x)$, which is an impulse train that samples the "envelope" $\frac{1}{2}G(u) = \frac{1}{4}\text{sinc}^2(\frac{1}{2}u)$ at impulse intervals of $f = \frac{1}{2}$. Thus, although the raw output data that is obtained by the DFT in both cases is identical, in the first case it will only undergo the standard amplitude scaling correction of Chapter 4 (multiplication by Δx), while in the second case it will also undergo the impulse amplitude correction of Appendix A (division of the impulse amplitudes by $R = 2$). Note that since this output array looks "continuous", unlike its clearly impulsive counterparts in Figs. D.10(a),(b), its interpretation as a continuous CFT or as a discrete impulse train (Fourier series) only depends on whether we consider the input array as a sampled version of $g(x)$ or as a sampled version of its periodic extension $p(x)$. In the second case, we must remember to apply the impulse amplitude correction of Appendix A (division by R), in accordance with Eq. (D.17). ∎

Just as in the case of Fig. D.8, it is interesting to note that in the periodic case, too, the very same N-element input arrays of Fig. D.10 can be also interpreted in a different way. This is shown in Figs. D.11(a)–(c). Fig. D.11(a) is identical to Fig. D.10(a), and shows the same periodic triangular signal $p(x)$ within a range of length $R = 8$, covering 4 periods of $p(x)$. Fig. D.11(b), although it shows exactly the same DFT input and output arrays as Fig. D.10(b), is interpreted here as a two-fold dilated version of $p(x)$, namely $p(x/2)$, within the same range length of $R = 8$ (which covers now only two periods of $p(x/2)$). Similarly, Fig. D.11(c), although it shows exactly the same discrete data as Fig. D.10(c), is interpreted here as a four-fold dilated version of $p(x)$, namely $p(x/4)$, within the same range length of $R = 8$ (which covers this time only one period of $p(x/4)$). Note that in the case of Fourier series, the dilation theorem (i.e. the spectral-domain behaviour when the original periodic function undergoes a dilation) is not the same as in the case of CFT (point 2 in Sec. 2.4.1), and it does *not* involve an amplitude change in the Fourier series coefficients [Kammler07 p. 187]. In the discrete DFT spectra of Fig. D.11 this is taken care of by the amplitude corrections of Chapter 4 (the multiplication by Δx) and of Appendix A (the division by R), since in all rows of Fig. D.11 both Δx and R remain unchanged. Note that in Fig. D.10, both Δx and R vary in each row; but since they vary proportionally (because $R = N\Delta x$, and N remains unchanged in all rows), the total amplitude correction (multiplication by Δx and division by R) remains the same in the three rows, just as in Fig. D.11.

Of course, if $g(x)$ is not periodic, then the height of its CFT *does* depend on the dilation factor of the given function, as shown in Fig. D.9. Furthermore, because the DFT spectrum is no longer impulsive, the impulse amplitude correction of Appendix A (the division of the impulse amplitudes in the DFT spectrum by R) should not be applied, and therefore the amplitude correction of the DFT output depends on Δx rather than on the ratio $\Delta x/R$ as in the periodic case.

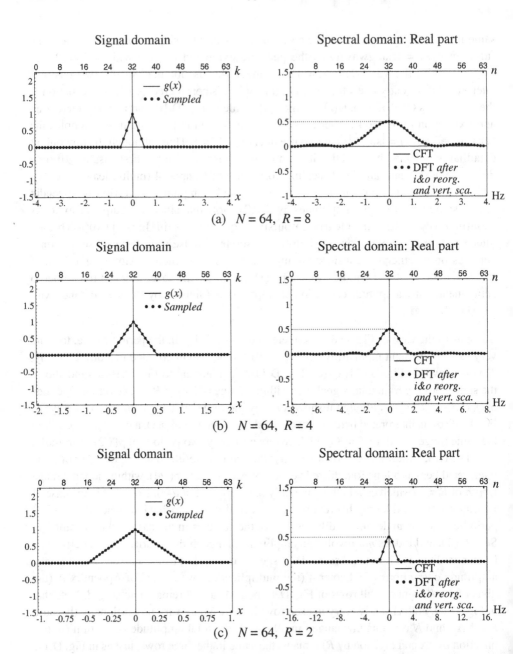

Figure D.8: The continuous function $g(x) = \mathrm{tri}(2x)$ and its CFT $G(u) = \frac{1}{2}\mathrm{sinc}^2(\frac{1}{2}u)$, represented by continuous lines, and their discrete counterparts that are obtained by sampling $g(x)$ and applying DFT (including the corrections of Chapters 3 and 4), represented by black dots. All the rows of the figure show the same functions $g(x)$ and $G(u)$, although the sampling range length R and the sampling step Δx in each row are different (note that $N = 64$ in all cases): (a) $R = 8$, $\Delta x = 8/64 = 1/8$; (b) $R = 4$, $\Delta x = 4/64$ $= 1/16$; (c) $R = 2$, $\Delta x = 2/64 = 1/32$.

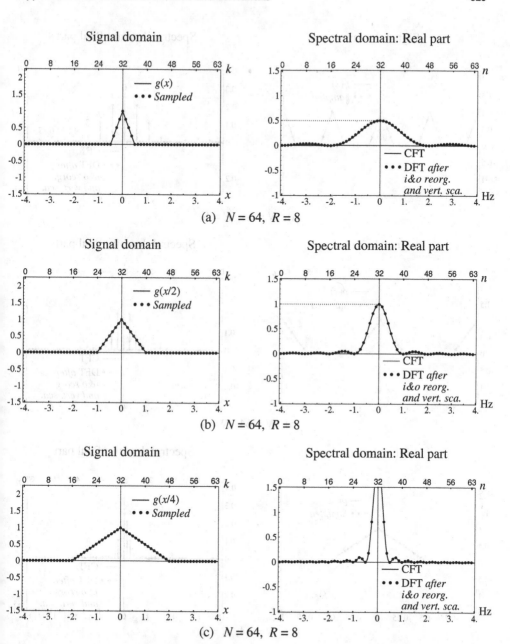

Figure D.9: A different interpretation of the very same discrete input arrays as in Fig. D.8, where the sampling range length and the sampling step Δx remain identical in the three rows ($R = 8$, $\Delta x = 8/64 = 1/8$), but each row represents a differently dilated version of $g(x)$, and its corresponding CFT: (a) $g(x) = \text{tri}(2x)$; $G(u) = \frac{1}{2}\text{sinc}^2(\frac{1}{2}u)$. (b) $g(x/2) = \text{tri}(x)$; $2G(2u) = \text{sinc}^2(u)$. (c) $g(x/4) = \text{tri}(x/2)$; $4G(4u) = 2\text{sinc}^2(2u)$. Although the discrete input array in each row is identical to its counterpart in Fig. D.8, the resulting discrete spectra do match the different CFTs thanks to the standard amplitude scaling correction of Chapter 4, which includes a multiplication by Δx.

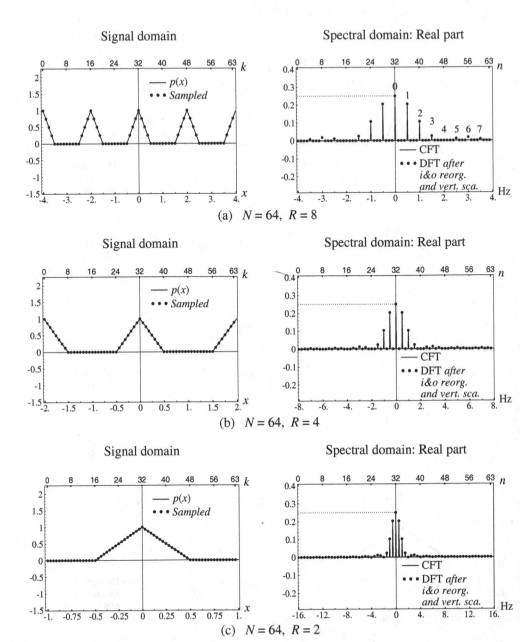

Figure D.10: The continuous periodic function $p(x) = \text{triangwave}(2x)$ (a periodic extension of $g(x) = \text{tri}(2x)$ of Fig. D.8) and its impulsive CFT $P(u)$, represented by continuous lines, and their discrete counterparts that are obtained by sampling $p(x)$ and applying DFT (including the corrections of Chapters 3, 4 and Appendix A), represented by black dots. All the rows of the figure show the same functions $p(x)$ and $P(u)$, although the sampling range length R and the sampling step Δx in each row are different (note that $N = 64$ in all cases): (a) $R = 8$, $\Delta x = 8/64 = 1/8$; (b) $R = 4$, $\Delta x = 4/64 = 1/16$; (c) $R = 2$, $\Delta x = 2/64 = 1/32$.

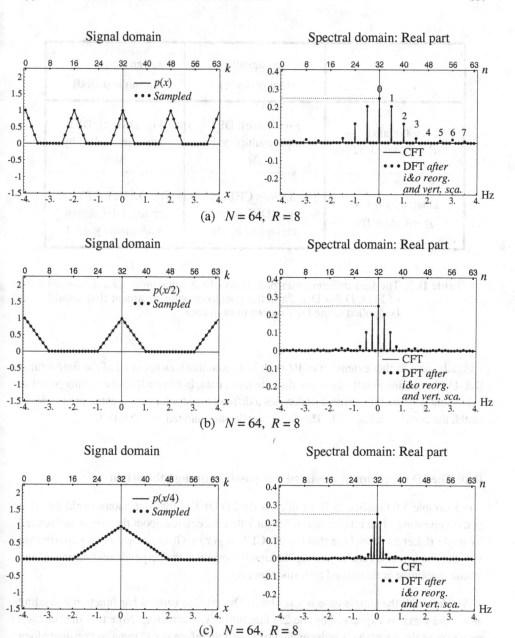

Figure D.11: A different interpretation of the very same input and output DFT arrays as in Fig. D.10, where the sampling range length remains identical in the three rows ($R = 8$) but each row represents a differently dilated version of $p(x)$, and its corresponding impulsive CFT: (a) $p(x) =$ triangwave($2x$); (b) $p(x/2) =$ triangwave(x); (c) $p(x/4) =$ triangwave($x/2$). Note that unlike in Figs. D.8–D.9, all the spectra in Figs. D.10–D.11 have the same amplitude.

	Given signal: $g(x)$ (not periodic)	Given signal: $p(x)$ (a single period)
Not dilated; $R = 2$, $\Delta x = 1/32$	Fig. D.8(c): DFT output multiplied by Δx	Fig. D.10(c): DFT output multiplied by $\Delta x/R$
Dilated by $a = 4$; $R = 8$, $\Delta x = 1/8$	Fig. D.9(c): CFT scaled by a; DFT output multiplied by Δx	Fig. D.11(c): CFT not scaled; DFT output multiplied by $\Delta x/R$

Table D.3: The four different interpretations of the same input data in row (c) of Figs. D.8 – D.11, and the postprocessing treatment that should be applied to the DFT output in each case.

Finally, as a further extension of Remark D.4, note that row (c) in all of the four figures D.8–D.11 contains exactly the same discrete input data, but in each case it is interpreted in a different way, and therefore it undergoes a different postprocessing treatment in order to match the corresponding CFT. This is succinctly summarized in Table D.3.

D.10 The 2D spectrum of a rotated bar passing through the origin

In Example 5.6 (see Sec. 5.5) we discuss the 2D DFT of the continuous-world function $g(x,y)$ consisting of an infinite bar of height 1 that is centered about the origin and rotated by angle θ. Let us show here that the 2D CFT of $g(x,y)$, $G(u,v)$, consists of a continuous straight line-impulse (blade) passing through the origin at the perpendicular direction, and whose amplitude is modulated by a sinc function.

We start with the simple case where $\theta = 0$. The infinite vertical bar function extending along the y axis is expressed by the 2D function $g(x,y) = \text{rect}(x)$. Note that the function rect(x) over the x,y plane is independent of y, and therefore $g(x,y)$ remains constant along the y direction. Similarly, the function sinc(u) over the u,v plane is independent of v, and therefore it remains constant along the v direction; therefore, in order to obtain a sinc-shaped blade (line impulse) along the u axis we have to multiply sinc(u) by $\delta(v)$.

Now, because $G(u,v) = \text{sinc}(u)\delta(v)$ is a product of a function of u and a function of v (both of which being considered throughout the u,v plane), we can apply the separable product theorem (see point 10 in Sec. 2.4.2 or [Bracewell95 p. 166]): Knowing that

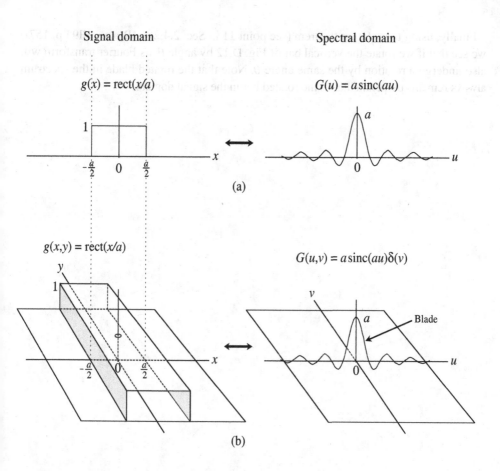

Figure D.12: (a) The 1D Fourier transform of the 1D function $g(x) = \text{rect}(x/a)$ is $G(u) = a\,\text{sinc}(a\,u)$. (b) Therefore, the 2D Fourier transform of the straight bar $g(x,y) = \text{rect}(x/a)$ along the y axis of the x,y plane is a vertical sinc-shaped blade (line impulse) $G(u,v) = a\,\text{sinc}(a\,u)\delta(v)$ extending along the u axis of the u,v plane.

$$\text{rect}(x) \leftrightarrow \text{sinc}(u)$$

and
$$1 \leftrightarrow \delta(v)$$

are 1D Fourier transform pairs [Bracewell86 pp. 412, 415], we obtain that

$$g(x,y) = \text{rect}(x) \leftrightarrow G(u,v) = \text{sinc}(u)\delta(v)$$

is a 2D Fourier transform pair. More generally, for any 2D function independent of y $g(x,y) = g_1(x)$ we have $G(u,v) = G_1(u)\delta(v)$ [Bracewell95 p. 166; Bracewell86 p. 253]. For example, as shown in Fig. D.12, if $g(x,y) = \text{rect}(x/a)$ then $G(u,v) = a\,\text{sinc}(a\,u)\delta(v)$.

Finally, using the rotation theorem (see point 11 in Sec. 2.4.2 or [Bracewell95 p. 157]) we see that if we rotate the vertical bar of Fig. D.12 by angle θ, its Fourier transform will also undergo a rotation by the same angle θ. Note that the rotated blade in the spectrum always remains perpendicular to the rotated bar in the signal domain.

Appendix E
Glossary of the main terms

E.1 About the glossary

Because the DFT has a very wide range of applications in many different fields of science and technology (see end of Sec. 1.2), it is not surprising that the terminology used in the vast DFT literature is far from being consistent and uniform. In many cases different authors use different terms for the same entities, and what is even worse, the same terms are often used in different meanings by different authors. As a few examples among many others, let us mention here the various Fourier theorems (or "rules"; see Sec. 2.4), whose names greatly vary between different sources. For instance, the *inversion* rule [Kammler07 p. 199] is called *symmetry* rule in [Brigham88 p. 107], while in [Bracewell86 pp. 364–365] it is called *reciprocity* and the term *symmetry* is reserved to properties related to oddness and evenness [Bracewell86 p. 366]. Similarly, the *reflection* rule [Kammler07 p. 199] is called in [Bracewell86 p. 366] *reversal* while in [Nussbaumer82 p. 82] the term being used is *symmetry* (again!). Even the *aliasing* artifact is often called *foldover*, and the *leakage* artifact is sometimes called *ringing* [Brigham88 pp. 94, 106].

Obviously, in such an interdisciplinary domain as the Fourier theory it would be quite impossible to adopt a universally acceptable standardization of the terms, because of the different needs and traditions in the various fields involved.[1] Nevertheless, even without having such far-reaching pretensions, we were obliged to make our own terminological choices in a systematic and coherent way, in order to avoid confusion and ambiguity in our own work. We tried to be consistent in our terminology throughout this work, even if it forced us to assign to some terms a somewhat different meaning than one would expect (depending on his own background, of course). A typical example is the systematic terminology we adopted for the different classes of continuous and discrete signals. Here, too, the terms in the literature largely vary between the different sources (compare for instance [Oppenheim99 pp. 8–9], [Bracewell95 p. 80], [Smith03 p. 11], [Mitra93 pp. 57–60], [Gomes97 pp. 16, 112], [PhysicsForums12], [Goodman05 p. 284], [Zayed93 p. 4], etc.).

We included in the present glossary all the terms for which we felt a clear definition was desirable to avoid any risk of ambiguity. Note, however, that this glossary is not ordered

---

[1] For example, the term *density* has very different meanings in almost any imaginable field of science or technology. A non-exhaustive list, just for the sake of illustration, may include: density of matter (in physics), density of population (in statistics), probability density (in probability), spectral density (in spectral analysis), density of a set (in mathematic topology), ink or colour density (in printing and colorimetry), etc. Another similar example is the term *phase*. It should be also noted that terminology sometimes tends to change over the years or according to fashions. As an example, the term *function convolution* in modern literature appears in older publications as *function composition* [Zygmund68 p. 36], the *resultant* of two functions, or even using the German term *Faltung* [Hardy68 p. 10].

alphabetically; rather, we preferred to group the various terms according to subjects. We hope this should help the reader not only to clearly see the meaning of each individual term by itself, but also to put it in relation with other closely related terms (which would be completely dispersed throughout the glossary if an alphabetical order were preferred). Note that terms in the glossary can be found alphabetically through the general index at the end of the book.

E.2 Terms in the signal domain

signal —

A function of one or more variables such as time, distance, 2D or MD space, etc., which represents a certain phenomenon of interest (physical, electrical, astronomical, statistical, etc.). For example, the temperature measured at a given point as function of time (either continuous or discrete) is a 1D signal; and any gray-level image or even colour image is a 2D signal. More precisely, a gray-level image is a *scalar-valued signal* of two variables, $z = g(x,y)$, while a colour image is a *vector-valued signal* of two variables, for example $(r,g,b) = \mathbf{g}(x,y)$. Note that we often use the terms "signal" and "function" interchangeably as synonyms.

digitization (or *digitizing*) —

The process of converting a continuous signal (such as voltage, sound, a 2D analog image, etc.) into digital form (see, for example, [Russ95 p. 16]). This process consists of two steps, which can be performed in any order: *discretization* (sampling), and *quantization*. Note that each of these steps involves some loss of information. The resulting digital signal is therefore said to be a *digital approximation* (or a *digital version*) of the underlying continuous signal. It should be noted, however, that in some references (such as [Jähne95 pp. 19, 40–41]) the term digitization is used in the sense we reserve here to the term *discretization*.

discretization (or *sampling*) —

The process of recording the values of a continuous signal (such as voltage, sound, a 2D analog image, etc.) at a given finite or countable set of points, usually with regular intervals, through the signal's domain of definition (for example, along the time axis, within space, etc.). Each single recording is called a *sample*. For example, in order to display an analog image on a monitor, the image must be first discretized (scanned or sampled) into a matrix of $M{\times}N$ pixels (samples). Discretization affects the *domain* of the given signal, and converts it from a continuous set into a discrete set. Note that discretization involves some loss of information. A sampled signal is therefore said to be a *discrete approximation* (or a *discrete version*) of the underlying continuous signal.

quantization —

The process of rounding off the values of a continuous signal (or of the samples of a continuous signal) into a finite or countable set of predefined values, usually with regular intervals (such as integer numbers, decimals, or any other values resulting from the finite precision of the measuring or computing equipment being used, for instance an 8-bit sensor or a 16-bit processor). As a concrete example, gray-level images that are displayed on a monitor are usually quantized into 256 gray levels, so that the value of the image at each of its discrete elements (pixels) can be stored in one byte (8 bits). Quantization affects the *range* of the given signal, and converts it from a continuous set into a discrete set. Note that quantization involves some loss of information. A quantized signal is therefore said to be a *quantized approximation* of the underlying continuous signal.

continuous signal[2] (or function; not to be confused with *analog signal*) —
Unfortunately, this term is used in the literature in two different but equally useful and legitimate meanings (see, for example, [Gomes97 pp. 16, 109, 110] or [Glassner95 pp. 128–129]). In mathematical branches such as analysis or topology, a continuous function is a continuous-valued function of continuous variable(s); the value of such a function varies continuously (i.e. without interruptions) when its independent variable(s) vary continuously. However, in digital signal processing (and in our context, too) the same term is being used in a much wider sense: A signal is said to be continuous if it has a continuous domain and a continuous range; but this does not yet imply that the signal has no interruptions. In other words, all we require from a signal in order to be continuous in this wider sense is that its variables as well as its amplitude be defined on continuous sets, rather than being limited to a certain number of discrete values. In most cases the intended meaning of the term should be understood from the context. But when a clear distinction between these two meanings is desired, one can use for the less restrictive meaning the longer but more precise term *continuous-domain continuous-range signal*, or alternatively *continuous-world signal*. Note that when referring to the restrictive meaning, we often say so explicitly.

continuous-domain signal (or function; also called *continuous-time signal*) —
A signal having a continuous domain. Such signals are often called in the literature *continuous-time signals* (see, for example, [Oppenheim99 p. 8] or [Glassner95 Sec. 4.2.1]); but this term may be misleading in 2D or MD signals, or even in 1D signals whose independent variable is not time but any other entity (such as distance, etc.).

continuous-range signal (or function; also called *continuous-amplitude signal* or *continuous-valued signal*) —
A signal having a continuous range.

---

[2] According to another convention, which is not compatible with our terminology, the term *discrete* implies a discrete domain; the term *continuous* implies a continuous domain; *digital* implies a discrete range; and *analog* implies a continuous range. In this convention, all four combinations are possible: a signal can be discrete and digital; discrete and analog; continuous and digital; or continuous and analog [PhysicsForums12].

digital signal[2] (or function) —

A discrete-valued signal of discrete variable(s). A digital signal has a discrete domain and a discrete range. A continuous signal that undergoes digitization, i.e. both *discretization* (sampling) and *quantization*, is a digital signal (see, for example, [Oppenheim99 pp. 8–9] or [Gomes97 p. 16]). Note that after discretization and quantization a digital signal may also undergo a third stage of *encoding*, in which it is converted into a certain encoded form such as a binary signal (a sequence of zeroes and ones) or any other encoded form that may be needed for its processing [Hayes12 pp. 101, 106; Gomes97 pp. 13, 133–137].

discrete-domain signal (or function; also called *discrete-time signal* or simply *discrete signal*[2]) —

A signal having a discrete domain. Such signals are often called in the literature *discrete-time signals* (see, for example, [Oppenheim99 p. 8] or [Glassner95 Sec. 4.2.2]); but this term may be misleading in 2D or MD signals, or even in 1D signals whose independent variable is not time but any other entity (such as distance, etc.). While many discrete-domain signals are obtained by *discretization* (sampling) of original continuous signals, not all discrete-domain signals are obtained in this manner. Trivial examples are the number of cars sold per day, the annual number of births per year in a given country, etc.

discrete-range signal (or function; also called *discrete-amplitude signal*, *discrete-valued signal* or *quantized signal*) —

A signal having a discrete range. A continuous signal that undergoes *quantization* is a discrete-range signal.

continuous-domain discrete-range signal (or function) —

A signal having a continuous domain but a discrete range. A continuous signal that undergoes *quantization* but not *discretization* (sampling) is a continuous-domain discrete-range signal. For example, a staircase function having a finite or countable number of steps (see, for example, Fig. 8.12(c)) is a continuous-domain discrete-range function.

discrete-domain continuous-range signal (or function) —

A signal having a discrete domain but a continuous range. A continuous signal that undergoes *discretization* (sampling) but not *quantization* is a discrete-domain continuous-range signal. Note that most of the mathematical developments in the field of digital signal processing, including the DFT theory, deal in fact with this class of signals, although in practice such signals are represented in digital computers by their quantized approximations, i.e. by digital signals (see, for example, [Mitra93 p. 58]).

discontinuous signal (or function) —

A signal having one or more discontinuities (interruptions). Note that any signal having a discrete domain or a discrete range is necessarily discontinuous, but a discontinuous signal does not necessarily have a discrete domain or a discrete range. For example, the signal:

$$g(x) = \begin{cases} x & x \leq 0 \\ x+2 & x > 0 \end{cases}$$

is discontinuous, but it is neither a discrete-domain signal nor a discrete-range signal. This is, indeed, an example of a signal which is neither strictly continuous (in the topological sense) nor discrete or digital; but since it still subsists in the continuous world (i.e., it is defined on a continuum and its range is basically a continuum), we can say it is a *continuous-world signal*. Note that signals such as $g(x)$, staircase functions, and histograms (see, for example, Fig. 8.12(c)) are discontinuous even if they can be graphically drawn by a continuous, uninterrupted line, i.e. without raising and shifting the pen. The reason is that instantaneous vertical jumps are always considered as discontinuities, since by definition a function cannot have more than one value (scalar or vectorial) at any given point. Hence, the function contains only one single point along the vertical jump, even if the jump is represented graphically by a vertical line.

continuous-world signal (or function) —
A signal which subsists in the continuous world, but still may have one or more discontinuities. For example, the signal $g(x)$ given above is neither strictly continuous (in the topological sense) nor discrete or digital, but since it still subsists in the continuous world (i.e., it is defined on a continuum and its range is basically a continuum), we say it is a *continuous-world signal*. This term is a convenient softening of the term *continuous signal* (in the strict topological sense), since it still allows for the existence of some discontinuities, as often happens in the real world. This term can be used as a synonym to the term *continuous signal* (in the wider sense), whenever we wish to avoid confusion with the strict (topological) sense of the term. Note that a continuous-world signal may be discrete; for instance, there exist continuous-world signals that are purely impulsive (see, for example, remark (i) in Sec. 2.5).

discrete-world signal (or function) —
A signal which subsists in the discrete world, as opposed to a continuous-world signal. Theoretically, a discrete-world signal is a *discrete-domain function* (i.e. a function having a discrete domain); but when processed by computer, it is in practice a *digital function* (since its range, too, is discrete). Note that a purely impulsive signal may be considered, depending on its nature, as a discrete-world signal or as a continuous-world signal (see, for example, remark (i) in Sec. 2.5): If the signal is undefined between its impulse locations, its domain is discrete and the signal is considered as a discrete-world signal. But if the signal takes the value zero throughout the intervals between its impulse locations, then its domain is continuous, and the signal is considered as a continuous-world signal. Similarly, a continuous-domain discrete-range signal such as a staircase fonction (see, for

example, Fig. 8.12(c)) is considered as a continuous-world signal, again, because its domain is continuous.

real-world signal (or entity) —
A signal or entity that is realizable in the real world (unlike conceptual or theoretical entities such as signals based on Cantor sets [Weisstein99 p. 187], Penrose stairs [Weisstein99 p. 1335], etc.). This definition is admittedly vague, just like that of the term *physically realizable* (as explained, for example, in [Bracewell86 p. 271]).

analog signal[2] (or function; not to be confused with *continuous signal*) —
A signal representing some real-world (physical, electrical, etc.) data that varies with respect to some variable such as time, distance, height, spatial location, etc. Examples are the temperature measured at a given location as function of time; the barometric pressure measured at a given time as function of the altitude; the temperature at a given time as function of the location (within a 2D plane or a 3D volume); etc. Theoretically, any real-world signal is analog, including digital signals (since digital signals, too, are actually represented by a varying voltage or current, and they are only *interpreted* as digital signals). From this point of view, it can be said that a signal is always analog, but the information it carries may be interpreted, depending on the case, as continuous or digital (in the former case small fluctuations in the signal are meaningful, whereas in the latter they are not taken into consideration).[3] The term "analog signal" originally comes from the fact that such a signal is *analogous* to the physical phenomenon it comes to express; for example, a signal representing barometric pressure varies up and down (in terms of voltage, current, etc.) *in analogy* with the original pressure it comes to represent.

period (or *repetition-period* of a function p) —
A number $T \neq 0$ such that for any $x \in \mathbb{R}$, $p(x+T) = p(x)$. In the case of a 2-fold periodic function $p(x,y)$, a double period (or period parallelogram) of $p(x,y)$ is any parallelogram A which tiles the x,y plane so that $p(x,y)$ repeats itself identically on any of these tiles (see also Sec. A.3.4 in Appendix A of [Amidror09]). If the function p is discrete, $\mathbb{R}$ is replaced by $\mathbb{Z}$. The MD generalization is similar, too.

period-vector (of a periodic function $p(x,y)$) —
A non-zero vector $\mathbf{P} = (x_0,y_0)$ such that for any $(x,y) \in \mathbb{R}^2$, $p(x+x_0,y+y_0) = p(x,y)$. If there exist two non-collinear vectors $\mathbf{P}_1$, $\mathbf{P}_2$ having this property, $p(x,y)$ is said to be 2-fold periodic; in this case, for any point $\mathbf{x} \in \mathbb{R}^2$ the points $\mathbf{x}$, $\mathbf{x}+\mathbf{P}_1$, $\mathbf{x}+\mathbf{P}_2$,

---

[3] Note, however, that the term analog is often used as the opposite of the term digital (for example, an analog display or watch as opposed to a digital one). This is also the case in the expressions *analog to digital conversion* and *digital to analog conversion*. Interestingly, in some references like [Oppenheim99 p.8] an analog signal is simply a synonym to continuous-domain signal, while in other references the term analog designates signals having continuous range (see the previous footnote). Finally, in yet other references like [Oppenheim75 p. 7] or [Mitra93 p. 57] the term analog refers to signals having both continuous domain and continuous range.

$\mathbf{x}+\mathbf{P}_1+\mathbf{P}_2$ define a period parallelogram of $p(x,y)$. If the function p is discrete, then $\mathbb{R}$ is replaced by $\mathbb{Z}$. The MD generalization is straightforward.

periodic function —

A function having a period. Note that a 2D function $p(x,y)$ can be *2-fold periodic* (such as $p(x,y) = \cos(x) + \cos(y)$) or only *1-fold periodic* (such as $p(x,y) = \cos(x)$). Similarly, an MD function can be *1-fold periodic, 2-fold periodic, ... , or M-fold periodic*. In the continuous world, periodic functions have impulsive spectra whose impulse locations are commensurable.

almost-periodic function —

See Appendix B in [Amidror09]. In the continuous world, almost-periodic functions have impulsive spectra whose impulse locations are not commensurable.

aperiodic function —

A function that is neither periodic nor almost periodic. In the continuous world, aperiodic functions have non-impulsive spectra.

grating (or *line-grating*) —

A pattern consisting of parallel lines. Unless otherwise mentioned it will be assumed that a grating is periodic and consists of equally wide parallel, straight lines that are separated by equal spaces. For example, a *binary grating* is a grating with a square-wave intensity profile consisting of white lines (with a constant value of 1) on a black background (whose value is 0).

E.3 Terms in the spectral domain

spectrum (or *frequency spectrum*) —

The frequency decomposition of a given function or signal, which specifies the contribution of each frequency to the function or signal in question. The frequency spectrum is obtained by taking the Fourier transform of the given function or signal.

continuous-world spectrum (not to be confused with *continuous spectrum*) —

This term refers to a spectrum in the continuous world, as opposed to a discrete-world spectrum. The spectrum obtained when applying a continuous Fourier transform to a continuous-world signal is a continuous-world spectrum; note, however, that this spectrum is not necessarily continuous (in the strict topological sense), and it can even be purely impulsive if the given continuous-world signal is periodic.

discrete-world spectrum (not to be confused with *discrete spectrum*) —

This term refers to a spectrum in the discrete world, as opposed to a continuous-world spectrum. The sequence obtained when applying DFT to a discrete-world signal is a discrete-world spectrum. Theoretically, a discrete-world spectrum is a *discrete-domain function* (i.e. a function having a discrete domain); but when processed by computer, it is in practice a *digital function* (since its range, too, is discrete). Note that a discrete-world spectrum is often a discrete approximation to an original continuous spectrum; but sometimes the continuous-world counterpart of a discrete-world spectrum is not continuous, and it can be even purely impulsive (if the given continuous-world signal is periodic).

continuous spectrum (not to be confused with *continuous-world spectrum*) —
A spectrum that is expressed by a continuous function (in the strict topological sense of this term). Note that continuous-world spectra are not necessarily continuous spectra. It should be also noted that a discrete (impulsive) continuous-world function may have a continuous spectrum; for example, the continuous-world spectrum of $g(x) = \frac{1}{2}\delta(x-f) + \frac{1}{2}\delta(x+f)$ is the continuous function $G(u) = \cos(2\pi f u)$ [Bracewell86 p. 413].

discrete spectrum (not to be confused with *discrete-world spectrum*) —
A spectrum that is expressed by a discrete function. Note that some continuous-world spectra are discrete; this is the case if the given continuous-world signal is periodic.

Hz (pronounced *Hertz*) —
A unit of frequency, named after the German physicist Heinrich Hertz. The unit Hz is defined in the literature as "cycles per second" (see, for example, [Briggs95 p. 18]). But for the sake of convenience we sometimes use this term more loosely as a *generic* frequency unit, i.e. as a compact shorthand notation for "cycles per signal-domain unit". This allows us to label all frequency axes (both in 1D and MD cases, including axes showing the DFT output array *after* the adaptations discussed in Sec. 4.3) in terms of Hz, rather than in terms of the more exact but too cumbersome "cycles per signal-domain unit". However, to avoid confusion, this abuse of language should not be adopted in real life, and the frequency unit Hz should only be used in its original, narrow sense, i.e. when the signal domain is indeed measured in seconds. For example, if the signal domain is measured in meters the spectral-domain frequency unit that one should use is "cycles per meter", and if the signal domain is measured in terms of minutes the spectral-domain frequency unit should be "cycles per minute".

visibility circle —
A circle around the spectrum origin whose radius represents the cutoff frequency, i.e., the threshold frequency beyond which fine detail is no longer detected by the eye. Obviously, its radius depends on several factors such as the viewing distance, the light conditions, etc. It should be noted that the visibility circle is just a first-order approximation. In fact, the sensitivity of the human eye is a continuous 2D bell-shaped function [Daly92 p. 6], with a steep "crater" in its center (representing frequencies which are too small to be perceived), and "notches" in

the diagonal directions (owing to the drop in the eye sensibility in the diagonal directions [Ulichney88 pp.79–84]). This term basically applies to the 2D case.

frequency vector —

A vector in the 2D (or MD) spectrum which represents the geometric location of a 2D (or MD) frequency in the spectrum (see, for example, Sec. 2.2 and Fig. 2.1 in [Amidror09]).

pulse (not to be confused with *impulse*) —

A 1D or MD mathematical function representing any phenomenon which only has non-zero values within a short, finite interval. For example, the function rect(ax) represents a square (or rectangular) pulse of height 1 that is centered about the origin of the x axis (see Fig. 3.5(a)). Its 2D counterpart rect(ax,ay) = rect(ax)rect(ay) represents a 2D pulse of height 1 that is centered over the origin of the x,y plane (see Fig. 3.6(a) or Fig. 6.8(a)).

impulse (not to be confused with *pulse*) —

A 1D or MD mathematical entity representing any phenomenon whose existence is concentrated in a single point, i.e. which has non-zero value at a single point (typically at the coordinate zero, i.e. at the origin). In the continuous world the impulse is also known as a *Dirac delta function*, and it has zero width and an infinite amplitude. In the discrete world the discrete impulse is also known as a *Kronecker delta function*, and it has the width of one discrete element and an amplitude of 1. See Sec. A.2 in Appendix A for further details.

DC impulse —

The impulse that is located on the spectrum origin. This impulse represents the frequency of zero, which corresponds to the constant component in the Fourier series decomposition of a periodic function. The amplitude of the DC impulse corresponds to the intensity of this constant component. This impulse is traditionally called the *DC impulse* because it represents in electrical transmission theory the direct current component, i.e., the constant term in the frequency decomposition of an electric wave. See, for example, [Smith03 pp. 152, 633].

compound impulse —

An impulse in the spectrum which is composed of several distinct impulses that happen to fall on the same location and hence "fuse down" into a single impulse. The amplitude of a compound impulse is the sum of the amplitudes (real or complex) of the individual impulses from which it is composed. See Sec. 6.4 in [Amidror09].

line-impulse —

A generalized 2D function which is impulsive along a 1D line through the plane, and null everywhere else. A line-impulse can be graphically illustrated as a "blade" whose behaviour is continuous along its 1D line support but impulsive in the perpendicular direction. Note that the amplitude of a line-impulse does not necessarily die out away from its center, and it may even rapidly oscillate between two constant values.

curvilinear impulse —

A generalized 2D function which is impulsive along a 1D curvilinear path through the plane, and null everywhere else. A curvilinear impulse can be graphically illustrated as a curvilinear "blade" whose behaviour is continuous along its 1D curvilinear support but impulsive in the perpendicular direction.

comb (or *impulse-comb, Dirac-comb, impulse-train*) —

An infinite train of equally spaced impulses located on a straight line in the spectrum. A 1-fold periodic function $p(x,y)$ is represented in the 2D spectrum by a comb centered on the spectrum origin. The step and the direction of this comb represent the frequency and the orientation of the periodic function; its impulse amplitudes, which are given by the Fourier series development of the periodic function, determine its intensity profile.

nailbed (or *impulse-nailbed*) —

An infinite 2D train of equally spaced impulses located in the spectrum on a dot-lattice (either square-angled or skewed). A 2-fold periodic function $p(x,y)$ is represented in the 2D spectrum by an impulse nailbed centered on the spectrum origin. The steps and the two main directions of this nailbed represent the frequency and the orientation of the two main directions of the function's 2D periodicity; the impulse amplitudes, which are given by the 2D Fourier series development of the periodic function, determine its intensity profile.

impulsive spectrum —

A spectrum which only consists of impulses, i.e., whose support consists of a finite or at most denumerably infinite number of points. All periodic and almost-periodic functions have impulsive spectra.

line-spectrum —

A spectrum which consists of line impulses (see, for example, Fig. 5.13).

wake —

A 2D continuous surface which trails off from an impulsive element (line-impulse, curvilinear impulse, etc.), gradually dying out as it goes away from the impulsive element in question. The amplitude of the wake may be considered as negligible with respect to that of its generating impulsive element. As an example, the spectrum of the cosinusoidal circular grating $\cos(2\pi f\sqrt{x^2+y^2})$ is a circular impulse ring of radius f, with a particular dipole-like impulsive behaviour on the perimeter of the circle and a negative, continuous wake which gradually trails off toward the center (see Example 10.6 in [Amidror09, Sec. 10.3] and Fig. 10.4(d) there).

Nyquist frequency —

We sometimes use this term to denote half of the sampling frequency, which is the highest frequency that discrete data (and the DFT spectrum) may contain [Smith03 pp. 42, 149]. Note, however, that we usually prefer to avoid using this term because of its ambiguity. As explained in [Smith03 pp. 41–42 and 638], it turns out that different authors use this term in four different meanings: the heighest frequency contained in a (band-limited) signal; twice this frequency; the sampling frequency; or half of the sampling frequency.

E.4 Miscellaneous terms

domain of a function (or *domain of definition*) —

The set of all values on which the function is defined, i.e. the set of values through which the independent variable(s) of the function may run. For example, the domain of the function $g(x) = x^2$ is $\mathbb{R}$, and the domain of its discrete counterpart $g(k) = k^2$ is $\mathbb{Z}$. Note, however, that we sometimes use the term "domain" in a wider sense, like in the terms *signal domain* and *spectral domain* of a Fourier transform.

range of a function —

The set of all values that the function can take when its independent variable(s) run throughout the domain of definition of the function. For example, the range of the function $g(x) = x^2$ is $\mathbb{R}^+ = [0...\infty)$, and the range of its discrete counterpart $g(k) = k^2$ is $\mathbb{Z}^+ = 0, 1, 2, ...$ Note, however, that we sometimes use the term "range" in a wider sense: For example, we may consider the range of the variable x (i.e. the set of values along which it may run); the frequency range of the DFT (which extends between minus half and plus half of the sampling frequency); the sampling range of a continuous function $g(x)$ (i.e. the range along the x axis within which the function has been sampled), etc. As illustrated by the last example, this wider sense of the term "range" may actually refer to the *domain* of a function (or a subset thereof) rather than to its *range*; but usually this ambiguity should not cause confusion.

point (not to be confused with *dot*) —

A single infinitely small geometric location within space, such as a point x along a line, a point (x,y) within a plane, etc. Note that outside the strict mathematical context the word "point" can be also used in a larger, figurative sense (such as a point in a list of items, stressing a point in a discussion, etc.).

dot (not to be confused with *point*) —

A graphical symbol that is used to represent a point pictorially. A dot may have any geometric shape (circular •, square ■, etc.).

vector (or *point* in a vector space) —

An element of the vector space in question ($\mathbb{R}^n$, the u,v plane, etc.). We always consider vectors as radius-vectors attached to the origin, and we do not distinguish between a *vector* and a *point* in the vector space (i.e. the end point or the head of the vector).

continuous —

An entity that is defined throughout a continuum such as $\mathbb{R}$, $\mathbb{R}^n$, an interval $[a...b]$ within $\mathbb{R}$ where $a < b$, etc., and has no jumps. Examples are a continuous line, a continuous set, a continuous signal or function (in the strict topological sense), etc.

discrete —

A subset D of $\mathbb{R}^n$ is called *discrete* if there exists a number $d > 0$ so that for any points $a,b \in D$ the distance between a and b is larger than d. Note, however, that the term *discrete* is also used as the opposite of *continuous*. For example, periodic functions have *discrete spectra*. See also the related terms *discrete signal* (in Sec. E.2), *discrete spectrum* and *discrete-world spectrum* (in Sec. E.3).

binary (grating, etc.) —

A structure which contains only two values, usually 0 and 1.

commensurable (or *commensurate*) —

Two vectors $v_1,v_2 \in \mathbb{R}^n$ (or real numbers in $\mathbb{R}$) are called *commensurable* if there exist non-zero integers m,n for which $nv_2 = mv_1$, i.e. $v_2 = \frac{m}{n}v_1$ (so that both v_1 and v_2 can be measured as integer multiples of the same length unit, say $\frac{1}{n}v_1$). Note that two numbers $x,y \in \mathbb{R}$ are commensurable *iff* their ratio x/y is rational. More generally, k vectors $v_1,...,v_k \in \mathbb{R}^n$ (or real numbers in $\mathbb{R}$) are called commensurable if they are linearly dependent over $\mathbb{Q}$ (which is identical to linear dependence over $\mathbb{Z}$). See also Problem 6-24.

incommensurable (or *incommensurate*) —

Two vectors $v_1,v_2 \in \mathbb{R}^n$ (or real numbers in $\mathbb{R}$) are called *incommensurable* if there do not exist non-zero integers m,n for which $nv_2 = mv_1$, i.e. $v_2 = \frac{m}{n}v_1$. Note that two numbers $x,y \in \mathbb{R}$ are incommensurable *iff* their ratio x/y is irrational. More generally, k vectors $v_1,...,v_k \in \mathbb{R}^n$ (or real numbers in $\mathbb{R}$) are called incommensurable if they are linearly independent over $\mathbb{Q}$ (which is identical to linear independence over $\mathbb{Z}$).

real-valued part; imaginary-valued part (of a complex-valued entity) —

If $c = a + bi$ is a complex-valued entity, a is called its *real-valued part*, and denoted by $\mathrm{Re}[c]$, and b is called its *imaginary-valued part*, and denoted by $\mathrm{Im}[c]$. Note that one also finds in the literature (and sometimes in the present book, too) the terms "real part" and "imaginary part"; but these terms may be misleading, because both a and b are indeed real-world entities (real numbers, integers, real-valued functions, etc.), and none of them is more real or more imaginary than the other. Similarly, one should use the term *complex-valued entity* rather than "complex entity", since such entities are not necessarily more complex than other ones.

phase (of a complex-valued entity) —

Any complex-valued entity c can be represented either by its *real-valued part* and its *imaginary-valued part*, i.e. by $c = \text{Re}[c] + \text{Im}[c]i$, or alternatively, using the polar notation, by its *magnitude* (also called *absolute value* or *modulus*) and its *phase* (also called *argument*), i.e. by $c = \text{Abs}[c] \cdot e^{i\,\text{Arg}[c]}$, where:

$$\text{Abs}[c] = \sqrt{\text{Re}[c]^2 + \text{Im}[c]^2} \quad \text{and} \quad \text{Arg}[c] = \arctan\frac{\text{Im}[c]}{\text{Re}[c]}$$

Note that $\text{Re}[c]$, $\text{Im}[c]$, $\text{Abs}[c]$ and $\text{Arg}[c]$ are all real-valued entities.

phase (of a periodic function, etc.) —

If $p(x)$ is a periodic function with period T, shifting it by T (or by any integer multiple of T) along the x axis leaves the function unchanged: $p(x + T) = p(x)$. But any other shift of $0 < a < T$ gives a shifted version of $p(x)$, $p(x + a)$, which is said to differ from $p(x)$ in its *phase*. For further details see [Amidror09, Sec. C.4 in Appendix C and Secs. 7.1–7.5 in Chapter 7].

Hermitian function (or sequence) —

A complex-valued function or sequence having a symmetric (i.e. even) real-valued part and an antisymmetric (i.e. odd) imaginary-valued part (see, for example, [Bracewell86 p. 16]). Thus, if $g(x)$ is Hermitian, then $g^*(x) = g(-x)$ and $g(x) = g^*(-x)$ [Bracewell86 p. 16]. The multidimensional generalization of a Hermitian function is discussed in Sec. 8.4; in particular, if $g(\mathbf{x})$ is Hermitian then $g^*(\mathbf{x}) = g(-\mathbf{x})$ and $g(\mathbf{x}) = g^*(-\mathbf{x})$. Hermitian functions (or sequences) are named after the French mathematician Charles Hermit. Note that in some references such entities are called *conjugate symmetric* or *conjugate even* (see, for example, [Briggs95 p. 76]), and the term "Hermitian" is reserved to the symmetry properties which prevail between the DFT and the IDFT [Briggs95 p. 74].

anti-Hermitian function (or sequence) —

A complex-valued function or sequence having an antisymmetric (i.e. odd) real-valued part and a symmetric (i.e. even) imaginary-valued part (see, for example, [Bracewell86 p. 15]). Thus, if $g(x)$ is anti-Hermitian, then $g^*(x) = -g(-x)$ and $g(x) = -g^*(-x)$ [Bracewell86 p. 22]. Note that in some references such entities are called *conjugate antisymmetric* or *conjugate odd* (see, for example, [Hayes12 p. 4]). The multidimensional generalization of an anti-Hermitian function is obtained in a similar way to that of a Hermitian function; in particular, if $g(\mathbf{x})$ is anti-Hermitian then $g^*(\mathbf{x}) = -g(-\mathbf{x})$ and $g(\mathbf{x}) = -g^*(-\mathbf{x})$.

separable function (of two or more variables) —

A function $f(x,y)$ of the variables x and y is said to be *separable* if it can be presented as (or separated into) a product of a function of x and a function of y:

$f(x,y) = g(x) \cdot h(y)$ [Gaskill78 pp. 16–17; Cartwright90 p. 117]. Note, however, that we use this term in a slightly larger sense: A 2D function $f(x,y)$ is separable if it can be presented as a product of two independent 1D functions. Therefore, although $f(x,y) = g(x) \cdot h(y)$ may no longer be separable (in the narrower sense) after it has undergone a rotation or a skewing transformation, we will still consider it as separable (with respect to the rotated or skewed axes x' and y': $f(x',y') = g(x') \cdot h(y')$). Note that in the MD case more possible variants exist, as illustrated for example by rules 10 and 10a in Sec. 2.4.3.

inseparable function (of two or more variables) —
A function that is not *separable*. For example, the function representing a square white dot is separable: $\text{rect}(x,y) = \text{rect}(x) \cdot \text{rect}(y)$, while the function representing a circular white dot is inseparable.

bounded function —
A function g is said to be bounded if there exists a finite, positive constant c for which $|g(x)| \le c$ for any x in the domain of definition of g. For example, the function $g(x) = \cos(x)$ is bounded (with $c = 1$). Note that a function which always has a finite value is not necessarily bounded; for example, the function $g(x) = x$ has a finite value at any point x in its domain of definition, but it is unbounded.

unbounded function —
A function that is not *bounded*. For example, the function $g(x) = 1/x$ is unbounded.

finite-duration function (also called *finite-length, support-limited, spatially-limited, time-limited* or *compactly-supported* function) —
A function g is said to be a finite-duration function if there exists a finite, positive constant c for which $g(x) = 0$ for $|x| > c$ (see, for example, [Brigham88 p. 86] or [Briggs95 p. 197]). For example, the function $g(x) = x$ for $x \in [0...1]$ and $g(x) = 0$ otherwise is a finite-duration function. A function $g(x)$ that is only defined on a finite interval such as $[a...b]$, where $a < b$, $a,b \in \mathbb{R}$, is also a finite-duration function [Chu08 p. 51]; an example of such a function is $g(x) = \arccos(x)$, which is only defined within the interval $[-1...1]$. The 2D or MD extensions of a finite-duration function are straightforward. Note that finite-duration functions are often called in the literature *time-limited functions*; but this term may be misleading in 2D or MD functions, or even in 1D functions whose independent variable is not time but any other entity (such as distance, etc.).

infinite-duration function (also called *infinite-length* function) —
A function that is not a *finite-duration function*. For example, the function $g(x) = \cos(x)$ is an infinite-duration function.

band-limited function —
A function g is said to be a band-limited function if its spectrum G is a finite-duration function. In other words, a function $g(x)$ is band limited if its Fourier

transform $G(u)$ is zero everywhere except over a finite interval of the variable u (see, for example, [Bracewell86 p. 189]). For example, the function $g(x) = \text{sinc}(x)$ is band limited since its spectrum, $G(u) = \text{rect}(u)$, is only non-zero over a finite interval. The 2D or MD extensions of a band-limited function are straightforward.

causal function —

A function that is identically zero for negative values of its independent variable. In other words, a function $g(x)$ is said to be causal if $g(x) = 0$ for all $x < 0$ [Marks09 p. 12]. For example, the function $g(x) = e^{-x} \text{step}(x)$ (see Fig. 3.3(a)) is causal. A short discussion on the origin of this term can be found in [Smith03 p. 130] or in [Gaskill78 pp. 140–141]. For our needs, an MD causal function is a function that is causal with respect to each of its M dimensions.

non-causal function —

A function that is not *causal* (see, for example, [Gaskill78 p. 141]).

transient function —

A function $g(x)$ is said to be transient if it dies out for large $|x|$ (see, for example, [Castleman79 p. 164]).

1D —

One dimensional (function, periodicity, spectrum, impulse comb, etc.). Strictly, a one dimensional entity is an entity that has only one dimension (such as a straight line). However, by abuse of language we often use this term to designate a 2D entity that only varies along one dimension, while its other dimension is constant. For example, we say that an impulse comb such as in Fig. C.4(b) is 1D, even though it subsists in the 2D u,v spectrum, because it varies only in one dimension, while in the orthogonal direction it remains constantly zero. Similarly, we often say that a 1-fold periodic structure such as a line grating (see, for example, Fig. C.4(a)) is a 1D structure, even though it actually spreads in the 2D x,y space, because it only varies along one dimension but remains constant along the orthogonal direction (i.e. along the individual grating lines). Such a 1D periodic structure is, in fact, a constant extension of a true one-dimensional structure (such as a square wave) into the second dimension. Note that we sometimes use the terms "1D periodic" and "1-fold periodic" interchangeably as synonyms.

2D —

Two dimensional (function, periodicity, spectrum, impulse nailbed, etc.). Strictly, any entity that has two dimensions, including a 1-fold periodic structure such as a line grating, is a 2D entity. However, by abuse of language we often use this term to designate a 2D entity that indeed varies along two dimensions. For example, we say that a function such as $g(x,y) = \text{rect}(x)\text{rect}(y)$ is 2D, because it varies along two dimensions. Note that we sometimes use the terms "2D periodic" and "2-fold periodic" interchangeably as synonyms.

moiré effect (or *moiré phenomenon*) —

A visible phenomenon which occurs when repetitive structures (such as line-gratings, dot-screens, etc.) are superposed or sampled. It consists of a new pattern which is clearly observed in the superposition or in the sampled structure, although it does not appear in any of the original structures. The moiré effect is represented in the spectrum of the superposition by new low frequencies which do not exist in any of the individual spectra (the spectra of the original structures).

pseudo moiré —

A visible moiré-like phenomenon which occurs in some cases when repetitive structures are superposed or sampled. Unlike a true moiré effect, a pseudo moiré is only a modulation artifact, and it is not accompanied by corresponding low frequencies in the spectrum of the superposition. For example, the new structure which becomes visible in the sum of two cosines having similar frequencies is a pseudo moiré (see Sec. D.3 in Appendix D).

sub-Nyquist artifact —

The term we use for a pseudo moiré effect that may occur when sampling a periodic signal whose frequency is typically (but not exclusively) just below the Nyquist frequency (i.e. just below half of the sampling frequency). See Sec. 8.6 and Problem 8-14.

singular moiré (or *singular state, singular superposition*) —

A configuration of the superposed or sampled structures in which the period of the moiré in question becomes infinitely large (i.e., its frequency becomes 0), and hence it can no longer be seen in the superposition. See Sec. 2.9 in [Amidror09].

macrostructure, microstructure (within a layer superposition or a sampled layer) —

The superposition of two or more layers (gratings, screens, etc.) may generate new structures which appear in the superposition but not in the original layers. Similar structures may also appear when sampling an original layer. These new structures can be classified into two categories: the *macrostructures*, i.e., the moiré effects proper, which are much coarser than the detail of the original layers; and the *microstructures*, i.e., the tiny geometric forms which are almost as small as the periods of the original layers, and are normally visible only from a close distance or through a magnifying glass (see Chapter 8 in [Amidror09]).

jaggies (or *staircasing effects*) —

The term we use for the jagged edges that often appear when discretizing 2D shapes, usually (but not exclusively) along borders with sharp 0/1 or black/white transitions. Jaggies tend to appear especially along slanted or curved edges. See Sec. 8.5.

wraparound —

See Remark C.10 in Appendix C and Problem 8-17.

operator —

An operator is a mathematical entity that gets as its input one or more functions and returns as its output a function. For example, $S[g(x),h(x)] = g(x) + h(x)$ is a *binary* operator (i.e. an operator that acts on two functions) which returns for any two given functions a third function, namely, their sum; $D[g(x)] = g'(x)$ is a *unary* operator (i.e. an operator that acts on a single function) which returns for any given function its derivative; and $F[g(x)] = G(u)$ is the unary operator that gives for any (suitable) function $g(x)$ its spectrum (Fourier transform) $G(u)$.

transform (not to be confused with *transformation*) —

A transform is a unary operator, i.e. a mathematical entity that maps any input function into an output function (including generalized functions such as impulses, etc.). For example, the continuous Fourier transform returns for any (suitable) function $y = g(x)$ another function, $w = G(u)$, which is called the spectrum of the original function. Note, however, that this definition is too wide, since it includes operators that are not usually considered as transforms such as $D[g(x)] = g'(x)$ (the operator which returns for any given function its derivative) or $P[g(x,y)] = g(x)$ (the operator which returns for each 2D function its 1D restriction to the x axis). Strictly speaking, the term transform is reserved to integral transforms (such as the Fourier transform), as well as their discrete counterparts (like the DFT). Thus, while every transform is a unary operator, not all unary operators are transforms. See also Sec. G.4 in [Amidror07] and in particular Footnote 2 there concerning some terminological subtleties. Note that a transform does not necessarily work for any possible input function; the set of functions for which it is defined is called its domain of definition (much like the domain of definition of a function).

transformation (or *mapping*; not to be confused with *transform*) —

A transformation is a function which maps a space such as $\mathbb{R}$, $\mathbb{R}^2$, $\mathbb{Z}$, $\mathbb{Z}^2$, etc. into another such space or into itself. Thus, for each element of the original space it returns an element of the target space. For example, the function $g(x) = 2x$ that maps any point x to $2x$ is a transformation from $\mathbb{R}$ onto $\mathbb{R}$; and the transformation $g(x,y) = x$ which projects any point (x,y) to the x axis is a transformation from $\mathbb{R}^2$ onto $\mathbb{R}$. More specifically, a *coordinate transformation* is a continuous transformation from a given space (such as $\mathbb{R}^2$, etc.) onto itself, which maps lines parallel to the main axes into distinct, continuous and not self-intersecting lines or curves (see also Appendices B–D in [Amidror07]). For example, a rotation by angle θ is a coordinate transformation in $\mathbb{R}^2$.

functional —

A functional is a mathematical entity whose input is a function but whose output is a number. Thus, a functional returns a number for any given function within its domain of definition. For example, the definite integral between 0 and 1 is a functional that returns for any suitable function $g(x)$ the area enclosed between the curve of the function, the x axis, and the two vertical lines $x = 0$ and $x = 1$.

rotation-invariant (transform, etc.; not to be confused with *rotation-preserving*) —
  A transform is said to be *rotation-invariant* (or *invariant under rotations*) if
  applying an arbitrary rotation transformation to any input of the transform does
  not affect the output of the transform. More generally, a transform is said to be
  invariant under a given coordinate transformation if applying the given
  coordinate transformation to any input of the transform does not affect the output
  of the transform. For example, the magnitude of the Fourier transform (as well as
  the power spectrum) are invariant under shifts in the spatial domain (i.e. they are
  shift-invariant), since shifts only affect the *phase* of the spectrum but not its
  magnitude (or power).

rotation-preserving (transform, etc.; not to be confused with *rotation-invariant*) —
  A transform is said to be *rotation-preserving* (or to *preserve rotations*) if applying
  an arbitrary rotation transformation to any input of the transform affects the output
  of the transform by exactly the same rotation. In other words, this happens if the
  transform P in question commutes with rotation: $P[R[g(x,y)]] = R[P[g(x,y)]]$,
  namely, if the transform of a rotated function equals the rotated transform of
  the original function. More generally, a transform is said to *preserve a given
  coordinate transformation* if applying the given coordinate transformation to any
  input of the transform results in applying the same coordinate transformation to
  the output of the transform. The coordinate transformation is then said to be
  preserved by the transform in question. For example, the 2D CFT preserves any
  rotation transformation (i.e. it is a *rotation-preserving* transform), since rotating
  the CFT input $g(x,y)$ by angle θ results in an equal rotation of the spectrum $G(u,v)$
  (thanks to the rotation theorem; see, for example, [Bracewell95 p. 157]). Note that
  some references use in this context the term *invariant*, saying that the 2D CFT is
  rotation-invariant, or *invariant under rotation transformations* (see, for example,
  [Prestini04 p. 229] or [Granlund95 p. 235]); but this abuse of language may cause
  confusion with the original meaning of the term *invariant*.

sinc (or *cardinal sine*) —
  A very important function in Fourier theory, defined as (see [Bracewell86 p.62]):

$$\text{sinc}(x) = \begin{cases} \dfrac{\sin(\pi x)}{\pi x} & x \neq 0 \\ 1 & x = 0 \end{cases}$$

  Confusingly, however, many references use a slightly different definition:

$$\text{sinc}(x) = \begin{cases} \dfrac{\sin(x)}{x} & x \neq 0 \\ 1 & x = 0 \end{cases}$$

  (see, for example, [Briggs95 pp. 96–97]). The integral of the sinc function between
  0 and x (using the *second* definition) is known as the *sine integral* function Si(x)
  [Bracewell86 p. 62]; see also Sec. D.5 in Appendix D. Note, however, that some
  references use the integral of the *first* definition of the sinc function, but usually
  without calling it Si(x) (see, for example, [Kammler07 p. 44]).

List of the main relations

The following list summarizes the main relations between the signal and frequency domain parameters in the 1D and MD cases. Each relation is accompanied by its equation number in the text.

1. Relations between Δx, Δu, R, F and N (see also Table 7.1 in Chapter 7):

$$R = N\Delta x \tag{4.2}$$

$$F = N\Delta u \tag{7.1}$$

$$\Delta u = \frac{1}{N\Delta x} \quad \text{or} \quad \Delta x\, \Delta u = 1/N \tag{4.5}$$

$$RF = N \tag{7.2}$$

$$\Delta u = 1/R \tag{7.3}$$

$$\Delta x = 1/F \tag{7.4}$$

The 2D or MD counterpart of each of these relations is simply obtained by applying the given 1D relation in each dimension separately.

2. Relations involving the sampling frequency f_s and the DFT minimal and maximal frequencies:

$$f_s = \frac{1}{\Delta x} \tag{4.6}$$

$$f_s = N\Delta u = F \tag{7.5}$$

$$f_{\min} = -\tfrac{1}{2}f_s \qquad f_{\max} = \tfrac{1}{2}f_s - \Delta u \tag{4.10}$$

The 2D or MD counterpart of each of these relations is obtained by applying the given 1D relation in each dimension separately.

3. If the signal $g(x)$ is periodic with period P and frequency f then we have, in addition:

$$Pf = 1 \tag{Sec. C.2}$$

and its discrete counterpart:

$$pq = N \tag{C.2}$$

Note that the 2D or MD counterparts of these two relations are *not* a coordinate-wise extension of the 1D relations, and they are given, as explained in Appendix C, by:

$$\mathbf{P}_i \cdot \mathbf{f}_j = \begin{cases} 1 & \text{if } i = j \\ 0 & \text{if } i \neq j \end{cases} \tag{C.7}$$

$$\mathbf{p}_i \cdot \mathbf{q}_j = \begin{cases} N & \text{if } i = j \\ 0 & \text{if } i \neq j \end{cases} \tag{C.11}$$

where $i,j = 1,...,n$ (n being the number of dimensions enjoying periodicity, $1 \leq n \leq M$), and where $\mathbf{P}_1,...,\mathbf{P}_n \in \mathbb{R}^M$, $\mathbf{f}_1,...,\mathbf{f}_n \in \mathbb{R}^M$, $\mathbf{p}_1,...,\mathbf{p}_n \in \mathbb{R}^M$ and $\mathbf{q}_1,...,\mathbf{q}_n \in \mathbb{R}^M$.

In the 1D case we also have the following connections between the discrete-world parameters p and q and their continuous-world counterparts P and f:

$$p = F/f = FP = \frac{N}{R} P \tag{C.4}$$

$$q = R/P = Rf = \frac{N}{F} f \tag{C.4}$$

$$p = P/\Delta x \tag{C.6}$$

$$q = f/\Delta u \tag{C.5}$$

The 2D or MD counterpart of each of these relations is obtained by applying the corresponding 1D relation to each dimension separately (note that the vectors $\mathbf{p}_i$ and $\mathbf{P}_i$ are collinear, just like the vectors $\mathbf{q}_i$ and $\mathbf{f}_i$):

$$\mathbf{p}_i = F\mathbf{P}_i = \frac{N}{R} \mathbf{P}_i \tag{C.14}$$

$$\mathbf{q}_i = R\mathbf{f}_i = \frac{N}{F} \mathbf{f}_i \tag{C.15}$$

$$\mathbf{p}_i = \mathbf{P}_i/\Delta x \tag{C.12}$$

$$\mathbf{q}_i = \mathbf{f}_i/\Delta u \tag{C.13}$$

where $i = 1,...,n$.

Note that in these MD relations it is assumed that the same parameters Δx, Δu, R, F and N are used in all of the dimensions; if this is not the case, we may simply append the index i to each of these parameters, too.

List of notations and symbols

This list consists of the main symbols used in the text. They appear here with a very brief description and a reference to the page in which they are first used or defined. Obvious symbols such as '+', '–', etc. have not been included.

Symbol	Short description	Page
x, y	Variables (coordinates) of the continuous signal domain	17
u, v	Variables (coordinates) of the continuous spectral domain	17
$\mathbf{x}$	Vector variable in the MD continuous signal domain	17
$\mathbf{u}$	Vector variable in the MD continuous spectral domain	17
$g(x)$	1D continuous-world function, input of the 1D CFT	15
$G(u)$	1D continuous-world function, output of the 1D CFT	15
$g(x,y)$	2D continuous-world function, input of the 2D CFT	17
$G(u,v)$	2D continuous-world function, output of the 2D CFT	17
$g(\mathbf{x})$	MD continuous-world function, input of the MD CFT	17
$G(\mathbf{u})$	MD continuous-world function, output of the MD CFT	17
k, l	Variables (coordinates) of the discrete signal domain	18, 20
n, m	Variables (coordinates) of the discrete spectral domain	18, 20
$\mathbf{k}$	Vector variable in the MD discrete signal domain	20
$\mathbf{n}$	Vector variable in the MD discrete spectral domain	20
$g(k)$	1D discrete-world function, input of the 1D DFT	18
$G(n)$	1D discrete-world function, output of the 1D DFT	18
$g(k,l)$	2D discrete-world function, input of the 2D DFT	20
$G(m,n)$	2D discrete-world function, output of the 2D DFT	20
$g(\mathbf{k})$	MD discrete-world function, input of the MD DFT	20

$G(\mathbf{n})$	MD discrete-world function, output of the MD DFT	20
$g(x_k)$	1D discrete-world function (alternative notation)	7
$g(x_k,y_l)$	2D discrete-world function (alternative notation)	7
$g(\mathbf{x_k})$	MD discrete-world function (alternative notation)	7
$\widetilde{g}$	Discrete approximation of the continuous-world function g	19, 20
$\widetilde{G}$	Discrete approximation of the continuous-world spectrum G	19, 20
R	Length of the sampling range in the input signal	70, 187
F	Length of the frequency range in the DFT spectrum	158, 189
M	Input / output space dimension of the CFT or DFT	17
N	Input / output array length (number of elements) in the 1D DFT	17
$N{\times}N$	Input / output array size (number of elements) in the 2D DFT	29, 47
$N{\times}...{\times}N$	Input / output array size (number of elements) in the MD DFT	69
P	Period of a 1D periodic function	74, 273
f	Frequency of a 1D periodic function	74, 273
$\mathbf{P}$	Period of an MD periodic function	281
$\mathbf{f}$	Frequency of an MD periodic function	281
f_1	The first frequency above 0 in the DFT spectrum	46, 72
f_s	The sampling frequency of the DFT input signal	72
f_G	The highest frequency in the spectrum $G(u)$	92
ω	Angular frequency	16
$\Delta x, \Delta y$	Sampling interval along the x or y axis	18, 72
$\Delta u, \Delta v$	Frequency step along the u or v axis	18, 72
p	Period of a 1D periodic function in terms of array elements	74, 273
q	Frequency of a 1D periodic function in terms of array elements	74, 273
$\mathbf{p}$	Period of an MD periodic function in terms of array elements	170, 286
$\mathbf{q}$	Frequency of an MD periodic function in terms of array elements	170, 286

$\mathbb{Z}$	The set of all integer numbers (positive, negative, and 0)	8
$\mathbb{Q}$	The set of all rational numbers	346
$\mathbb{R}$	The set of all real numbers	283
$\mathbb{Z}^M$	The M-dimensional integer space	29
$\mathbb{R}^M$	The M-dimensional Euclidean space	25
$a...b$	The number range from a to b in $\mathbb{R}$	48, 68
$[a...b]$	The number range from a to b in $\mathbb{R}$ (alternative notation)	348
$[a...b)$	The number range from a up to (but not including) b in $\mathbb{R}$	187
$k,...,l$	The number range from k to l in $\mathbb{Z}$	17
Re[ ]	The real-valued part of a complex entity	22, 202
Im[ ]	The imaginary-valued part of a complex entity	22, 202
Abs[ ]	The magnitude of a complex entity	202
Arg[ ]	The phase of a complex entity	202
$a{\cdot}b$	Multiplication	196, 276
$\mathbf{v}{\cdot}\mathbf{w}$	Scalar product of two vectors	17, 20
$\mathbf{v}{\times}\mathbf{w}$	Vector product of two vectors	285
$\mathbf{v}/\mathbf{w}$	Element-wise vector division	20
$|a|$	The absolute value of the number a (real or complex)	22
$|\mathbf{v}|$	The length ($=$ Euclidean norm) of the vector $\mathbf{v}$	282, 285
$\mathbf{v}{\perp}\mathbf{w}$	$\mathbf{v}$ is perpendicular to $\mathbf{w}$	285
$\text{proj}(\mathbf{v})_{\mathbf{w}}$	The projection of $\mathbf{v}$ on $\mathbf{w}$	282
$G(u) = \mathcal{F}[g(x)]$	$G(u)$ is the Fourier transform of $g(x)$	92, 143
$g(x) \leftrightarrow G(u)$	$g(x)$ and $G(u)$ are a Fourier pair	35, 333
$\sim a$	Approximately a	318
$\sim$	Is Fourier series of ...	321
$\approx$	Approximately equal	19, 20

List of abbreviations

References

[Abdou82]   I. E. Abdou and K. Y. Wong, "Analysis of linear interpolation schemes for bi-level image applications," *IBM Jour. of Research and Development*, Vol. 26, No. 6, 1982, pp. 667–680.

[Amidror97]   I. Amidror, "Fourier spectrum of radially periodic images," *Jour. of the Optical Society of America A*, Vol. 14, No. 4, 1997, pp. 816–826.

[Amidror07]   I. Amidror, *The Theory of the Moiré Phenomenon, Volume II: Aperiodic Layers.* Springer, Dordrecht, 2007.

[Amidror09]   I. Amidror, *The Theory of the Moiré Phenomenon, Volume I: Periodic Layers.* Springer, London, 2009 (second edition).

[Amidror09a]   I. Amidror and R. D. Hersch, "The role of Fourier theory and of modulation in the prediction of visible moiré effects," *Journal of Modern Optics*, Vol. 56, 2009, pp. 1103–1118.

[Bagchi99]   S. Bagchi and S. K. Mitra, *The Nonuniform Discrete Fourier Transform and its Applications in Signal Processing.* Kluwer, Boston, 1999.

[Bergland69]   G. D. Bergland, "A guided tour of the fast Fourier transform," *IEEE spectrum*, Vol. 6, 1969, pp. 41–52; also reprinted in *Digital Signal Processing*, L. R. Rabiner and C. M. Rader (Eds.), IEEE Press, NY, 1972, pp. 228–239.

[Birkhoff77]   G. Birkhoff and S. Mac Lane, *A Survey of Modern Algebra.* Macmillan Publishing, NY, 1977 (fourth edition).

[Boff86]   K. R. Boff, L. Kaufman and J. P. Thomas, *Handbook of Perception and Human Performance (Vol. 2)*, John Wiley & sons, NY, 1986.

[Bovik05]   A. Bovik (Ed.), *Handbook of Image and Video Processing.* Elsevier, Amsterdam, 2005 (second edition).

[Bracewell86] R. N. Bracewell, *The Fourier Transform and its Applications.* McGraw-Hill Publishing Company, Reading, NY, 1986 (second edition).

[Bracewell95] R. N. Bracewell, *Two Dimensional Imaging.* Prentice Hall, NJ, 1995.

[Briggs95]   W. L. Briggs and V. E. Henson, *The DFT: An Owner's Manual for the Discrete Fourier Transform.* SIAM, Philadelphia, 1995.

[Brigham88]   E. O. Brigham, *The Fast Fourier Transform and Its Applications.* Prentice-Hall, NJ, 1988.

[Cartwright90] M. Cartwright, *Fourier Methods for Mathematicians, Scientists and Engineers.* Ellis Horwood, UK, 1990.

[Castleman79] K. R. Castleman, *Digital Image Processing.* Prentice Hall, NJ, 1979.

[Champeney73] D. C. Champeney, *Fourier Transforms and their Physical Applications*. Academic Press, London, 1973.

[Champeney87] D. C. Champeney, *Fourier Theorems*. Cambridge University Press, UK, 1987.

[Čížek] V. Čížek, *Discrete Fourier Transforms and Their Applications*. Adam Hilger, Bristol, 1986.

[Chu08] E. Chu, *Discrete and Continuous Fourier Transforms: Analysis, Applications and Fast Algorithms*. CRC, Boca Raton, 2008.

[Cleveland93] W. S. Cleveland, *Visualizing Data*. AT&T Bell Laboratories, NJ, 1993.

[Cleveland94] W. S. Cleveland, *The Elements of Graphing Data*. AT&T Bell Laboratories, NJ, 1994 (revised edition).

[Cochran67] W. T. Cochran *et al.*, "What is the fast Fourier transform?" *IEEE Transactions on Audio and Electroacoustics*, Vol. 15, June 1967, pp. 45–55; also reprinted in *Digital Signal Processing*, L. R. Rabiner and C. M. Rader (Eds.), IEEE Press, NY, 1972, pp. 240–250.

[Cooley65] J. W. Cooley and J. W. Tukey, "An algorithm for the machine calculation of complex Fourier series," *Mathematics of Computation*, Vol. 19, 1965, pp. 297–301.

[Cooley69] J. W. Cooley, P. A. W. Lewis and P. D. Welch, "The finite Fourier transform," *IEEE Transactions on Audio and Electroacoustics*, Vol. 17, June 1969, pp. 77–85; also reprinted in *Digital Signal Processing*, L. R. Rabiner and C. M. Rader (Eds.), IEEE Press, NY, 1972, pp. 251–259.

[Cooley70] J. W. Cooley, P. A. W. Lewis and P. D. Welch, "The fast Fourier transform algorithm: programming considerations in the calculation of sine, cosine and Laplace transforms," *Journal of Sound and Vibration*, Vol. 12, July 1970, pp. 315–337; also reprinted in *Digital Signal Processing*, L. R. Rabiner and C. M. Rader (Eds.), IEEE Press, NY, 1972, pp. 271–293.

[Cornsweet70] T. N. Cornsweet, *Visual Perception*. Harcourt Brace Jovanovich, Florida, 1970.

[Coulon84] F. de Coulon, *Theorie et Traitement des Signaux*. Presses Polytechniques Romandes, Lausanne, 1984 (in French).

[Crow77] F. C. Crow, "The aliasing problem in computer-generated shaded images," *Communications of the ACM*, Vol. 20, No. 11, 1977, pp. 799–805.

[Daly92] S. Daly, "The visible differences predictor: an algorithm for the assessment of image fidelity," *Human Vision, Visual Processing and Digital Display III*, SPIE proceedings, Vol. 1666, USA, 1992, pp. 2–15.

[Dutt93] A. Dutt and V. Rokhlin, "Fast Fourier transforms for nonequispaced data," *SIAM Journal on Scientific Computing*, Vol. 14, No. 6, November 1993, pp. 1368–1393.

[Erdélyi54] A. Erdélyi (Ed.), *Tables of Integral Transforms (Vol. 1)*. McGraw-Hill, NY, 1954.

[Fielding06] G. Fielding, R. Hsu, P. Jones and C. DuMont, "Aliasing and reconstruction distortion in digital intermediates," *SMPTE Motion Imaging Journal*, Vol. 115, No. 4, 2006, pp. 128–136.

[Foley90] J. D. Foley, A. van Dam, S. K. Feiner and J. F. Hughes, *Computer Graphics: Principles and Practice*. Addison-Wesely, Reading, Massachusetts, 1990 (second edition).

[Fourier22] J. B. J. Fourier, *Théorie Analytique de la Chaleur*. Didot, Paris, 1822 (in French).

[Fourier24]   J. B. J. Fourier, "Théorie du mouvement de la chaleur dans les corps solides," *Mémoires de l'Académie des Sciences de l'Institut de France*, Vol. 4, 1824, pp. 185–556 (in French).

[Fourier26]   J. B. J. Fourier, "Théorie du mouvement de la chaleur dans les corps solides," *Mémoires de l'Académie des Sciences de l'Institut de France*, Vol. 5, 1826, pp. 153–246 (in French).

[Gaskill78]   J. D. Gaskill, *Linear Systems, Fourier Transforms, and Optics*. John Wiley & Sons, NY, 1978.

[Glassner95]   A. S. Glassner, *Principles of Digital Inage Synthesis, Volume 1*. Morgan Kaufmann, San Francisco, 1995.

[Gomes97]   J. Gomes and L. Velho, *Image Processing for Computer Graphics*. Springer, NY, 1997.

[Gonzalez87]   R. C. Gonzalez and P. Wintz, *Digital Image Processing*. Addison-Wesley, Reading, Ma, 1987 (second edition).

[Goodman05]   J. W. Goodman, *Introduction to Fourier Optics*. Roberts & Company, Colorado, 2005 (third edition).

[Gradshteyn07] I. S. Gradshteyn and I. M. Ryzhik, *Table of Integrals, Series, and Products*. Academic Press, San Diego, 2007 (seventh edition).

[Granlund95]   G. H. Granlund and H. Knutsson, *Signal Processing for Computer Vision*. Kluwer, Dordrecht, 1995.

[Gröchenig01] K. Gröchenig, *Foundations of Time-Frequency Analysis*. Birkhäuser, Boston, 2001.

[Hardy68]   G. H. Hardy and W. W. Rogosinski, *Fourier Series*. Cambridge University Press, London, 1968 (reprint of third edition of 1956).

[Harris78]   F. J. Harris, "On the use of windows for harmonic analysis with the discrete Fourier transform," *Proceedings of the IEEE*, Vol. 66, 1978, pp. 51–83.

[Harris98]   J. W. Harris and H. Stocker, *Handbook of Mathematics and Computational Science*. Springer, NY, 1998.

[Hayes12]   M. H. Hayes, *Digital Signal Processing*. Schaum's Outline Series, McGraw-Hill, NY, 2012 (second edition).

[Heideman85]   M. T. Heideman, D. H. Johnson and C. S. Burrus, "Gauss and the history of the fast Fourier transform," *Archive for History of Exact Sciences*, Vol. 34, No. 3, 1985, pp. 265–277. An earlier version was published in *IEEE ASSP Magazine*, Vol. 1, No. 4, 1984, pp. 14–21.

[Henson92]   V. E. Henson, "DFTs on irregular grids: The anterpolated DFT," Technical Report NPS-MA-92-006, Naval Postgraduate School, Monterey, CA, 1992. Available online at http://www.dtic.mil/dtic/tr/fulltext/u2/a255187.pdf (accessed Oct. 2, 2012).

[Herivel75]   J. Herivel, *Joseph Fourier: The Man and the Physicist*. Clarendon Press, Oxford, 1975.

[Jähne95]   B. Jähne, *Digital Image Processing: Concepts, Algorithms and Scientific Applications*. Springer, Berlin, 1995 (third edition).

[Jain89]   A. K. Jain, *Fundamentals of Digital Image Processing*. Prentice Hall, NJ, 1989.

[Kammler07]   D. W. Kammler, *A First Course in Fourier Analysis*. Cambridge University Press, Cambridge, 2007 (revised edition).

[Komrska79]   J. Komrska, "Symmetry of Fraunhofer diffraction expressed by series expansions of Fourier transforms," *Optica Acta*, Vol. 26, No. 2, 1979, pp. 173–195.

[Kreyszig93]   E. Kreyszig, *Advanced Engineering Mathematics*. John Wiley & Sons, NY, 1993 (seventh edition).

[Lay03]   D. C. Lay, *Linear Algebra and its Applications*. Addison-Wesley, Boston, 2003 (third edition).

[Loan92]   C. van Loan, *Computational Frameworks for the fast Fourier transform*. SIAM, Philadelphia, 1992.

[Mackay91]   A. L. Mackay, *A Dictionary of Scientific Quotations*. Institute of Physics, Bristol, 1991 (second edition).

[Mallat91]   S. Mallat, *A Wavelet Tour of Signal Processing*. Academic Press, San Diego, 1991 (second edition).

[Manolakis11] D. Manolakis and V. Ingle, *Applied Digital Signal Processing*. Cambridge University Press, UK, 2011.

[Maplesoft]   DiscreteTransforms/FourierTransform – Maple® Help. http://www.maplesoft.com/support/help/Maple/view.aspx?path=DiscreteTransforms/FourierTransform (accessed January 19, 2012).

[Marks91]   R. J. Marks II, *Introduction to Shannon Sampling and Interpolation Theory*. Springer, NY, 1991. http://marksmannet.com/RobertMarks/REPRINTS/1999_Introduction ToShannonSamplingAndInterpolationTheory.pdf (accessed May 30, 2012).

[Marks09]   R. J. Marks II, *Handbook of Fourier Analysis and its Applications*. Oxford University Press, NY, 2009.

[Marvasti01]   F. Marvasti (Ed.), *Nonuniform Sampling: Theory and Practice*. Kluwer, NY, 2001.

[Matlab02]   *Signal Processing Toolbox for Use With Matlab®: User's Guide*. The MathWorks Inc., MA, 2002 (fourth printing).

[Mitchell88]   D. P. Mitchell and A. N. Netravali, "Reconstruction filters in computer graphics," Proceedings of SIGGRAPH 1988, *Computer Graphics*, Vol. 22, No. 4, 1988, pp. 221–228.

[Mitra93]   S. K. Mitra and J. F. Kaiser, *Handbook for Digital Signal Processing*. Wiley, NY, 1993.

[Morrison94]   N. Morrison, *Introduction to Fourier Analysis*. Wiley, NY, 1994.

[Nussbaumer82] H. J. Nussbaumer, *Fast Fourier Transforms and Convolution algorithms*. Springer, Berlin, 1982 (second edition).

[Nuttall81]   A. H. Nuttall, "Some windows with very good sidelobe behavior," *IEEE transactions on Acoustics, Speech and Signal Processing*, Vol. 29, 1981, pp. 84–91.

[Oberhettinger90] F. Oberhettinger, *Tables of Fourier Transforms and Fourier Transforms of Distributions*. Springer, Berlin, 1990.

[Oppenheim75] A. V. Oppenheim and R. W. Schafer, *Digital Signal Processing*. Prentice Hall, London, 1975.

[Oppenheim99] A. V. Oppenheim and R. W. Schafer with J. R. Buck, *Discrete-Time Signal Processing*. Prentice Hall, London, 1999 (second edition).

[Papoulis62]   A. Papoulis, *The Fourier Integral and its Applications*. McGraw-Hill, NY, 1962.

[Papoulis68] A. Papoulis, *Systems and Transforms with Applications in Optics*. McGraw-Hill, NY, 1968.

[Park09] W. Park *et al.*, "Accurate image rotation using Hermite expansions," *IEEE Transactions on Image Processing*, Vol. 18, No. 9, September 2009, pp. 1988–2003.

[Petersen62] D. P. Petersen and D. Middleton, "Sampling and reconstruction of wave-number-limited functions in N-dimensional Euclidean spaces," *Information and Control*, Vol. 5, 1962, pp. 279–323.

[PhysicsForums12] Physics Forums, 2012, "Analog and digital, continuous and discrete." http://www.physicsforums.com/showthread.php?t=479118 (accessed October 2, 2012).

[Poularikas96] A. D. Poularikas (Ed.), *The Transforms and Applications Handbook*. CRC, Boca Raton, 1996.

[Pratt91] W. K. Pratt, *Digital Image Processing*. John Wiley & Sons, NY, 1991 (second edition).

[Press02] W. A. Press, S. A. Teukolski, W. T. Vetterling and B. P. Flannery, *Numerical recipes in C++*. Cambridge University Press, UK, 2002 (second edition).

[Prestini04] E. Prestini, *The Evolution of Applied Harmonic Analysis: Models of the Real World*. Birkhäuser, Boston, 2004.

[Ramirez74] R. W. Ramirez, "The fast Fourier transform's errors are predictable, therefore manageable," *Electronics*, Vol. 47, June 1974, pp. 96–102.

[Ramirez85] R. W. Ramirez, *The FFT: Fundamentals and Concepts*. Prentice Hall, NJ, 1985.

[Rao98] R. M. Rao and A. S. Bopardikar, *Wavelet Transforms: Introduction to Theory and Applications*. Addison-Wesley, MA, 1998.

[Rosenfeld82] A. Rosenfeld and A. C. Kak, *Digital Picture Processing (Vol.1)*. Academic Press, Florida, USA, 1982 (second edition).

[Rubinstein88] R. Rubinstein, *Digital Typography: An Introduction to Type and Composition for Computer System Design*. Addison-Wesely, Reading, Massachusetts, 1988.

[Russ95] J. C. Russ, *The Image Processing Handbook*. CRC Press, Boca Raton, 1995 (second edition).

[Smith96] C. C. Smith, J. F. Dahl and R. J. Thornhill, "The duality of leakage and aliasing and improved digital spectral analysis techniques," *Journal of Dynamic Systems, Measurement and Control*, Vol. 118, 1996, pp. 741–747.

[Smith03] S. W. Smith, *Digital Signal Processing: A Practical Guide for Engineers and Scientists*. Newnes, USA, 2003.

[Smith07] J. O. Smith III, *Mathematics of the Discrete Fourier Transform (DFT) with Audio Applications*. W3K Publishing, WA, 2007 (second edition).

[Spiegel68] M. R. Spiegel, *Mathematical Handbook of Formulas and Tables*. Schaum's Outline Series, McGraw-Hill, NY, 1968.

[Stark79] H. Stark, "Sampling theorems in polar coordinates," *Jour. of the Optical Society of America*, Vol. 69, No. 11, November 1979, pp. 1519–1525.

[Stark82] H. Stark, "Polar sampling theorems of use in optics," *Applications of Mathematics in Modern Optics*, SPIE proceedings, Vol. 358, USA, 1982, pp. 24–30.

[Stein71]   E. M. Stein and G. Weiss, *Fourier Analysis on Euclidean Spaces*. Princeton University Press, NJ, 1971.

[Strang93]   G. Strang, "Wavelet transforms versus Fourier transforms," *Bulletin of the American Mathematical Society (New Series)*, Vol. 28, No. 2, April 1993, pp. 288–305.

[Stucki79]   P. Stucki (Ed.), *Advances in Digital Image Processing*, Plenum Press, NY, 1979.

[Thornhill83]   R. J. Thornhill and C. C. Smith, "Time aliasing: A digital data processing phenomenon," *Journal of Dynamic Systems, Measurement and Control*, Vol. 105, 1983, pp. 232–237.

[Ulichney88]   R. Ulichney, *Digital Halftoning*. MIT Press, USA, 1988.

[Vygodski73]   M. Vygodski, *Aide-Mémoire de Mathématiques Supérieures*. Mir, Moscow, 1973 (French translation).

[Wandell95]   B. A. Wandell, *Foundations of Vision*. Sinauer, Massachusetts, 1995.

[Weaver89]   H. J. Weaver, *Theory of Discrete and Continuous Fourier Analysis*. Wiley, NY, 1989.

[Weisstein99]   E. W. Weisstein, *CRC Concise Encyclopedia of Mathematics*. CRC, Boca Raton, 1999.

[Williams00]   G. L. Williams, "Sub-Nyquist distortions in sampled data, waveform recording, and video imaging," NASA Technical Memorandum TM-2000-210381, Ohio, 2000. Available online at http://ntrs.nasa.gov/archive/nasa/casi.ntrs.nasa.gov/20000120584_2000177329.pdf (accessed Oct. 2, 2012).

[Wikipedia12]   Wikipedia®, 2012, "Aliasing." http://en.wikipedia.org/wiki/Aliasing (accessed October 2, 2012).

[Wolberg90]   G. Wolberg, *Digital Image Warping*. IEEE Computer Society Press, California, 1990.

[Wolfram96]   S. Wolfram, *The Mathematica® Book*. Cambridge University Press, UK, 1996 (third edition).

[WolframAlpha09] WolframAlpha® LLC, 2009. http://www.wolframalpha.com/input/?i=limit (accessed August 24, 2011).

[Wright07]   H. Wright, *Introduction to Scientific Visualization*. Springer, London, 2007.

[Zayed93]   A. I. Zayed, *Advances in Shannon's Sampling Theory*. CRC, Boca Raton, 1993.

[Zwillinger96]   D. Zwillinger (Ed.), *CRC Standard Mathematical Tables and Formulae*. CRC, Boca Raton, 1996 (30th edition).

[Zygmund68]   A. Zygmund, *Trigonometric Series (Vol. 1)*. Cambridge University Press, US, 1968 (corrected reprint of second edition of 1959).

Index

Page numbers followed by the letter "g" indicate entries in the glossary (Appendix E).

Printed in the United States
By Bookmasters